ADVANCED PULVERIZED COAL INJECTION TECHNOLOGY AND BLAST FURNACE OPERATION

ACKNOWLEDGEMENT

The editors gratefully acknowledge financial contribution by the Industry Club of Japan.

ADVANCED PULVERIZED COAL INJECTION TECHNOLOGY AND BLAST FURNACE OPERATION

Edited by

Kuniyoshi Ishii
Hokkaido University, Japan

Research Group of Pulverized Coal Combustion in Blast Furnace,
Ironmaking 54th Committee,
Japan Society for the Promotion of Science (JSPS)
and
Technical Division of High-Temperature Processes, ISIJ

PERGAMON

ELSEVIER SCIENCE Ltd
The Boulevard, Langford Lane
Kidlington, Oxford OX5 1GB, UK

First edition 2000
Transferred to digital printing 2006

Library of Congress Cataloging in Publication Data
A catalog record from the Library of Congress has been applied for.

British Library Cataloguing in Publication Data
A catalogue record from the British Library has been applied for.

ISBN: 0 08 043651 X

∞ The paper used in this publication meets the requirements of ANSI/NISO Z39.48-1992 (Permanence of Paper).
Printed and bound by Antony Rowe Ltd, Eastbourne

Chairman:

Kuniyoshi Ishii, Hokkaido University

Editorial board:

Tatsuro Ariyama, NKK Corp.
Tadashi Deno, Nippon Steel Corp.
Masayuki Fukui, Nakayama Steel Works
Yoshimasa Kajiwara, Sumitomo Metal Industry
Yoshiaki Kashiwaya, Hokkaido University
Kanji Takeda, Kawasaki Steel Corp.
Yoshimichi Takenaka, Kobe Steel Ltd.
Kazuyoshi Yamaguchi, Nippon Steel Corp.

Members:

Tomohiro Akiyama, Miyagi National College of Technology
Takashi Ando, Idemitsu Kosan Co. Ltd.
Toshihiko Ide, Godo Steel Corp.
Yoshiaki Iguchi, Nagoya Institute of Technology
Takanobu Inada, Sumitomo Metal Industry
Masahiro Ishigaki, Tohoku University
Yuji Iwanaga, Sumitomo Metal Industry
Tsunao Kamijo, Kobe Steel Ltd.
Eiki Kasai, Tohoku University
Masahiro Kawakami, Toyohashi University of Technology
Isao Kobayashi, Kobe Steel Ltd.
Mamoru Kuwabara, Nagoya University
Takatoshi Miura, Tohoku University
Shigekatsu Mori, Nagoya University
Tetsuya Nagasaka, Tohoku University
Ichirou Naruse, Toyohashi University of Technology

Hiroshi Nogami, Tohoku University
Masayoshi Sadakata, Tokyo University
Yasushi Sasaki, Hokkaido University
Masakata Shimizu, Kobe Steel Ltd.
Shin-ich Suyama, Sumitomo Metal Industry
Noboru Taguchi, Akita University
Katsuhiro Tanaka, Nisshin Steel Co., Ltd.
Masanori Tokuda, Tohoku University
Yukio Tomita, Nishin Steel Corp.
Hiromitsu Ueno, Nippon Steel Corp.
Tateo Usui, Osaka University
Junichiro Yagi, Tohoku University
Yasumasa Yamashita, National Institute for Resources and Environment

The late members:

Kazutomo Ohtake, Toyohashi University of Technology
Takeaki Murayama, Kyusyu University

Preface

A huge amount of pig iron is made as the crude materials of high quality steel and a few hundred million ton of coal is used in the world every year. Although coal is mainly used as coke, extensive replacement of coke by injected pulverized coal is a major challenge for blast furnace operation to decrease the cost and reduce coke demand. According to theoretical investigations and actual industrial operation, it seems clear that coal combustion is hardly achieved in the raceway at high injection rate. Consequently unburned fine materials are blown with the gas into the bosh and dead man areas with possible detrimental effects on gas flow and permeability of the coke column. The capacity of the furnace to consume these particles by solution loss is probably one of the limitations to coal injection. It is therefore important to understand the physicochemical and aerodynamic behavior of fines including the change of in-furnace phenomena.

As engineers and researchers in the field of ironmaking in Japan are well aware of the importance of PCI operation, the serious scientific study on PCI had been carried out in a decades ago in parallel with many practical investigations. Thus, the frontier has been opened in the established domain of metallurgy. Exploiting this opportunity, the committee on pulverized coal combustion and in-furnace reaction in BF was set up in May 1993 as a cooperative research of the Japan Society for the Promotion of Science (JSPS) and the Iron and Steel Institute of Japan (ISIJ). Consisting of total thirty-four members and advisors drawn from the fields of academy and industry, this committee collected, discussed, and evaluated numerous papers during its four year commission.

Particular attention was paid to the interpretation of findings drawn from the combustion experiments and results of applying to the live furnace, and furnace performance. The results of these intense research activities are presented here in the hope that they will serve not only as a source of enrichment to the professional knowledge of researchers and operators, but also as a textual material for graduate students in the field of metallurgy, chemical engineering , chemistry, mechanical engineering, and so on.

A lot of original papers are included within ten chapters as outlined in the following.

Chapter 1 is devoted to the short view on the development of PCI techniques in Japan, to grasp the whole view of this book.

Chapter 2 deals with the combustion of single particle of pulverized coal to basically understand the character of coal combustion, while Chapter 3 discusses the behavior of group of pulverized coal particles, focusing the progress in recent laboratory scaled experiments.

Chapter 4 deals with combustion behavior of PC in the practical BF to elucidate the effect of combustion condition on the combustibility and at least BF performance.

In Chapter 5, to quantitatively deal with the combustion of PC in blowing pipe and near the raceway, mathematical treatment is described on basis of chemical engineering. The performances of high efficiency lance developed by applying these methods are shown in Chapter 6.

Chapter 7 deals with the changes in in-furnace phenomena according with high rate PCI such as burden distribution, flows and pressure drop of gas, descent of liquid and solid materials in the upper and lower parts of the furnace, heat-loss from the furnace wall, replacement ratio and related phenomena in the dropping zone, raceway and the deadman.

Chapter 8 deals with the generation of fines increased with increasing PCR, and its effect on furnace operation. In Chapter 9 burden properties needed in the high rate PCI operation are discussed for the both of agglomerated iron ore and coke.

Chapter 10 describes the factors to rule the upper limit of PCI. This will present a number of views regarding future blast furnace technology.

I wish to express my sincere thanks to all members and advisors for their eager efforts. My special appreciation goes to Prof. Masanori Tokuda, the former chairman of the 54th committee of JSPS, and Mr. Takashi Sumigama, the former chairman of the Ironmaking Committee of ISIJ. Finally, I also want to thank deeply the late lamented, Prof. K. Ohtake and Prof. T.Murayama. Their names will be remembered forever in the annals of ironmaking research.

October 2000

Kuniyoshi Ishii
GROUP CHAIRMAN

Contents

CHAPTER 3

Tatsurou Ariyama

CHAPTER 4

Tsunao Kamijou, Masakata Shimizu

CHAPTER 5

Hiroshi Nogami, Kanji Takeda

CHAPTER 6

Tatsurou Ariyama

CHAPTER 7

Kuniyoshi Ishii, Yoshiaki Kashiwaya

CHAPTER 8

Tomohito Akiyama, Yoshimasa Kajiwara

CHAPTER 9

Masahiro Kawakami, Kazuyoshi Yamaguchi

CHAPTER 10

Tadashi Deno

CHAPTER 1

Introduction - High rate PCI operation in Japan

1.1 Outline of high rate PCI operation in Japan

The first pulverized coal injection (PCI) in Japan was installed at Oita 1 BF of Nippon Steel in 1981 after the 2nd oil crisis, and subsequently the pulverized coal injection operation has become more widely used. In 1998, PCI operation is under being conducted at all of 30 blast furnaces in Japan. The average injection rate in 1998 was 131.1kg/thm and especially four blast furnaces have conducted high rate PCI over 200kg/thm; Fukuyama 3 BF (266kg/thm) of NKK, Kakogawa 1 BF (254kg/thm), Kobe 3 BF (203kg/thm), and Kakogawa 3 BF(203kg/thm) of Kobe Steel as shown in Table 1-1.

The purpose of high rate PCI is to reduce the pig iron cost through the utilization of non-coking coal and to extend the available coke oven life. Figure 1-1 shows the age distribution of coke oven in Japan. Figure 1-2 shows the relationship between coke supply and coke demand in Japan based on the assumption that a coke oven can last 35 years on an average. Since the average age of coke oven in Japan is currently about 25 years as shown in Fig.1-1, it is estimated that coke supply will reduce to half in 2005 on the assumption of 35 years coke oven life. Otherwise, as shown in Fig.1-2, if the pulverized coal injection rate (PCR) reaches 200kg/thm at all blast furnaces, the required coke demand can be remarkably reduced. From this reason, high rate PCI can be considered to be an attractive technology in ironmaking process, and the development of high rate PCI is now being actively pursued in Japan.

The PCR in Japan until 1997 is shown in Fig.1-3. Actually, the injection rate is increasing year by year, and as described above, high rate PCI operation over 200kg/thm was carried out at some blast furnaces. However, there are still some problems to be solved in high rate PCI. For example, the relationship between fuel rate and productivity is shown in Fig.1-4. The increase in fuel rate would bring about a decrease in productivity. The increase in fuel rate could reduce the benefit acquired from the high rate PCI. It is necessary to lower the coke rate without raising the fuel rate sharply, aiming at increasing PCR. Thus, many fundamental researches on high rate PCI operation and PC combustion are being promoted in this area. In this report, recent activities on high rate PCI in Japan are described.

Table 1-1 Results of PCI operation in Japan

PCI Operation in Japan (≥130Kg/thm)

Company		NKK	Kobe	NKK	Kobe	Kobe
Blast furnace		Fukuyama3	Kakogawa1	Fukuyama4	Kakogawa3	Kobe3
Charging device		Bell+MA	Bell+MA	Bell+MA	Bell+MA	Bell-less
Inner volume	m^3	3223	4550	4288	4550	1845
Month	y.m	'98.6	'98.3	'9410	'98.3	'98.3
Productivity	$t/d/m^3$	1.90	1.88	1.97	1.88	2.14
Coke rate	Kg/thm	289.0	291.2	314	291.2	294.1
PC rate	Kg/thm	266.0	254.4	218.0	203.3	203.0
O_2 enrichment	%	4.80	4.10	2.70	4.10	2.80
Blast temperature	°C	1220	1233	1238	1233	1200
Blast moisture	g/Nm^3	32.0	17.0	27.0	17.0	13.0
Slag rate	Kg/thm	266	265	274	265	279
Pellet ratio	%	15.5	35.0	25.0	35.0	0.0
Type of lance		Oxy-coal	Single	Excentric Double	Single	Single

Company		NSC	Nisshin	NSC	NSC	Hokkai
Blast furnace		Kimitsu3	Kure1	Kimitsu2	Oita1	Muroran2
Charging divice		Bell-less	Bell+MA	Bell-less	Bell+MA	Bell-less
Inner volume	m^3	4063	2150	3273	4884	2296
Month	y.m	'93.11		'96.2	'96.3	'95.9
Productivity	$t/d/m^3$	1.86	2.19	2.23	2.13	2.28
Coke rate	Kg/thm	318	328.0	323.3	313.8	325.4
PC rate	Kg/thm	181.0	180.0	174.6	165.3	165.2
O_2 enrichment	%	3.67			2.30	1.77
Blast temperature	°C	1224			1258	1246
Blast moisture	g/Nm^3	26.0			10.0	13.7
Slag rate	Kg/thm	284		293	318	281
Pellet ratio	%	11.6			4.2	0.0
Type of lance		Single	Double	Single	Single	Single

Company		NSC	Nisshin	Sumitomo	NSC	Kawasaki
Blast furnace		Ooita2	Kure2	Kokura2	Tobata1	Mizushima3
Charging divice		Bell+MA	Bell-less	Bell+MA	Bell- less	Bell-less
Inner volume	m^3	5245	1650	1850	4407	4359
Month	y.m	'96.3		'93.3	'96.2	'94.4
Productivity	$t/d/m^3$	2.07	2.34	1.8	2.15	1.8
Coke rate	Kg/thm	344.0	340.5	348	372.1	372
PC rate	Kg/thm	159.7	159.0	154.1	142.2	138.0
O_2 enrichment	%	2.30	3.24	1.00	2.80	1.90
Blast temperature	°C	1247	1200	1196	1221	1110
Blast moisture	g/Nm^3	11.0	25.6	15.0	26.3	38.0
Slag rate	Kg/thm	320	288	315	296	325
Pellet ratio	%	4.6	6.3	11.7	5.7	0.0
Type of lance		Single	Double	Single	Single	Single

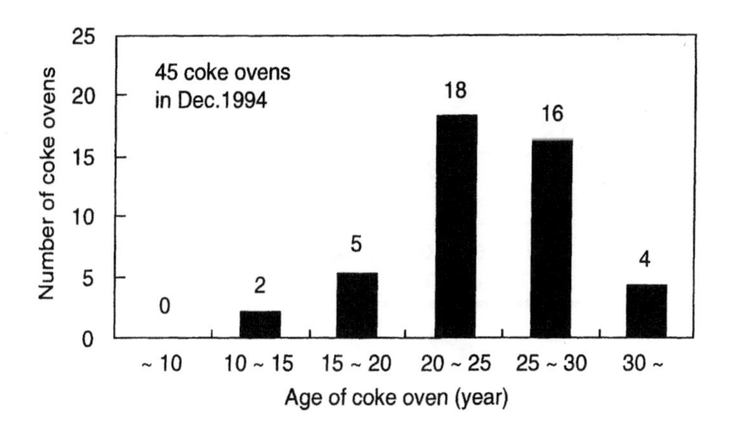

Fig.1-1 Age distribution of coke oven in Japan

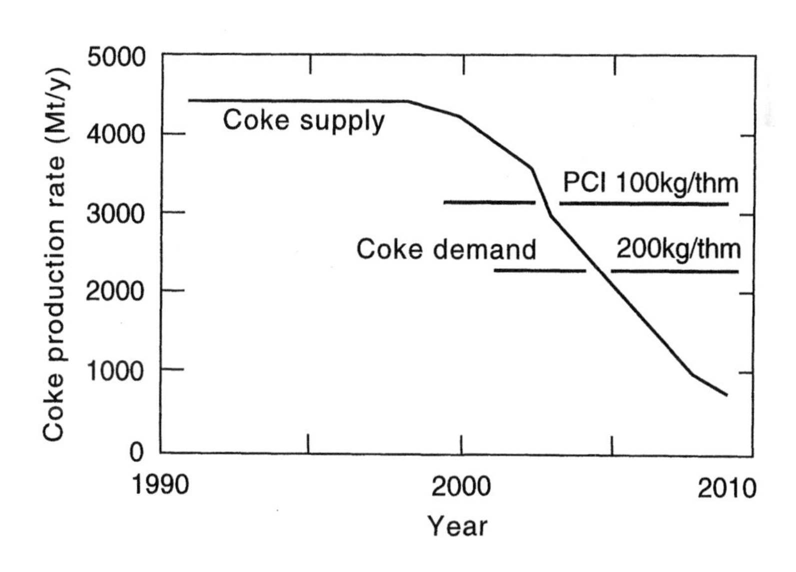

Fig.1-2 Prediction of coke supply and demand in Japan

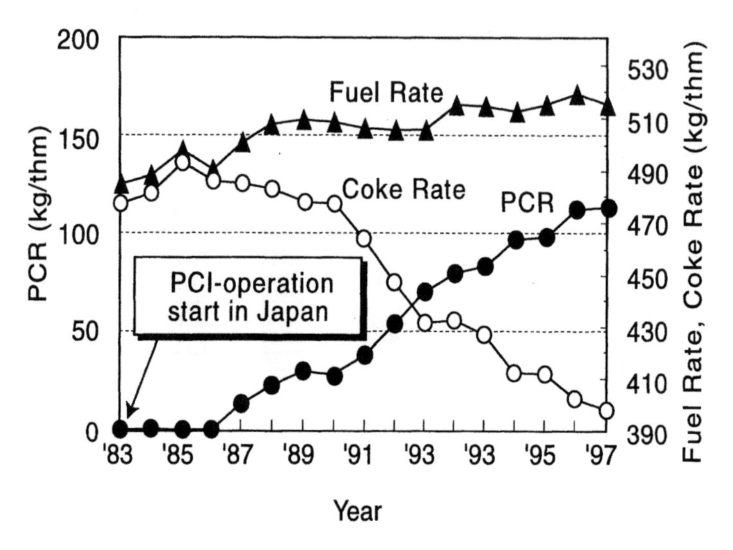

Fig.1-3 Change in fuel rate, coke rate and PCR in Japan

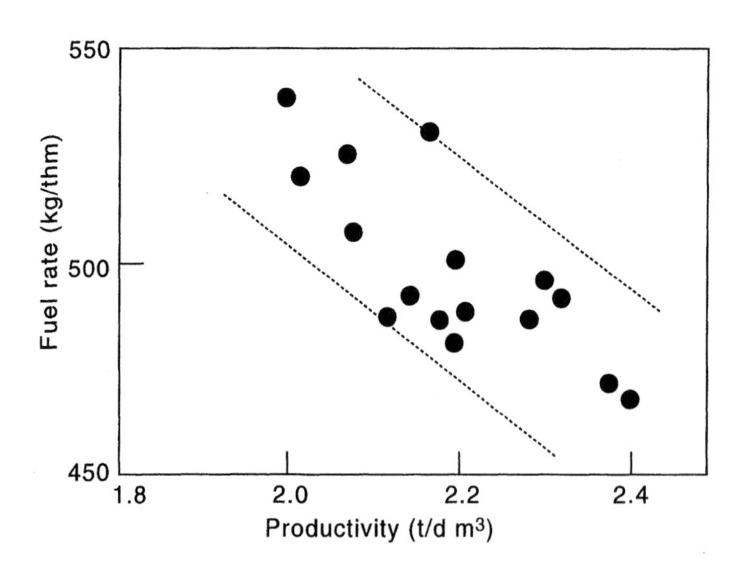

Fig.1-4 Relationship between productivity and fuel rate in Japan

1.2 Actual situation of high rate PCI

Figure 1-5 [1)] shows the relationship between the coke rate and PCR in the representative countries. It indicates that the replacement ratio has a tendency to be lowered when the PCR becomes over 180kg/thm.

From the overall operating results obtained until now, the relationship of fuel rate with PCR, excess air ratio(stoichiometric oxygen ratio) (μ) and Ore/Coke (O/C) are represented as shown in Fig. 1-6. From Fig.1-6, the following points can be summarized at high rate PCI operation.

(1) When PCR is over about 180kg/thm, the fuel rate gradually increases accordingly as the pulverized coal rate becomes higher.

(2) There is a limit to the stoichiometric oxygen ratio with μ of about 0.7 supposed to be the lower limit.

(3) The upper limit of the Ore/Coke seems to be 5.5.

The increase in the fuel rate at high rate PCR over 180kg/thm could be related with

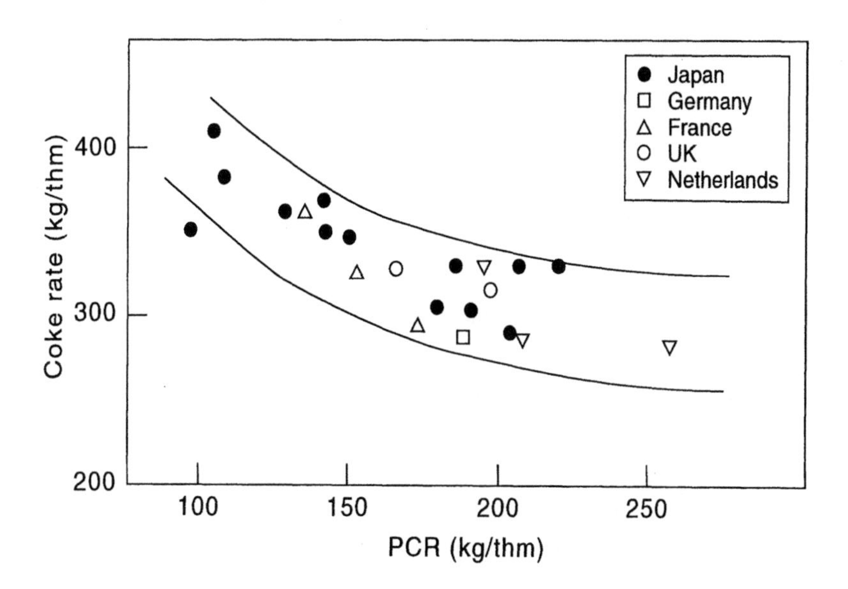

Fig.1-5 Relationship between PCR and coke rate

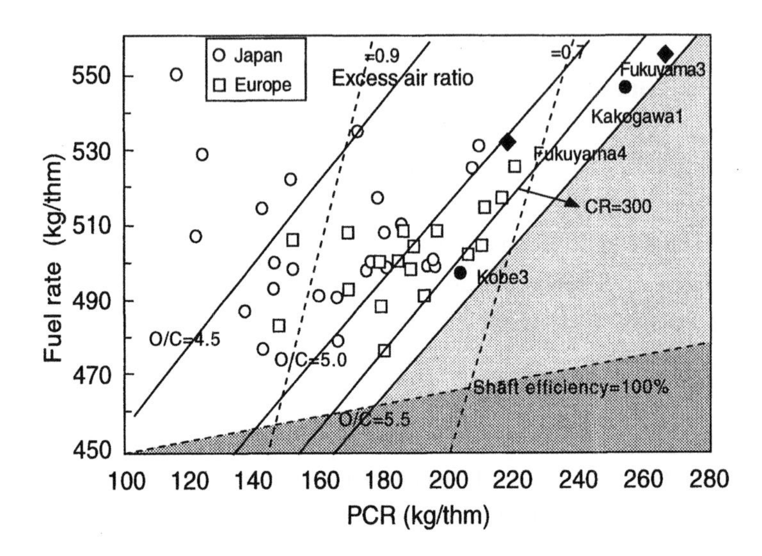

Fig.1-6 Relationships of fuel rate with PCR, excess air ratio and O/C

lowering in combustion efficiency or an increase in the heat loss due to the change of in-furnace conditions. These changes can be given as a tendency of replacement ratio (Coke rate/PCR). Figure 1-7 [2] shows a change in the replacement ratio at Kakogawa No.2BF of Kobe Steel. Then, a figure obtained in Fig.1-7 shows the corrected PCR based on a calorific value. Moreover, the relation between PC/oxygen ratio and the combustion efficiency can be shown in Fig.1-8. The PC/oxygen ratio has an great influence on the combustion efficiency.

The influence of PCI has been studied from the various viewpoints, such as burden quality, burden charging conditions, heat flux ratio at the shaft, blast conditions (flame temperature and stoichiometric oxygen ratio) and solution loss reaction of coke, etc. As to the burden quality, the actual blast furnace conditions should be referred. Generally, as the Ore/Coke ratio becomes high in high rate PCI operation, it is required to secure proper gas permeability and liquid permeability. Consequently, it has been tried until now to improve the melting property of sinter. Figures 1-9 and 1-10 show the correspondence of Al_2O_3 and that of MgO contained in the sinter to PCR. The contents of these components are decreased under the high rate PCI because they influence the reducibility and melting property of sinter. Decreasing the contents of these components leads to the decrease of slag volume produced, as shown in Fig. 1-11 , thus contributing

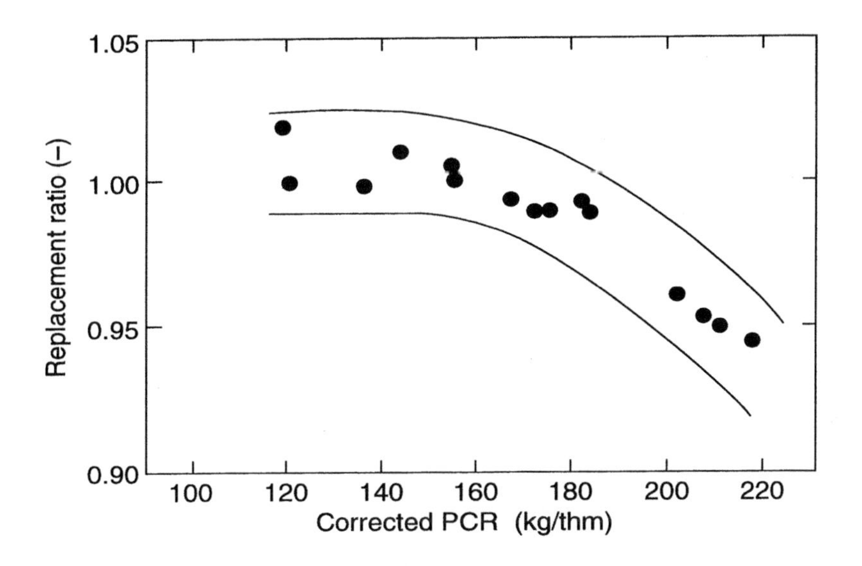

Fig.1-7 Relationship between corrected PCR and coke replacement ratio

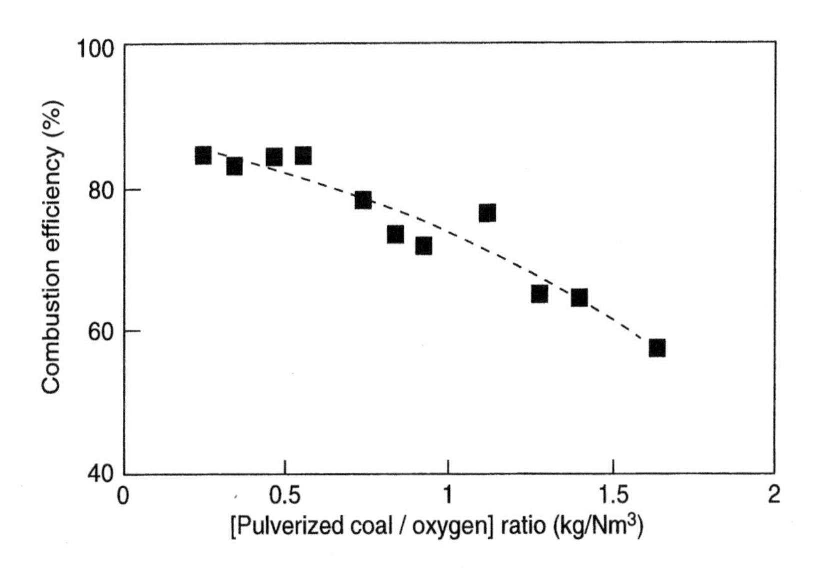

Fig.1-8 Influence of pulverized coal to oxygen ratio on combustion efficiency

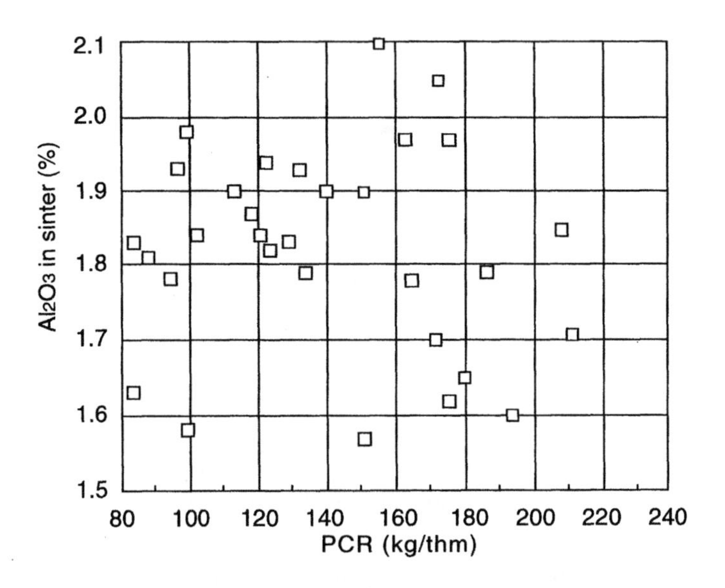

Fig.1-9 Variation of Al_2O_3 in sinter with increase of PCR

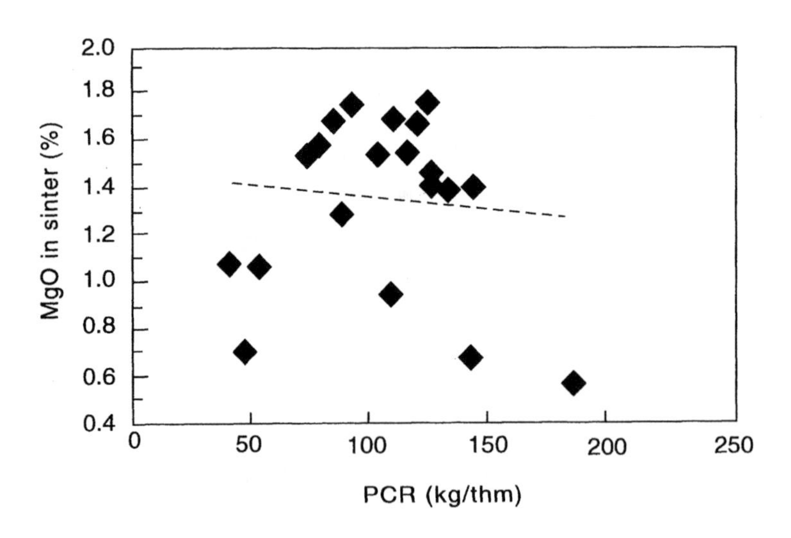

Fig.1-10 Change in MgO in sinter with increase of PCR

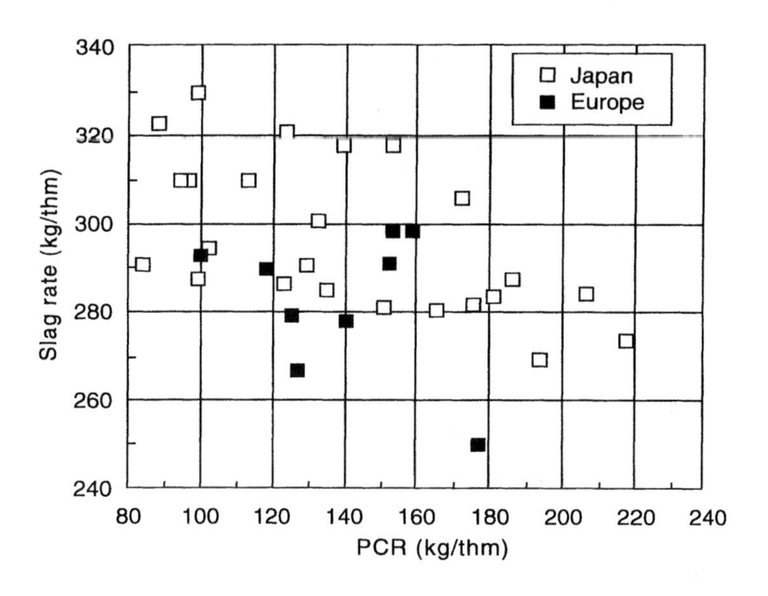

Fig.1-11 Relationship between PCR and slag rate

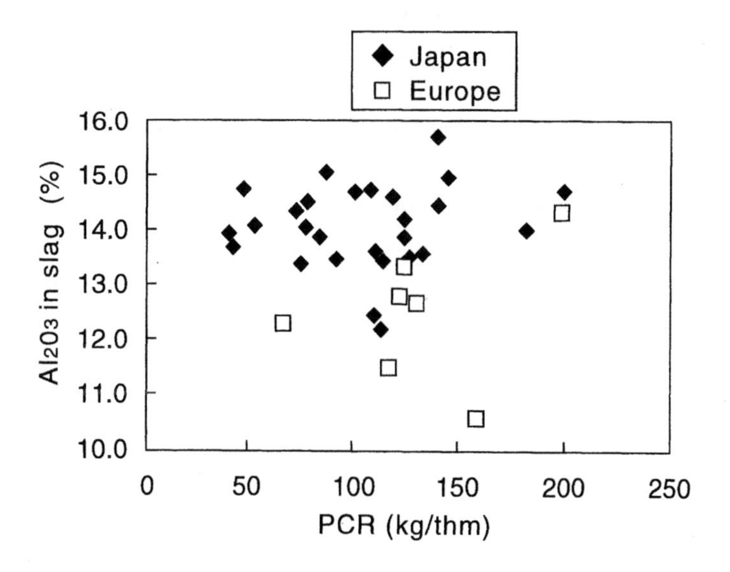

Fig.1-12 Relationship between PCR and Al_2O_3 content in slag

to a reduction in fuel rate as well as an improvement in gas and liquid permeability.

In Europe, most of blast furnaces are operated in such raw ore conditions that the high rate PCI operation can be carried out easily, since the slag volume is small, and Al_2O_3 content in slag is low, as shown in Figs.1-11 and 1-12 . Also, coke strength can be regarded as important factor, and a coke quality in Europe is generally higher by 1 to 2% than that of coke in Japan at DI of 150. However, to establish the high rate PCI operation in japan, the improvement of burden qualities accorded with Japanese conditions should be considered.

1.3 Subjects on high rate PCI

The subjects to be studied regarding high rate PCI can be represented as shown in Fig.1-13 [3] based on operating results and the current research activities.

As mentioned above, as to the factors causing a change in the in-furnace phenomena at high rate PCI, the following items are noted; increase of Ore/Coke, decrease in theoretical flame temperature and decrease in stoichiometric oxygen ratio and lowering of heat flux ratio (heat flux ratio can be defined as the heat capacity of solid to the heat capacity of gas). Subsequently, these phenomena result in various changes in the blast furnace, as shown in Fig.1-13. Actually, the following points are characterized; (1) rise in pressure drop at tuyere and in furnace, (2) increase of heat loss due to intensified peripheral gas flow and lowering of heat flux ratio and (3) generation of unburnt pulverized coal due to reduction in combustion rate. The overall phenomena are described below.

1.3.1 Increase in permeability resistance in furnace

Generally, increase of pressure drop at high rate PCI has a relation with; 1) decreased coke layer thickness at high Ore/Coke (decrease of coke rate), 2) rise of furnace heat level and rise of gas temperature due to lowering of heat flux ratio, 3) increase in solution loss reaction of coke near central area in accordance with Ore/Coke distribution, 4) degradation of coke due to increase in burden load and 5) the fact that generation and accumulation of coke fines ranging from the middle area to central area of the furnace varies a packing condition in the deadman and obstructs the gas permeability.

1.3.2 Increase in heat loss

In line with the increase in PCR, since a coke combustion rate in the raceway is reduced, a burden descending velocity at the upper part is decreased. The decrease in the descending velocity causes the heat flux ratio to be lowered, which leads to where the furnace internal temperature rises and the temperature rising at the furnace top also going up, which puts the furnace in an unstable state.

From the change of the peak point of CO_2 gas concentration in the raceway, which is regarded as being a maximum combustible point, it is estimated that a maximum temperature zone is coming nearer the tuyere nose at high rate PCI . Thus it can be

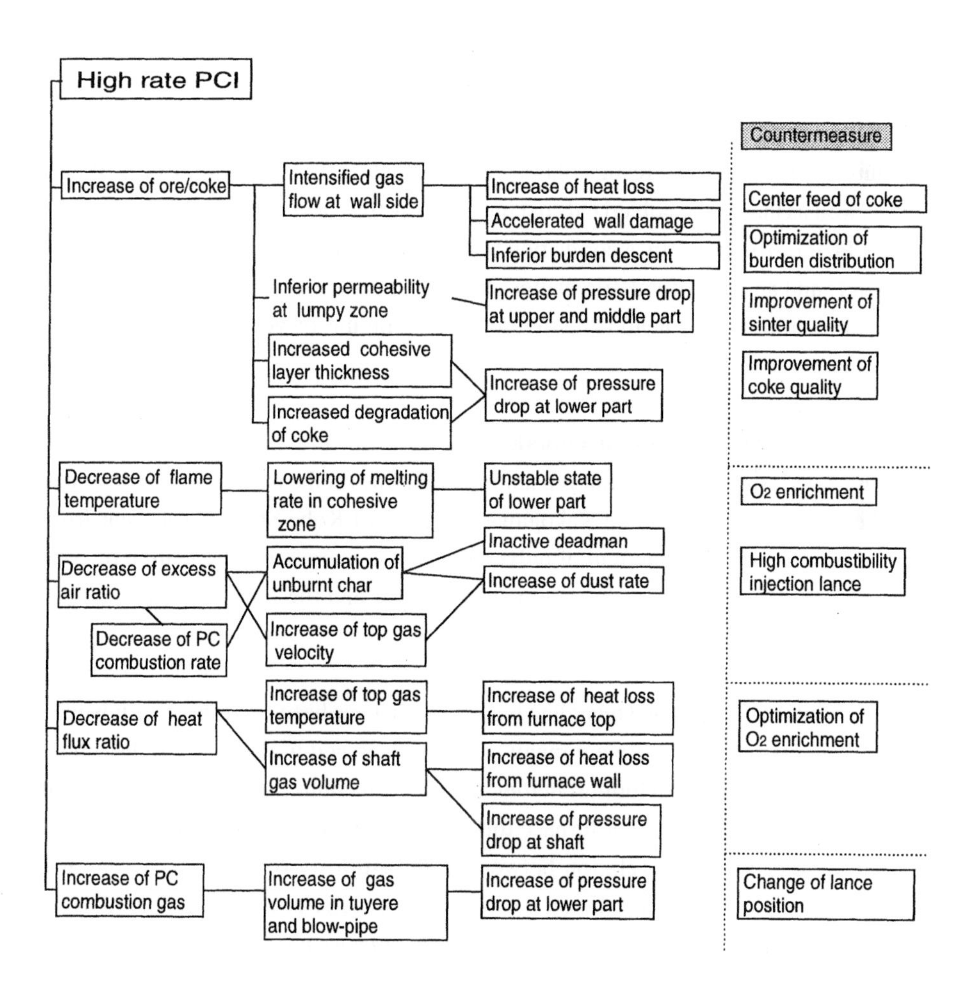

Fig.1-13 Subjects and countermeasure at high rate PCI

assumed that this will result in a heat loss at the lower part of the blast furnace probably being increased.

1.3.3 Accumulation of unburnt char

From the observation of the dry dust at the furnace top, it is found that C content in the dry dust at the furnace top remarkably increased when an stoichiometric oxygen ratio is coming to 0.7. Especially, the amount of unburnt char in dust of less than 0.2mm in particle size sharply has increased. It suggests that combustion rate in the raceway can be suppressed due to insufficient oxygen supply.

1.3.4 Influence of coal brand on combustion of PC in raceway

Figure 1-14 [4)] shows the results obtained by estimating the effect of pulverized coal brand as well as that of the pulverized coal injection rate by a test furnace (VMR=VM(%) in coal x PCR) . As shown in a correlation between a maximum combustible point (CO_2 peak point) and amount of volatile matter charged, a tendency by which the maximum combustible point comes nearer the tuyere as the amount of volatile matter charged becomes larger is recognized.

1.3.5 Accumulation of fines in deadman

Figure 1-15 [5)] shows the relationship among PCR, fines at the middle area of deadman and coke particle diameter at the Kakogawa No. 1 BF of Kobe Steel. As is clear from this relationship, when the amount of pulverized coal is increased, degradation of coke occurs ,the amount of fines at the deadman increases, and the coke particle diameter decreases.

The subjects to be examined regarding high rate PCI have been cited above. However, it is to be understood that there are many subjects remained with regard to the whole blast furnace phenomena including; burden distribution control, relation between heat flux ratio and gas temperature at furnace top, relation between ore layer thickness and coke slit at cohesive zone, relation between combustibility of pulverized coal and consumption of unburnt char, relation between coke reactivity and coke degradation, inactivity of the deadman, and so on. Moreover, as to precise understanding combustion behavior of pulverized coal, a study based on the blast furnace raceway condition characterized by rapid heating over 2000°C should be carried out.

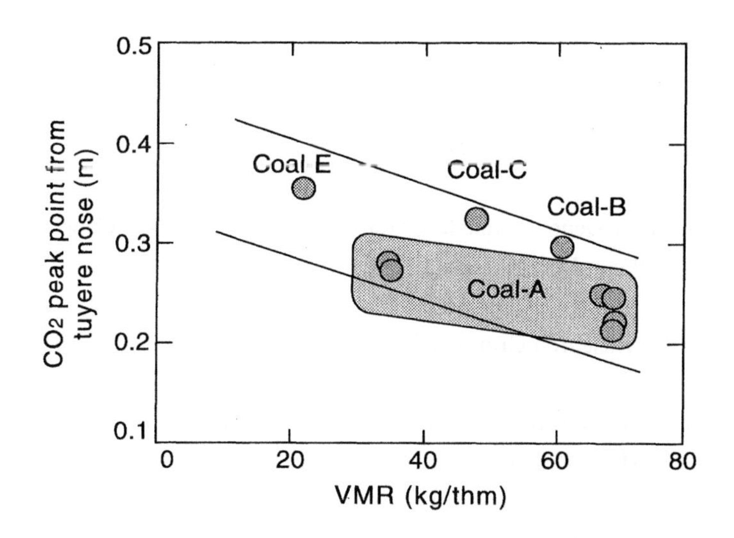

Fig.1-14 Relationship between VM in pulverized coal and CO_2 peak point in raceway

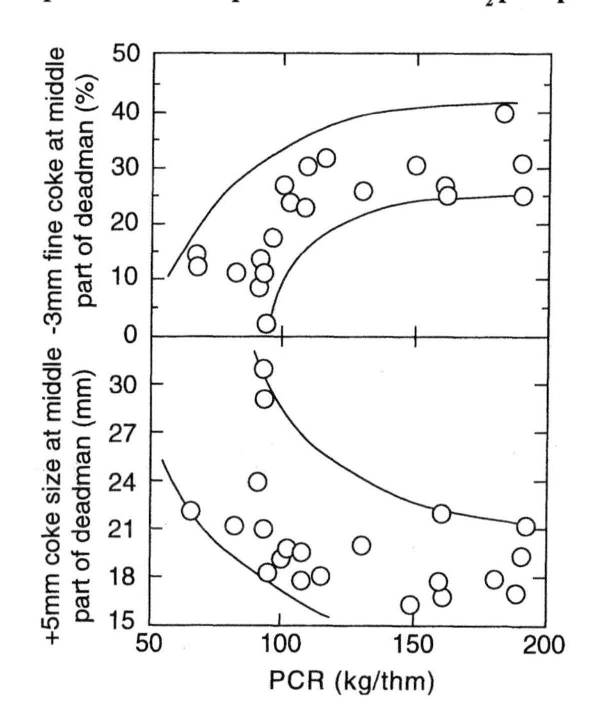

Fig.1-15 Influence of PCR on coke size and coke fines in deadman

REFERENCES

1) Y. Okuno and T.Deno: Private letter referred from Proc.1st Int. Cong. Science and Tech. Ironmaking, ISIJ, Sendai, (1995), 444.
2) S.Inaba:the146th,Nishiyama-Memorial-Lecture, ISIJ, (1993), 203.
3) M. Shimizu:Research Group of Pulverized Coal Combustion in Blast Furnace-1(1995),JSPS54.
4) T.Ariyama, M.Sato,R.Murai,K.Miyagawa,K.Nosawa and T.Kamijo:Tetsu-to-Hagane,**81**(1995),1114.
5) M.Shimizu,A.Kasai,T.Kamijo,H.Iwakiri,R.Ito and M.Atsushi : JSPS 54th committee, No.54-2037 (1995)

CHAPTER 2

Characteristics of pulverized coal combustion

A large number of types of coal imported into Japan from many foreign countries. This situation suggests that the coal used in Japan is in great variety, and this requires that the combustion technologies have to be developed to apply to any coal to attain high combustion efficiency and low environmental pollution.

Pulverized coal combustion has been adopted widely in large-scale boilers mainly in electric power stations. Recently, pulverized coal is also injected into blast furnaces, so-called PCI (Pulverized Coal Injection) technology has been introduced in the steel industries of some developed countries since the cost of coal has become more reasonable than that for oil. In PCI technology, however, a preheated air of about 1500K of gas temperature is used as an oxidizer, and the gas temperature is much higher than that in ordinary pulverized coal combustion boilers. Additionally, coal essentially contains ash consisting of inorganic constituents and is incombustible. The ash or char particles may sometimes hinder the performance of blast furnaces due to increases of the pressure drop, particulate emission and so forth resulting in the aggravation of gas and liquid permeability and burden descent as described in Chapter 1.

In order to solve these problems, it is necessary to clarify the fundamental combustion phenomena of pulverized coal under the operating conditions in the blast furnace, both experimentally and numerically. In this chapter, fundamentals on coal, including coal properties, characteristics on ordinary pulverized coal combustion, in which such phenomena as volatile matter evolution and combustion, ignition, char combustion, burnout and ash formation and its behaviors are separately discussed. The combustion behaviors of pulverized coal under high temperature condition obtained by a lab-scale furnace and so on will also be described. The information may be of help to elucidate the practical behaviors in the blast furnaces.

2.1 Classification of coal

Combustion of coal is a complex process. Consequently, it is necessary to assess a coal consignment by means of easily analyzed parameters which reflect its thermal value and any tendencies relating to process efficiency or the environment. The following four systems of classification of coals by rank are internationally standardized [1].

ASTM D 388: Classification of Coals by Rank

International Classification of Hard Coals by Type

International System for Classifying Brown Coals and Lignites

ISO 2950: Brown Coals and Lignites - Classification by Types on the Basis of Total Moisture Content and Tar Yield

Table 2-1[1], taken from the standard of ASTM D 388, defines the limits of the various criteria used in this classification system. This standard classifies coals according to the

Table 2-1 ASTM D 388 classification of coals by rank

Class / Group	Fixed carbon limits (%) (dry,mineral-matter-free base)		Volatile matter limits (%) (dry,mineral-matter-free base)		Calorific value limits (Btu/lb) (moist mineral-matter-free base)		Agglomerating character
	≧	<	>	≧	≧	<	
1.Anthracitic							
1.Meta-anthracite	98	-	-	2	-	-	nonagglomerating
2.Anthracite	92	98	2	8	-	-	nonagglomerating
3.Semianthracite	86	92	8	14	-	-	nonagglomerating
2.Bituminous							
1.Low volatile bituminous coal	78	86	14	22	-	-	commonly agglomerating
2.Medium volatile bituminous coal	69	78	22	31	-	-	commonly agglomerating
3.High volatile A bituminous coal	-	69	31	-	14,000	-	commonly agglomerating
4.High volatile B bituminous coal	-	-	-	-	13,000	14,000	commonly agglomerating
5.High volatile C bituminous coal	-	-	-	-	11,500	13,000	commonly agglomerating
					10,500	11,500	agglomerating
3.Subbituminous							
1.Subbituminous A coal	-	-	-	-	10,500	11,500	nonagglomerating
2.Subbituminous B coal	-	-	-	-	9,500	10,500	nonagglomerating
3.Subbituminous C coal	-	-	-	-	8,300	9,500	nonagglomerating
4.Lignitic							
1.Lignitic A	-	-	-	-	6,300	8,300	nonagglomerating
2.Lignitic B	-	-	-	-	-	6,300	nonagglomerating

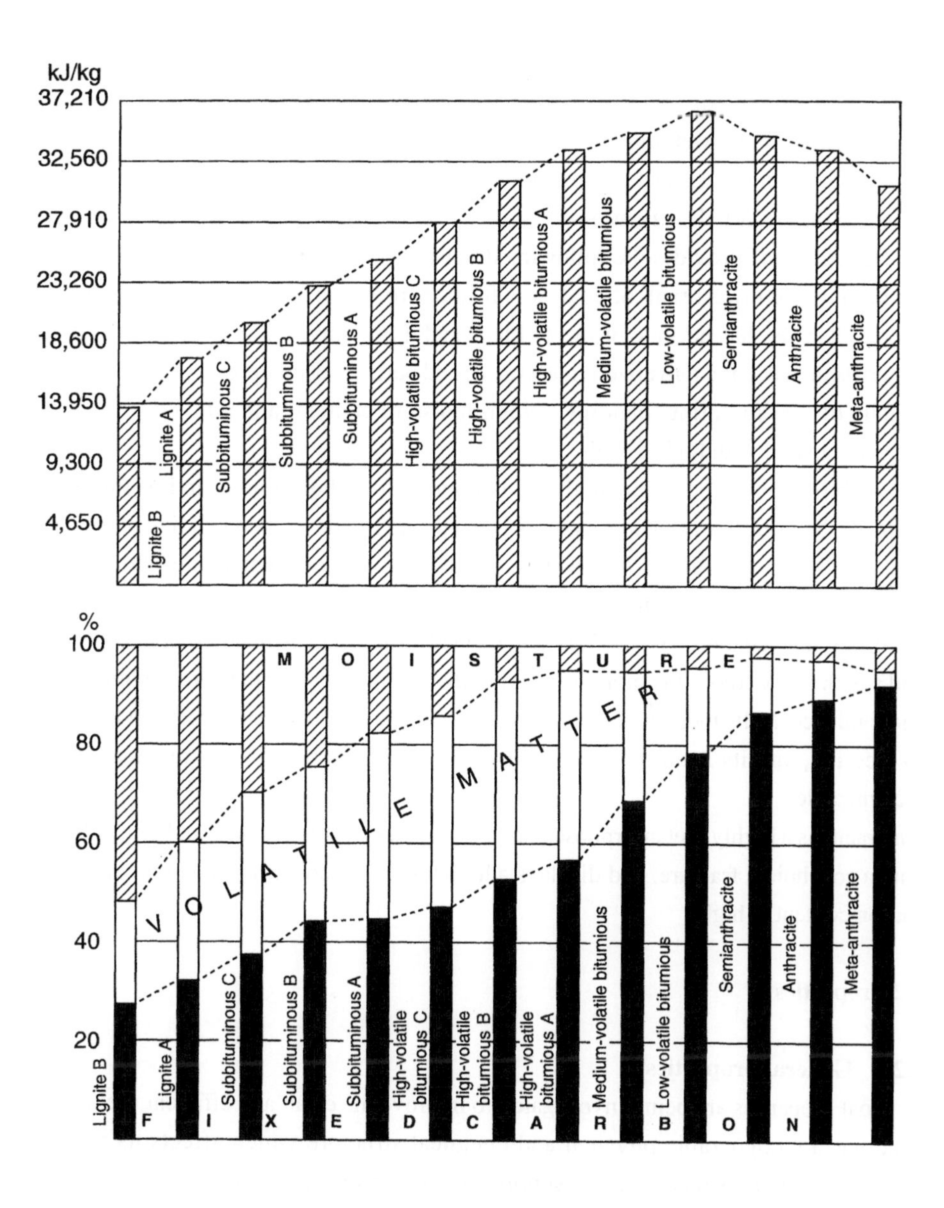

Fig.2-1 Schematic of variations of heating value and composition with a moisture calculated on the mineral-matter-free basis

degree of metamorphism or the progressive alternation in the natural series from lignite to anthracite. Fixed carbon and heating value, calculated on the mineral-matter-free basis, have been chosen as the criteria of rank. The higher rank coals are classified according to fixed carbon on the dry mineral-matter-free basis; the lower rank coals according to heating value on the wet mineral-matter-free basis. Figure 2-1[2] also shows the heating value and moisture of 12 coal groups, ranging from soft lignite to very hard metaanthracite, based on the ASTM classification.

Hendrickson[3] provides the following description of some of three types of coal.

Lignite, the lowest rank of coal, was formed from peat, which was compacted and altered. Its color has become brown to black and it is composed of recognizable woody materials imbedded in pulverized (macerated) and partially decomposed vegetable matter. Lignite displays jointing, banding, a high moisture content, and a low heating value when compared with higher coals

Subbituminous coal is difficult to distinguish from bituminous and is dull, black colored, shows little woody materials, is banded, and has developed bedding planes. The coal usually splits parallel to the bedding. Although it does not contain moisture, it is still of relatively low heating value.

Bituminous coal is dense, compacted, banded, brittle, and displays columnar cleavage and a dark black color. It is more resistant to disintegration in air than are subbituminous and lignite coals. Its moisture content is low, volatile matter content is variable from high to medium, and its heating value is high. Several varieties of bituminous coal are recognizable.

Anthracite is a highly metamorphosed coal, is jet black in color, is hard and brittle, breaks with a conchoidal fracture, and displays a high luster. Its moisture content is low and its carbon content is high.

2.2 Properties

2.2.1 General properties[4]

Coal properties are being investigated to improve the clean and efficient use of coal. Coal has a recognizable physical and chemical structure which affects its reactive processes. For cases where coal is supplied to power utilities subject to a lot of quality limitations, designated to minimize operational problems, a typical range of quality parameters is given in Table 2-2[4]. Each property has some technological relevance to the

pulverized coal combustion process:
i) Specific energy means the thermal value of each coal consignment. Ash content provides an estimate of the mineral matter content of a sample, thus indicating both the amount of inert material being processed through the combustor and the extent of the ash disposal required afterwards.
ii) Moisture and sulfur are also process contaminants; moisture affects the thermal value, total sulfur content gives some indication of the quality of sulfur oxides emitted and the likelihood of combustor corrosion problems.
iii) The volatile matter content is used to select the appropriate combustor operating conditions, which will minimize unburnt fraction and maintain good flame stability.
iv) Ash deposition is assessed from ash fusion temperatures, and is usually determined

Table 2-2 Coal combustion quality parameters

Quality parameter	Technological relevance
· Specific energy	- Thermal value
· Ash	- Process contaminant - Ash disposal
· Moisture	- Process contaminant
· Sulfur	- SOx emissions - Corrosion
· Volatile matter	- Unburnt fuel loss - Flame stability
· Ash fusion temperatures	- Slagging
· Hardgrove grindability index	- Grindability
· Sodium/chlorine	- Fouling

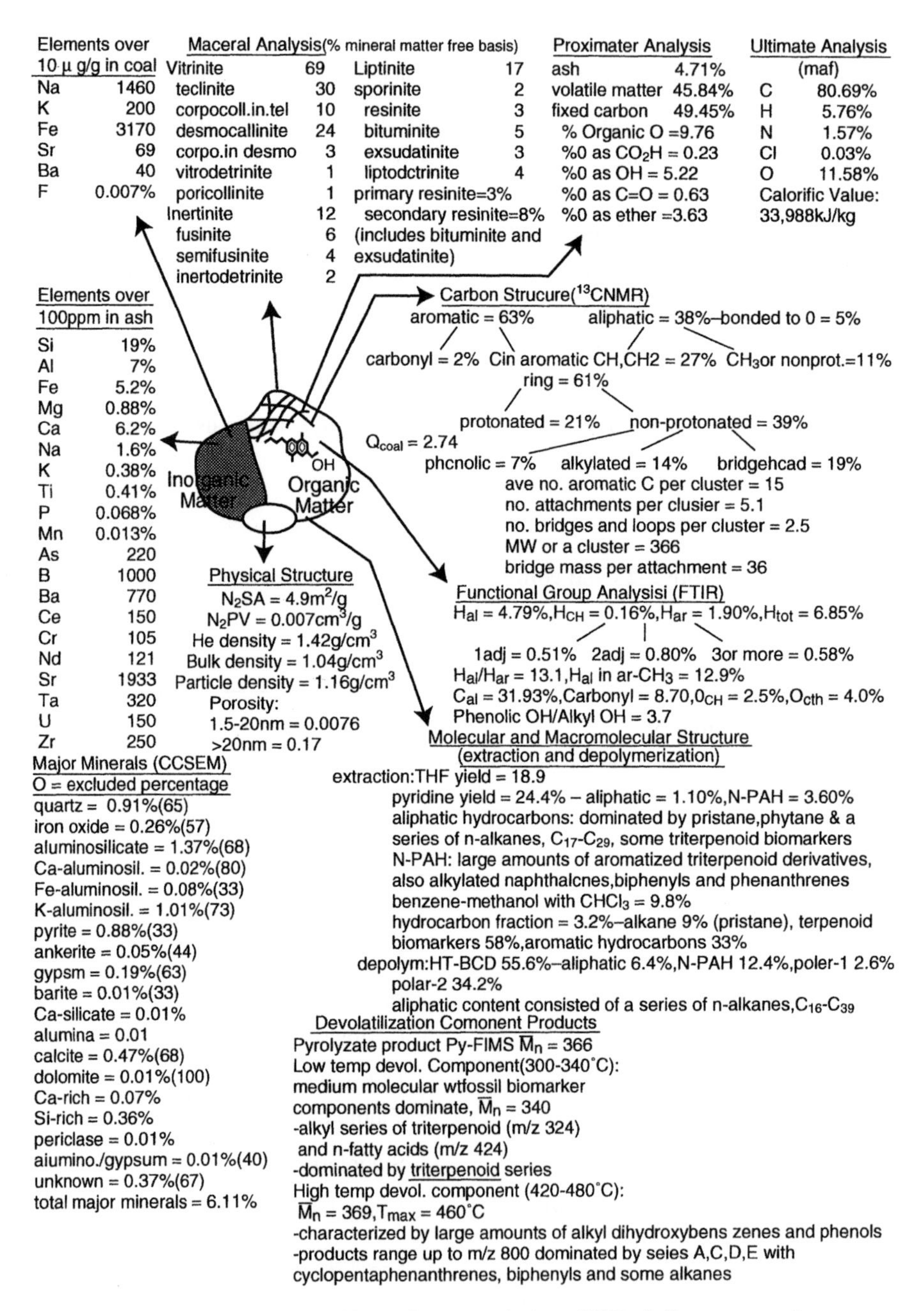

Fig.2-2 Summary of key characteristics of Blind Canyon coal

Table 2-3 Classfication of coal macerals

STOPES-HEERLEN SYSTEM		CLASSIFICATION IN U.S.A (SPACKMAN/SYSTEM)			
Maceral-Group	Maceral-Suite	Maceral-Group	Range of maximum reflectance (per cent) under oil	Other distinguishing characteristics	Macerals
Vitrinite	Vitrinite suite	Anthrinoid Group	2.50- 10.00	Opaque in transmitted light. Greysh-white in reflected light.	$A_{25}-A_{100}$
		Vitrinoid Group	0.40- 2.40	Translucent in transmmitted light - usually yellow,red or brown, Grey in reflected light.	V_4-V_{24}
		Xylinoid Group	0.10- 0.39	Translucent in transmitted light - usually buff, whitish-yellow to yellowish-brown. Dark grey in reflected light.	X_1-X_3
Exinite	Liptinite suite	Exinoid Group	0.05- 1.50	Coalified spore, pollen, cuticular or endodermal materials. Translucent in transmitted light - whitish-yellow, yellow, golden yellow or red. Black, dark grey to light grey in reflected light.	E_0-E_{15}
		Resinoid Group	0.05- 1.50	Coalified resins or other plant secretions or exudates. Translucent in transmitted light - whitish-yellow, yellow, golden yellow or red. Black, dark grey to light grey in reflected light.	F_0-R_{15}
Inertinite	Inertinite suite	Fusinoid Group	4.00- 10.00	Characteristics essentially those of fusinite (Stopes-Heerlen System).	$F_{40}-F_{100}$
		Semi-Fusinoid Group	0.20- 3.99	Characteristics essentially those of semifusinite (Stopes-Heerlen System).	SF_2-SF_{39}
		Micrinoid Group	0.20- 8.00	Characteristics essentially those of micrinite (Stopes-Heerlen System).	M_2-M_{80}

[a] Adapted from Spackman (1963).

under reducing conditions. Sometimes, limits on the elemental composition of the ash are also included, particularly in terms of sodium content.

v) Results from a standard grindability test (Hardgrove) are used to indicate possible problems in pulverizing a coal consignment.

Figure 2-2[5] shows an example of coal properties which are necessary to estimate combustion behaviors. Only expensive instruments may measure some of the properties in the figure. However, some of those properties are effective for predicting quality and/or the combustion behaviors in the furnace.

2.2.2 Macerals[6]

Macerals are one of the indexes to classify the physical structure of coal. Macerals are optically homogeneous, discrete, microscopic constituents of the organic fraction of coals. They are the building blocks of coals analogous to minerals of rocks. As shown in Table 2-3[6], macerals are classified into three major groups and many individual macerals on the basis of source mineral morphology, nature of formation, similarity of chemical composition, internal structures, level of reflectivity and degree of coalification.

Vitrinites are coalified woody tissues derived from stems, roots and vascular tissues of leaves. Vitrinite is the most abundant and most important maceral in coal. The macerals of the exinite group are derived from resinous and waxy material of plants, including resins, cuticles, spore and pollen exines and algal remains that constitute resinite, cutinite, sporonite and alginite. Inertinites are a group of macerals that are opaque to transmitted light and bright to extremely bright in incident light. They are derived presumably from charring of plant tissues. Certain inertinites may be the result of intensive biochemical processes.

2.2.3 Char structures[4]

Char microstructure should also be considered in order to estimate the ignition behaviors and burnout characteristics. Three main different char classifications have evolved from several studies[7-11]. At first sight these appear to be contradictory but on closer examination it is clear that the range of coal types employed in each investigation has influenced the choice of char types. During devolatilization many coal particles soften, develop fluidity and in some cases swell as volatile gases are rapidly evolved, thus producing and an open porous structure. Table 2-4 lists the criteria used to distinguish char types. The classification consists of two principal types, viz. fused and unfused, which are then subdivided into four and three classes, respectively.

2.2.4 Grindability[12)]

The grindability of coal and coke may be determined by various methods, the best-known of which are the "Ball mill grindability test" and "Hardgrove grindability test", which have been adopted as standard procedures. Dryden[13)] has investigated the relation between grindability and coal rank. As shown in Fig.2-3[13)], the hardgrove index reaches a maximum between 89 and 90% carbon. However, it is not only the strength but also the hardness that is at a minimum in this region as is shown by measurements of the Vickers

Table 2-4 Amalgamation of char-type classification

Nomenclature used by Oka et. al. Jones et. al., Street et. al. and Tsai & Scaroni			Nomenclature of Bailey et. al.[37]	Degree of anisotropy	Origin	
					Maceral	Rank
FUSED	Cenospheres	Thin -walled	Tenuispheres	**(Rv>0.8%) nil (R<0.8%)	Vitrinite(+ exinite)	Lrb/mrb/hrb
		Thick -walled	Crassispheres	*/**(RV>0.8%) nil (R<0.8%)	Vitrinite	Lrb/mrb/hrb
		Lacy (Network)	Tenuinetwork	nil	Vitrinite(+ inertinits) Or + mineral) Intranet(+exinete)	Sub/lrb
	Honeycomb		Inertoid	**	Intranet R1< threshold value (rank dependent)	Lrb/mrb
UNFUSED	Microdisrupted		Mineroid	Ni l / *	Intranet	sub
	Skeleton		Fusinoid	* / **	Fusinite	all
	Unfused block		Unfused block	* / **	Sclerotinite Oxidised vitrinte	all

Rv : mean random vitrinite reflectance
Ri : mean random inertinite reflectance

**highly
*weakly

sub : sub-bituminous
lrb : low-rank bituminous
mrb : midium-rank bituminous
hrb : high-rank bituminous

microhardness.

2.2.5 Ash constituents[14)]

Composition of the noncombustible fraction and the mode of distribution of the mineral species in coals are based on the results of chemical and mineralogical analyses. Tables 2-5 and 2-6 lists mineral species that have been identified in different coal deposits. Those inorganic coal components undergo complex chemical and physical transformations during combustion to produce the intermediate ash species, which consist of gases, liquids and solids. The partitioning of the inorganic components during combustion to form ash intermediates depends on the association and chemical characteristics of the inorganic components, the physical characteristics of the coal particles, the physical characteristics of the coal minerals and the combustion conditions.

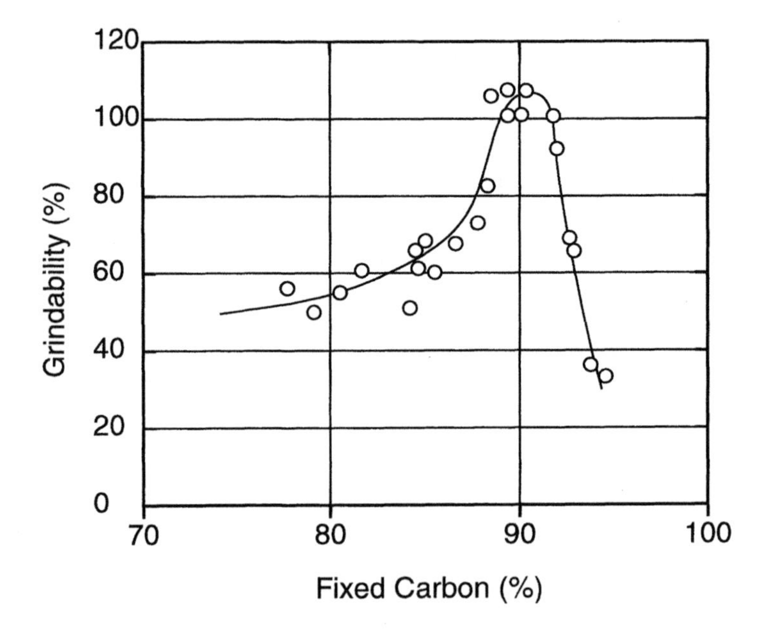

Fig.2-3 Hardgrove grindability index of coal

Table 2-5 Silicate and oxide mineral species in coal

Species	Chemical formula	Specific[a] gravity (kg m^{-3})	Melting point (K)
	Silica and silicates-common occurrence		
Quartz	SiO_2	2650	1983
Kaolinite	$Al_2O_3 \bullet 2SiO_2 \bullet 2H_2O$		2083
Muscovite	$K_2O \bullet 3Al_2O_3 \bullet 6SiO_2 \bullet 2H_2O$	2900	(Mullite)
Illite	As muscovite with Fe,Ca,and Fe		
Montmorillonite	$(1\text{-}X)Al_2O_3 \bullet X(MgO,Na_2O) \bullet 4SiO_2 \bullet nH_2O$		
Chlorite	$Al_2O_3 \bullet 5(FeO,MgO) \bullet 3.5SiO_2 \bullet 7.5H_2O$		
Orthoclase	$K_2O \bullet Al_2O_3 \bullet 6SiO_2$	2500	
Plagioclase	$Na_2O \bullet Al_2O_3 \bullet 6SiO_2$-Albite		
	$CaO \bullet Al_2O_3 \bullet 2SiO_2$-Anorthite		
	Silicates-less common occurrence		
Augite	$Al_2O_3 \bullet Ca(Mg,Fe,Al,Ti) \bullet 0.2SiO_2$		
Amphibole	Augite + Na,Fe,P	3100	
Biotite	$Al_2O_3 \bullet 6(MgO \bullet FeO) \bullet 6SiO_2 \bullet 4H_2O$	3100	
Granite	$Al_2O_3 \bullet 3(CaO,MgO,FeO,MnO) \bullet 3SiO_2$		
Epidote	$4CaO \bullet 3(Al,Fe)O_3 \bullet 6SiO_2 \bullet H_2O$	3350	
Kyanite	$Al_2O_3 \bullet SiO_2$	3350	2083(Mullite)
Sanidite	$K_2O_3 \bullet Al_2O_3 \bullet 6SiO_2$	2570	
Straurolite	$Al_2O_3 \bullet FeO \bullet 2SiO_2 \bullet H_2O$		
Tourmaline	$Na(Fe,Mn)_3 \bullet 3Al_2O_3 \bullet 6SiO_2 \bullet 3BO \bullet 2H_2O$	3100	
Zircon	$ZrO_2 \bullet SiO_2$	4500	2825
	Oxides and hydrated oxides		
Rutite	TiO_2[b]	4200	2100
Mangetite	Fe_3O_4	5140	1865
Hematite	Fe_2O_3	5200	1840
Limonite	$Fe_2O_3 \bullet H_2O$	4300	675[c]
Diaspore	$Al_2O_3 \bullet H_2O$	3400	425[c]

[a]The specific gravity of silicate minerals is in the range of 2500 to 3500 kg m^{-3};it increases with Al_2O_3/SiO_2 ratio and decreases with H_2O content. The silicates containing Na,K,CA,Mg,and Fe do not have a definite melting point temperature.

[b]With the exception of rutile, the oxide minerals rarely occur in coal.

[c]Denotes loss of water.

Table 2-6 Carbonate, sulfide, sulfate, phosphate and chloride minerals in coal

Species	Chemical formula	Specific gravity (kg m^{-3})	Melting/decomposition temperature (K)
Carbonates			
Calcite	$CaCO_3$	2710	1200[b]
Aragonite	$CaCO_3$	2710	1150[b]
Dolomite	$CaCO_3 \cdot MgCO_3$	2850	1050[b]
Ankerite	$CaCO_3 \cdot FeCO_3$		1000[b]
Siderite	$FeCO_3$	3830	800[b]
Sulfides			
Pyrite	FeS_2	5000	1075[b]
Marcasite	FeS_2	4870	1075[b]
Pyrrhotite	FeS_x	4600	1300
Chalcopyrite	CuFeS	4100	1300
Melnikovite	FeS_2+(As,FeS,H_2O)	~5000	1075[b]
Galena	PbS	7500	1370
Mispickel	$FeS_2 \cdot FeAs_2$	~5000	1075[b]
Sphalerite	ZnS		
Sulfates			
Barytes	$BaSO_4$	4500	1855
Gypsum	$CaSO_4 \cdot 2H_2O$	2320	1725
Kieserite	$MgSO_4 \cdot H_2O$	2450	1395[b]
Thenerdite	Na_2SO_4	2680	1157
Mirahilite	$Na_2SO_4 \cdot 10H_2O$	1460	1157
Melanterite	$FeSO_4 \cdot 7H_2O$	1900	755[b]
Keramolite	$Al_2(SO_4)_3 \cdot 16H_2O$	1690	945[b]
Jarosite	$K_2SO_4 \cdot xFe_2(SO_4)_3$	2500	900[b]
Phosphates			
Apatite	$Ca_5F(PO_4)_3$	3100	>1500
Evansite	$3Al_2O_3 \cdot P_2O_5 \cdot 18H_2O$	2560	>1775
Chlorides			
Halite	NaCl	2170	1074
Sylvite	KCl	1980	1043
Bischofite	$MgCl_2 \cdot 6H_2O$	1570	987

[a]Calcite, dolomite, ankerite, siderite, pyrite, barytes, and apatite minerals occur frequently. Gypsum and other sulfates are found mainly in low-lank and weathered coals;other mineral species are rarely found.

[b]Denotes the decomposition temperature.

2.3 Processes of pulverized coal combustion

Combustion phenomena of pulverized coal in the furnace are illustrated in Fig.2-4. As soon as coal particles are injected into the furnace, they are preheated radiatively by the flame in the downstream. Then, volatile matter is evolved, and ignition takes place on the particle or in the gas phase. Although depending on coal type, fixed carbon is also burnt during the volatile matter evolution. With the progress of combustion, particle temperature rapidly increases. After almost all the volatile matter is evolved, char combustion becomes dominant. The rate of char combustion is relatively slow compared to that of volatile matter evolution or combustion. Finally, ash is produced after the combustion completes.

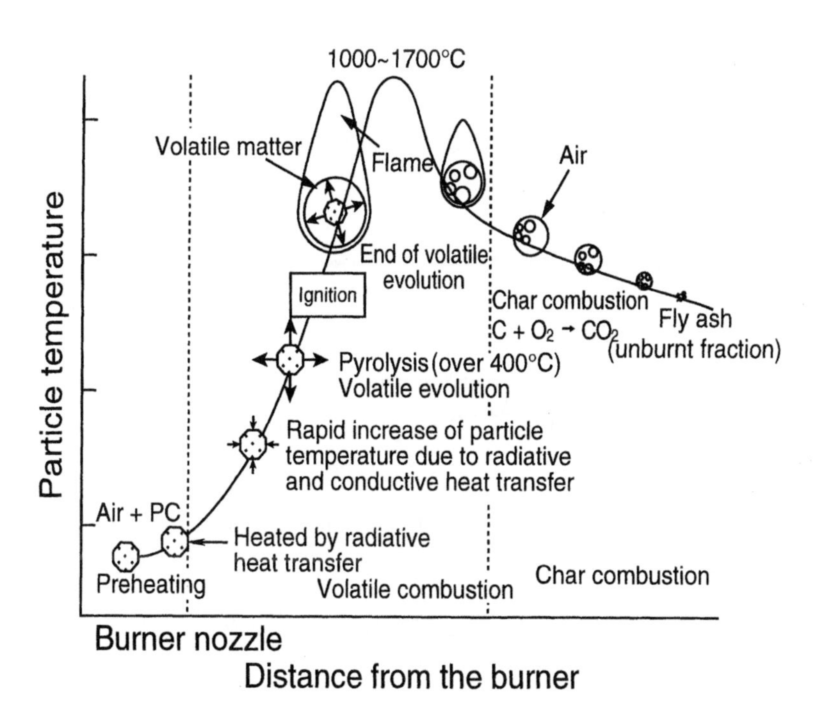

Fig. 2-4 Illustration of combustion phenomena of PC

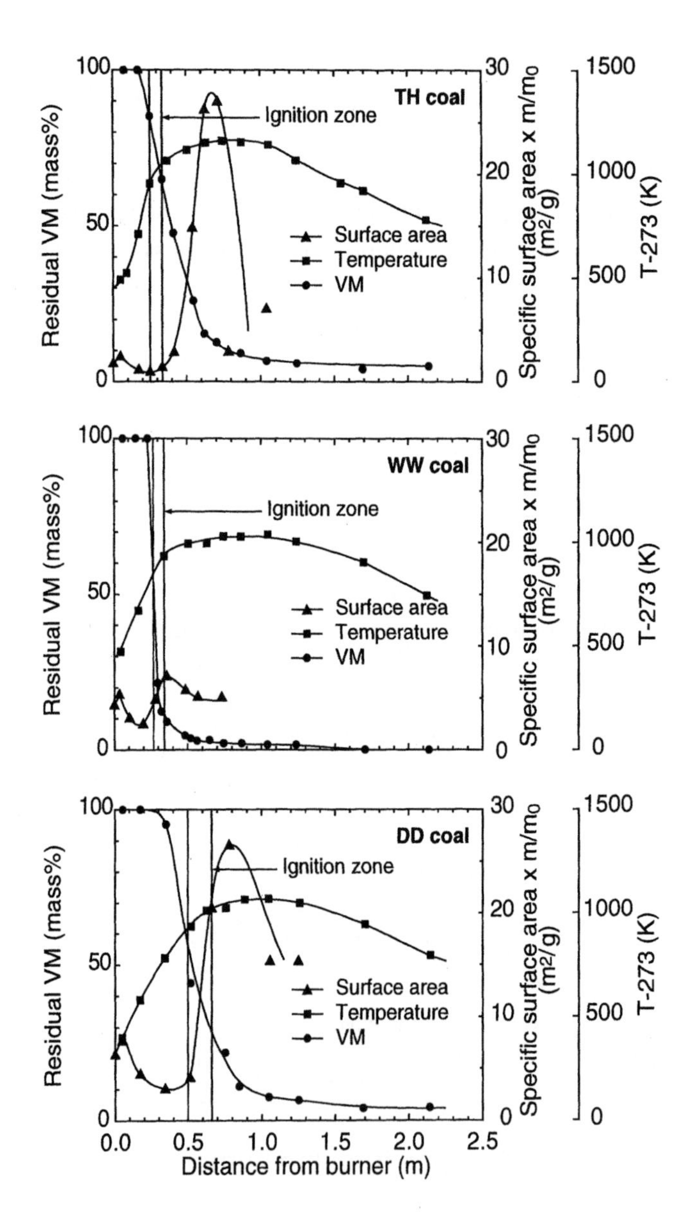

Fig.2-5 Changes of specific surface area and residual volatile matter for three types of coal along the central axis of swirling turbulent flow combustor

In order to understand such a series of phenomena as ignition, devolatilization, char combustion, burnout, pollutant emissions and ash behaviors or elucidate the respective detail mechanisms, it is necessary for a chain of phenomena to be divided into the individual phenomenon. The fundamentals on the individual phenomenon will be described in order as follows.

2.3.1 Ignition[15)]

Essenhigh et al.[16)] provide a review of ignition phenomena of coal. They suggested the ignition of coal particles could be a multi-staged process. A particle could ignite either homogeneously or heterogeneously. Small sized particles are heated up quickly and volatile matter evolves strongly and homogeneous ignition takes place. On the other hand, the temperature increase rates of large particles (larger than 100 μm) is slow (lower

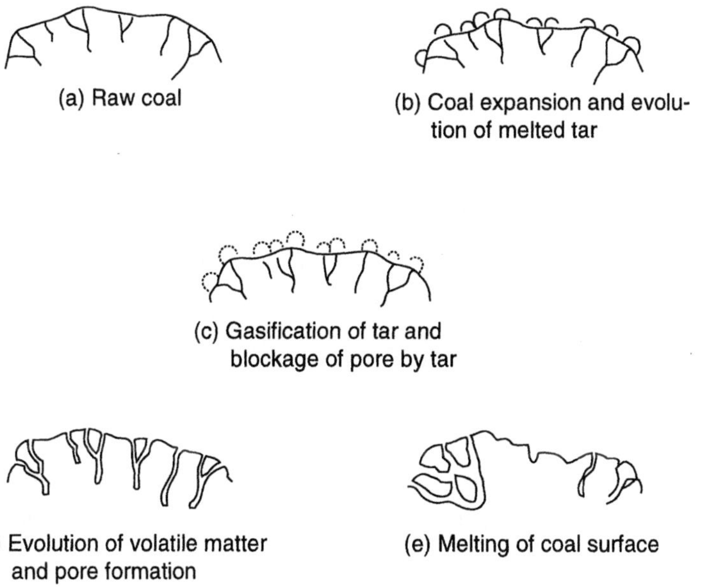

(a) Raw coal

(b) Coal expansion and evolution of melted tar

(c) Gasification of tar and blockage of pore by tar

(d) Evolution of volatile matter and pore formation

(e) Melting of coal surface

Fig.2-6 Structural change of coal surface during ignition process

than 100K/s) and heterogeneous ignition may selectively occur.

Ignition mechanisms of pulverized coal particles are important as they support the whole combustion history and the stability of the whole flame. Ignition and combustion are governed by the heat and mass transfer and chemical reactions. These phenomena are affected by the coal properties such as macerals, specific surface area, pore size, content of volatile matter and so forth. The internal structure of burning particles is also an important function to elucidate the ignition behavior.

Combustion histories of different pulverized coals were obtained by using a swirling, turbulent flow combustor and a drop tube furnace[17]. Figure 2-5[17] shows the change of specific surface area and residual volatile matter for three types of coal. These results may suggest a change of surface structure during the ignition and combustion as shown schematically in Fig.2-6[17]. The process can be explained as follows: soon after particles are introduced into furnace, raw coal (a) slightly increase their specific surface area by some expansion (b), next the area decreases rapidly through tar solidification in the pores or on the surface (c). After that, the area increases rapidly in the ignition zone and attains a maximum value as tar evaporates and new pores are formed by volatile matter evolution

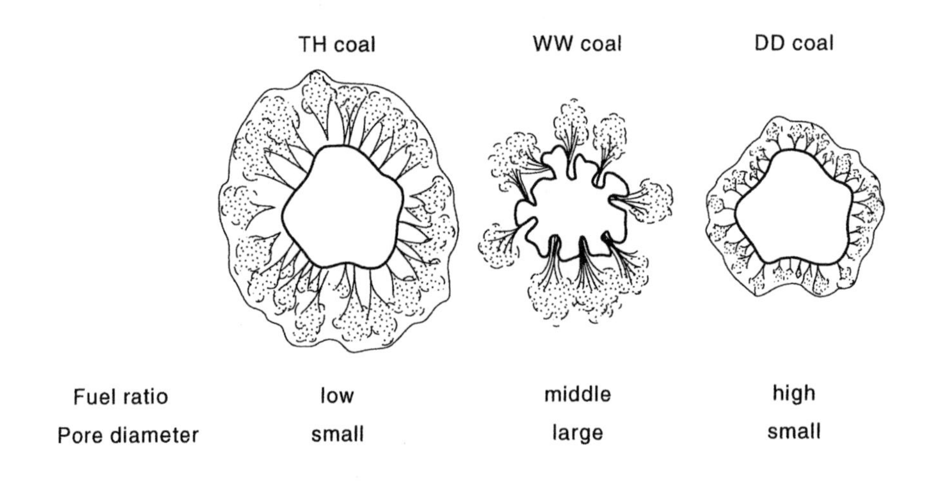

Fig.2-7 Models of flame structure around a coal particle

(d), and finally the area decreases as ash melts to fill the pores (e). The change of area is shown in Fig.2-5. This indicates that TH and DD coals show almost the same behavior, but WW coal differs significantly. In addition, from the observation of surface and char structuresTH and DD coals had many small pores on the surface and formed a network-type char, but WW coal had only large holes and formed a balloon-type char. These results suggest that there might occur a basic difference in flame structure around a coal particle as shown in Fig.2-7[17]. In case of TH and DD coals, volatile matter is evolved almost uniformly over the surface of coal particle, and an enveloping flame is formed surrounding the particle. Ignition may occur just on the surface because at that moment the surface is at the highest temperature. Once ignition occurs surface temperature increases rapidly and volatile matter near the surface evolves strongly and the flame is pushed into the gas phase. This phenomenon can be shown by theoretical calculation[18]. In the case of WW coal, volatile matter may spout out from several large holes. In this case, surface reaction could continue after the ignition because oxygen can diffuse to the surface. This case may result in higher surface temperature and ash near surface may melt and plug the pores, but the volatile matter evolves strongly from the dispersed holes and the evolution rate of volatile matter is the highest among these three coals.

2.3.2 Evolution of volatile matter

In pulverized coal combustion, volatile matter is evolved in about 100ms. Even if this process completes in a short period, evolution of volatile matter plays a role for flame stabilization and low NO combustion.

Generally, the evolution process of volatile matter depends on coal properties as described above. Very many kinds of evolution model have already been proposed. One simple modeling of volatile matter evolution is Eq.(2.1), based on the first-order reaction scheme.

$$\frac{dV}{dt} = ki\ (V^* - V) \qquad (2.1)$$

$$ki = A_0 \exp\left(-\frac{E}{RT}\right)$$

where V and V* indicate the fraction of volatile matter in burning particles and that at infinite time, respectively. E and A are activation energy and frequency factors, respectively.

2.3.3 Char combustion

In char combustion after evolving volatile matter, lump combustion in stoker boilers and coal stoves and so on, combustion behaviors inside a particle should be considered.

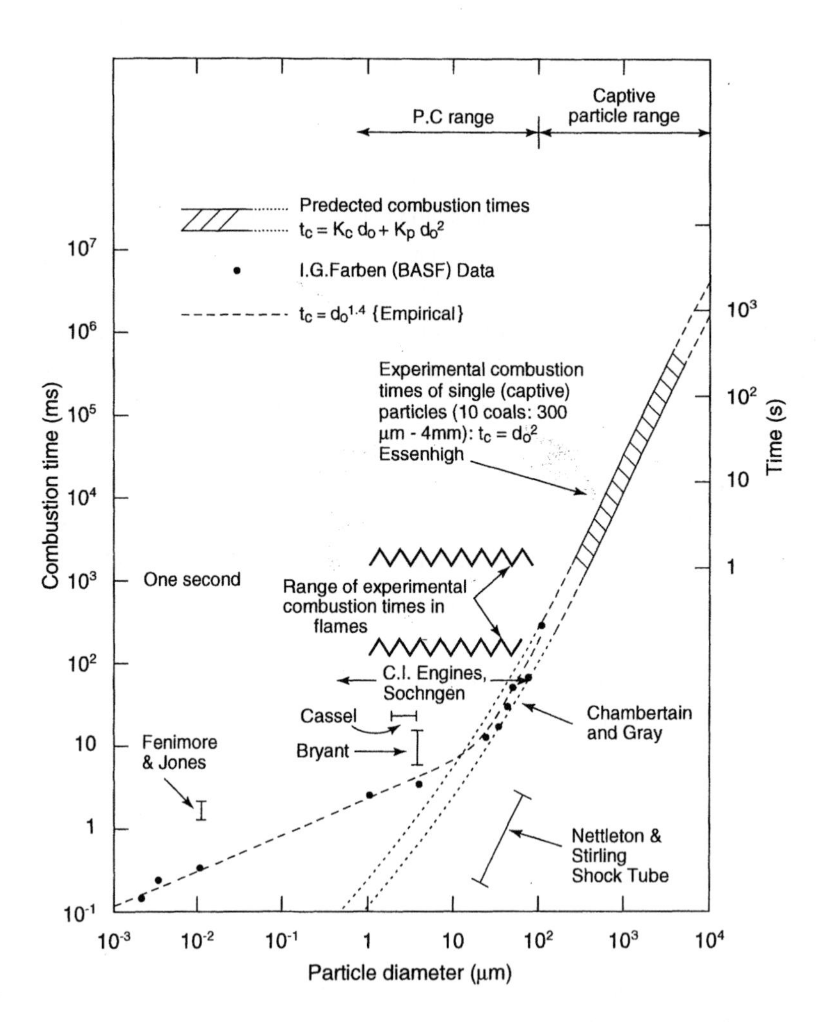

Fig.2-8 Relationship between burnout time and particle diameter of coal particle

It is also difficult to explain the detailed characteristics of combustion behavior there. Therefore, many kinds of modeling have been carried out. One simple modeling is "The Shell and Core reaction model". In this model, comparing the experimental result obtained with the result simulated, we can predict what the reaction behavior is controlled by. As a result, the reaction characteristics are classified by into three kinds of control stages, reaction control, diffusion control in boundary layer and diffusion control in ash layer.

2.3.4 Burnout

Relationship between burnout time and particle diameter of coal particles has been summarized by Essenhigh[18] as shown in Fig.2-8. Under pulverized coal combustion conditions, generally, the burnout time is proportional to the square of the particle diameter since pulverized combustion is controlled by diffusion of oxygen surrounding each particle.

2.4 Combustion characteristics of a single coal particle

Figure 3-14 shows a series of photographs of an instantaneous flame around a single particle after being ignited. These photographs are taken by a high-speed camera and show change of the flame pattern over time. They indicate how a single particle of the same coal containing volatile matter of 33.3% burns, corresponding to changes of the ambient oxygen concentration. Almost instantly upon ignition, a large flame of evolved volatile matter appear around the particle. The flame rapidly expands in size, and then shrinks and disappears, being continued by char combustion after a period of time as the volatile matter burns out. By focusing on the expanded flame, it is noted that the flame size decreases with an increase of the oxygen concentration. The combustion degree of a particle at a distance of 300mm is only 50% in an atmosphere containing 7% of oxygen, but reaches 70% in an atmosphere containing 25% of oxygen. The general combustion mechanism of a single particle can be considered as indicated in Fig.3-15. As shown in this figure, when oxygen concentration is high, the volatile matter evolved from the particle is consumed rapidly by combustion. The resultant flame is located in the immediate vicinity of the particle, further promoting its pyrolysis. When the oxygen concentration is low, as represented by Fig.3-15(a), the flame expands in size, and the generated heat tends to dissipate into the surrounding area. It may be surmised by applying these findings to particle group combustion, when the oxygen concentration is high, the individual flames tend to be compact, and the combustion of each particle is isolated. However, when the oxygen concentration is low, or the spatial concentration of particles is high, the flames tend to bulge and interfere with each other, which adversely affects the combustion rate.

2.5 Conclusion

Coal combustion technologies will become more significantly relate to energy utilization and environmental issues even in the near future. In particular, it will be necessary to develop super high efficient and clean technologies for coal combustion. In order to solve this problem, elucidation of detailed coal reaction phenomena independent of the coal type, facilities, operating conditions and so forth is required.

REFERENCES

1) C. Karr Jr. (ed.): "Analytical Methods for Coal Products", Academic Press, 1(1978), p.237.
2) P. Averitt: "Coal Resources of the United States, January 1, 1974", Geological Survey Bulletin 1412, U. S. Government Printing Office, Washington DC(1975).
3) T. A. Hendrickson (ed.): "Synthetic Fuels Data Handbook", Cameron Engineers, Inc., Denver, CO(1975).
4) J. F. Unsworth, D. J. Barratt, P. T. Roberts: "Coal Quality and Combustion Performance, an International Perspective", Coal Science and Technology 19, elsevier, (1991).
5) L. D. Smoot (ed.): "Fundamentals of Coal Combustion", Coal Science and Technology 20, elsevier, (1993), p.191.
6) C. Karr Jr. (ed.): "Analytical Methods for Coal Products", Academic Press, 1(1978), 8-13.
7) R. B. Jones, C. B. McCourt, C. Morley and K. King: Fuel, 64(1985), p.1460.
8) C.-Y. Tsai and A. W. Scaroni: Fuel, 66(1987), p.203.
9) P. Lightman and P. J. Street: Fuel, 40(1967), p.128.
10) N. M. Skorupska: Coal Combustibility, Ph.D. Thesis, Univ. of Newcastle-upon-Tyne, U.K., October(1987).
11) P. J. Street, R. P. Weight and P. Lightman: Fuel, 48(1969), p.343.

12) D. W. Van Krevelen: "Coal", elsevier, (1993), p.470.
13) I.G. C. Dryden: Fuel, 30(1951), p.217.
14) E. Raask: "Mineral Impurities in Coal Combustion", Hemisphere (1985).
15) K. Ohtake: Proc. of 3rd Int. Symp. on Coal Combustion, (1995), p.195.
16) R. H. Essenhigh, M. K. Misra and D. W. Shaw: Combust. and Flame, 77(1989), p.3.
17) Y. Yamamoto, T. Ohe, I. Naruse and K. Ohtake: Trans. Jpn. Soc. Mech. Eng., ser. B, 60(1994), p.649.
18) R. H. Essenhigh: Proc. of the 16th Symp. (Int.) on Combust., (1977), p.353.

CHAPTER 3

Combustion behavior of PC particle group

In a blast furnace, pulverized coal (PC) particles are injected into the hot blast from a lance with small diameter. The hot blast temperature is 1000 to 1200°C and the gas velocity of the hot blast is over 100-200 m/s in the blowpipe and tuyere. Under these conditions, the injected PC particles are instantaneously heated in the hot blast; subsequently, they ignite and start burning. At the same time, these pulverized coal particles are accelerated and dispersed by the high speed hot blast while burning. The combustion process of PC particle groups is based on that of a single coal particle. However, the movement of PC particle groups and their dispersing process significantly affects the combustion behavior. These all factors described above are therefore important

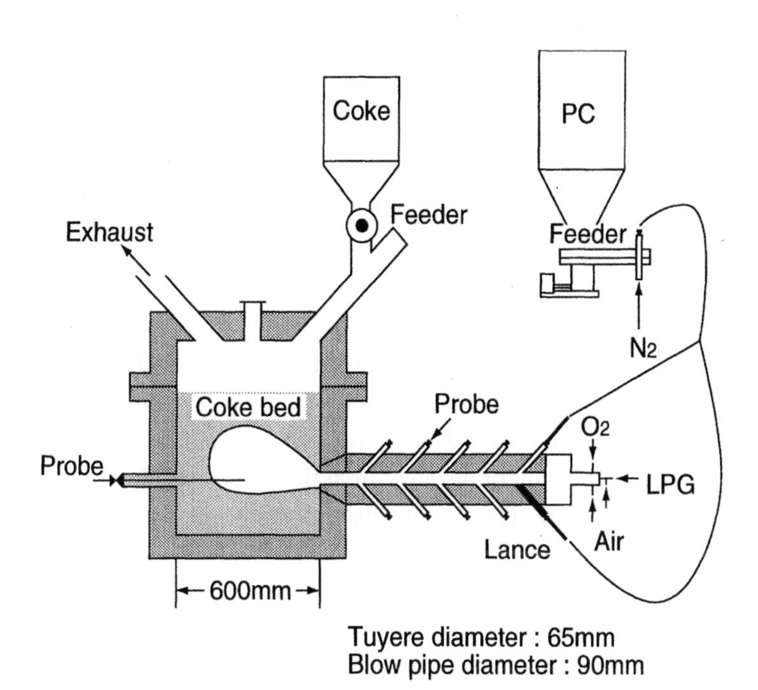

Fig.3-1 Hot model for pulverized coal combustion

for improving the combustibility of PC.

3.1 Movement of PC particle group based on direct observation

Direct observation of the movement of particle groups is difficult due to the high speed and high temperature condition of the gas flow; therefore, findings based on actual phenomenon are insufficient because of the lack of the site measurement data. A case is introduced here in which the behavior of particle groups injected from lance is directly observed by means of the hot model shown in Fig. 3-1.[1] This hot model has one tuyere and consists of a coke-packed bed and a blowpipe section.

The blowpipe used for the experiment is equipped with pairs of 40 mm diameter observation holes along both sides, as shown in Fig. 3-2. Each pair of observation holes are placed symmetrically across the axis to cancel the effect of the background. The movement of PC particle groups was photographed by a high-speed camera through these observation holes immediately after being injected from lances. The film speed was 5000 frames per second, and the exposure time of each frame was 2.5 μ s. The film was analyzed by image processor to divide the image brightness into 255 graduations. The conditions for this experiment were: a gas velocity in the blowpipe of 82 m/s, a gas temperature of 1200°C, and a PC injection rate of 200 kg/thm. Figure 3-3 shows one set of example photographs. One lance was placed above the blowpipe. The gas velocity in the injection lance was 15 m/s. These high-speed photographs that were image-processed at different levels of luminescent intensity show the high-luminescence flames clearly. Volatile matter emitted from burning PC particles decompose and generate soot, which emits light and forms high-luminescence flames. The photographic images show these combustion flames. The photographs show different images corresponding to the distance from the lance. At distances of 150 and 300 mm from the injection lance, the flames are located somewhat above the center of the view from the observation hole and do not spread over the whole field of view. This indicates that combustion proceeds in a dispersed manner within the cross section of the blowpipe. At distances of 600 and 900 mm from the lance, the combustion of PC spreads over the entire view. It is estimated that PC particles do not immediately disperse in the cross sectional direction on leaving the lance. From these observations, the flow pattern of PC particles may be illustrated as shown in Fig. 3-4; namely, PC particles are confined within a narrow stream by the high-speed hot blast and do not disperse immediately after leaving the lance. Generally, mathematical

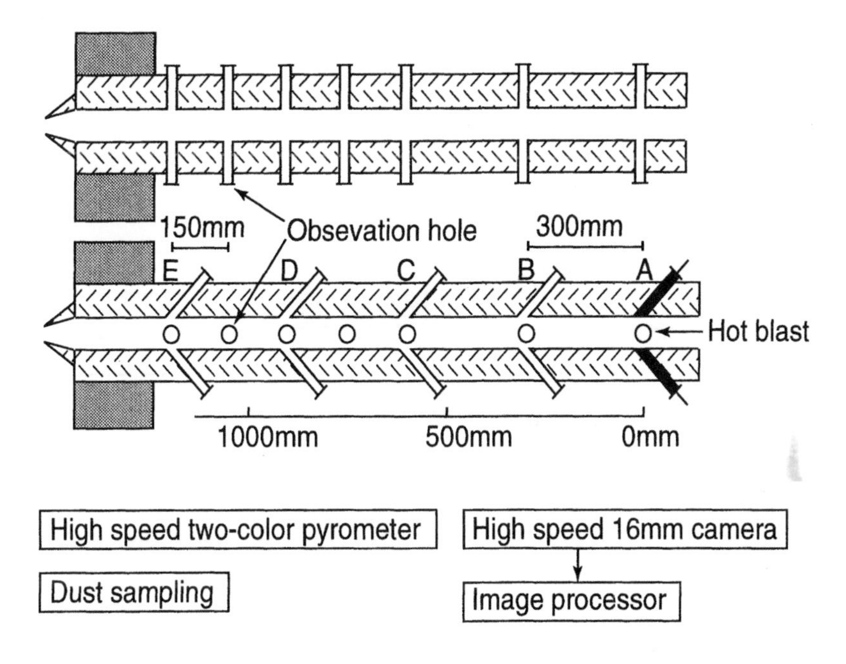

Fig.3-2 Structure of blow pipe and measuring device

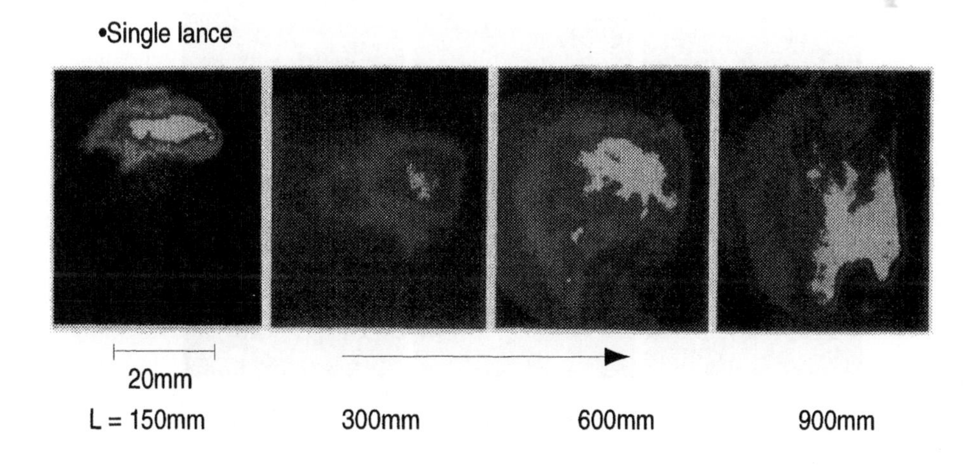

Fig.3-3 Combustion flame of pulverized coal particle group in the blowpipe

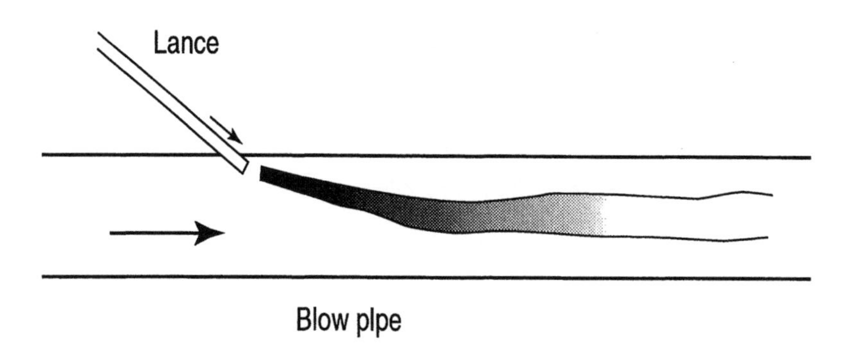

Fig.3-4 Flow pattern of pulverized coal particles injected into the blowpipe

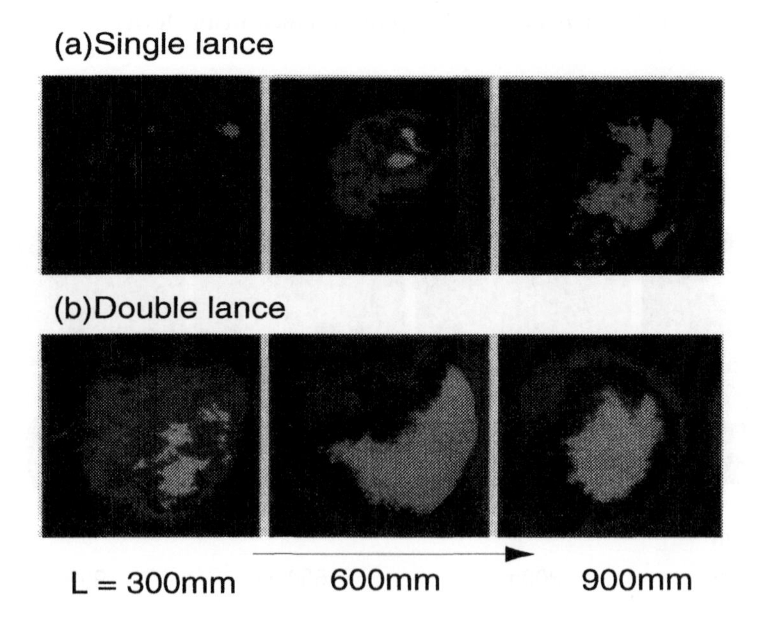

Fig.3-5 Combustion flame of pulverized coal paricle group

models for PC combustion have tended to assume a uniform distribution of PC particles in cross sectional area, but it may be noted from the above observation that these assumptions do not agree with the actual conditions.

Figure 3-5 compares the photographs of the combustion flames when double lance was used at the same experimental facility to those for a single lance. The combustion flames in the double lance expand to the entire view immediately after PC particles is injected from the lances. By using multiple injection points, a uniform flow of PC can be obtained across the entire cross section. PC particle groups do not easily disperse in the cross sectional area of the blowpipe and tuyere because they are restricted by the high-speed gas flow. Therefore, the movement of PC particles is significantly affected by their starting point, and their non-uniform distribution could be improved by an appropriate arrangement of lances. This provides an important view to achieve better combustibility of PC: improving the arrangement of lances can more effectively utilize the oxygen in the hot blast.

Table 3-1 Experimental conditions

• Experimental conditions of hot model.

Gas volume	350 Nm^3/h
Blast temperature	1200°C
Gas velocity in blow pipe	82 m/s
Coal injection rate	48 , 65 kg/h
Oxygen content	21%
Fuel equivalence ratio	1,03,0,78
PCR	150,200 kg/thm
Size distribution of PC	-74μm,80%

• Proximate analysis of coal (dry base)

VM (%)	FC (%)	Ash (%)
33.2	57.0	9.8

3.2 Combustion of PC in the blowpipe

3.2.1 Movement of PC particle group and its combustion behavior

Combustion efficiency and combustion temperature were measured using the hot model shown in Fig. 3-1. Table 3-1 gives the conditions for the experiment. The combustion efficiency was determined by the consumption rate of combustible matter in the coal, using the ash content as a basis. Figure 3-6 shows the measured combustion efficiency. The combustion efficiency increases rapidly within 600 mm from the injection point. In this range, combustion of the volatile matter released by rapid pyrolysis of PC probably plays the major role. In comparing the two types of lance, the combustion efficiency increases rapidly in the first half of the combustion process in the case of double lance and reaches approximately 60% at a distance of 600 mm. In the case of a single lance, in contrast, the combustion efficiency reaches only 40% at the same distance. After this distance, char takes the place of volatile matter in combustion, so the combustion

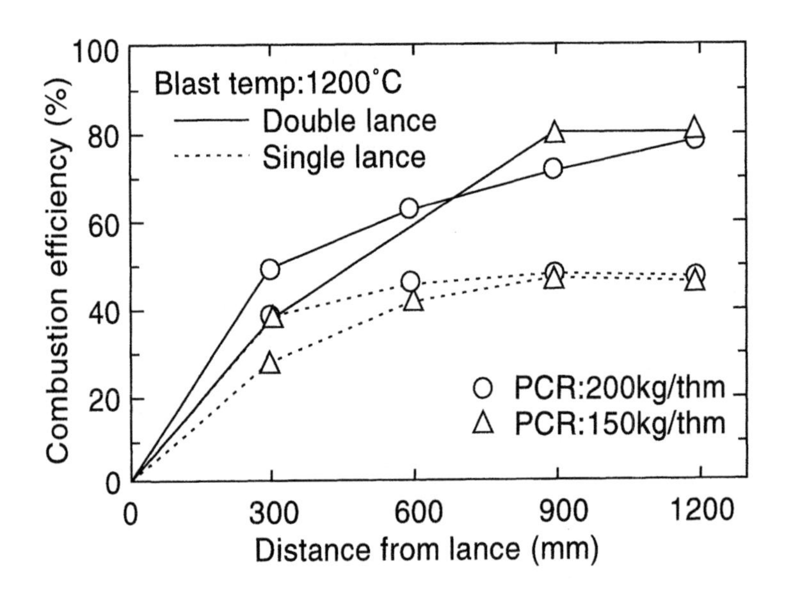

Fig.3-6 Combustion degree of pulverized coal

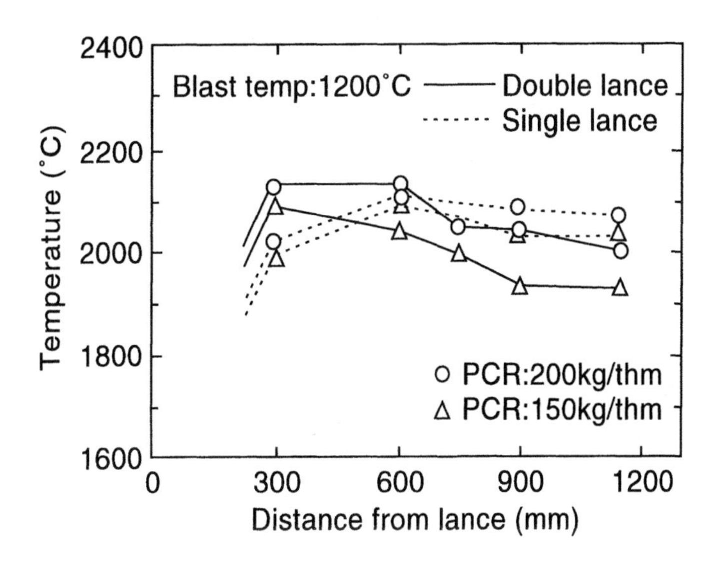

Fig.3-7 Temperature change of combustion flame measured by high-speed two-color pyrometer

efficiency increases much slower and in a similar manner for both cases. The effect of the injection rate is not clear in either case. The effect of the number of lances, double or single, is most noticeable in the range of 300 to 600 mm from the lance.

Figure 3-7 shows temperature measured by high-speed, two-color pyrometers (wavelengths of 0.85 and 1.0 μm, and response times of 0.5 ms). Then, it is considered that the observed temperature indicates temperature of the luminous flame observed in Figs. 3-3 and 3-4. For the double lance, the temperature reaches about 2100°C almost instantaneously at a point 300 mm from the lances, while for the single lance, the temperature rises slower and attains the maximum at a point 600 mm from the lance. The change in temperature corresponds to that in the combustion efficiency mentioned above. It is therefore evident that the difference in the flow pattern of PC particle groups, as observed by high-speed photograph, is the predominant factor governing combustibility in the range of 300 to 600 mm from the lance.

Furthermore, the gas compositions immediately downstream of the lance were

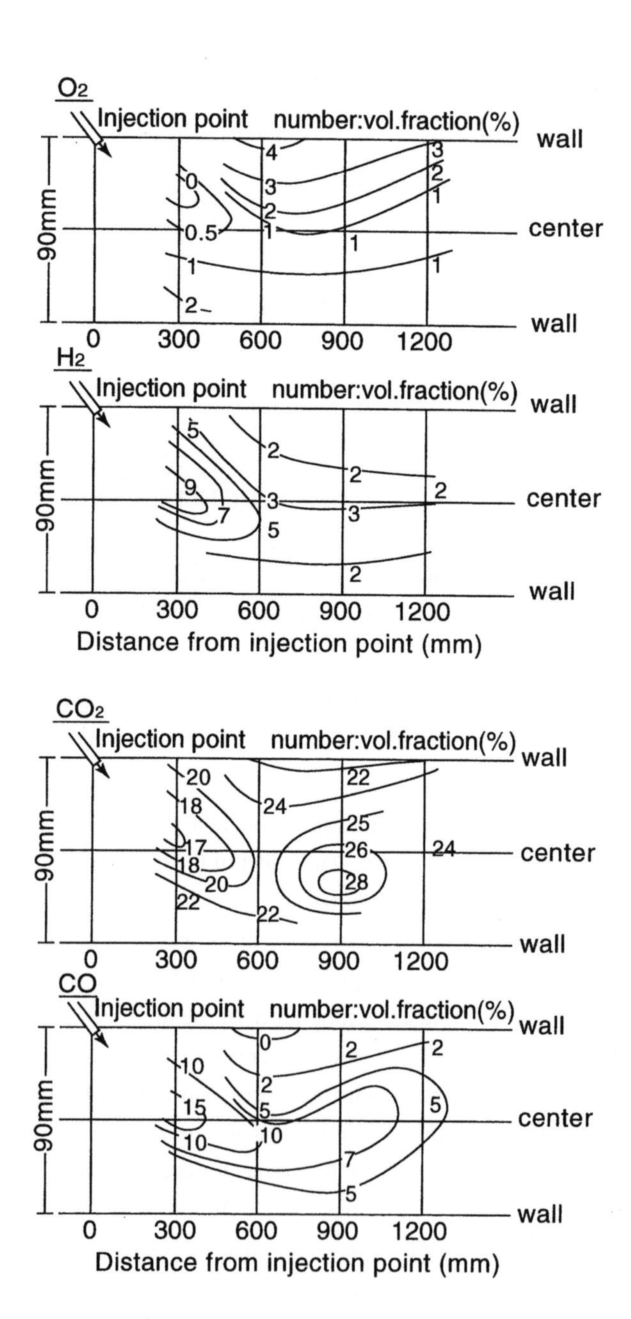

Fig.3-8 Gas concentration change nearby injection lance

analyzed at the same facility. The results are shown in Fig.3-8[2] These results indicate that the gas composition varies widely across the blowpipe immediately after the lance and demonstrate the large effect that flow patterns of PC particle groups have on combustion.

3.2.2 Effect of coal type

Kugisaki et al.[3] conducted an experiment using the experimental combustion apparatus shown in Fig. 3-9 to clarify the structural change of PC with combustion by injecting PC at a rate of 5 kg/h into vitiated air. Figure 3-10 shows the conversion efficiency of the volatile matter (VM) and fixed carbon (FC), and the changes in specific surface area when the hot blast temperature was varied. The coal samples used were PR coal (VM 39.2 %, FC 63.5 %, ash 2.5 %, moisture 4.8 %) and K-9 coal (VM 18.8 %, FC 70.2 %, ash 10.4%, moisture 0.6 %). It is noted from Fig. 3-10 that, in the case of PR coal, the rise in hot blast temperature increases the volatile matter emission rate, but does not affect the conversion of the fixed carbon. It is also noted that the specific surface area, which affects the flame pattern around the particle, increases as volatile matter is emitted. The increase in the hot blast temperature further increases the specific surface area. In

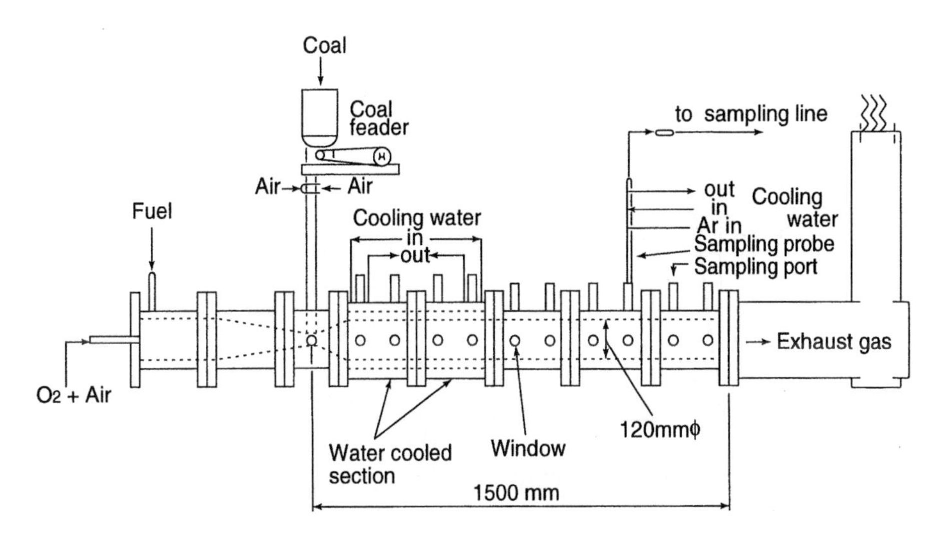

Fig.3-9 Experimental apparatus for high temperature combustion

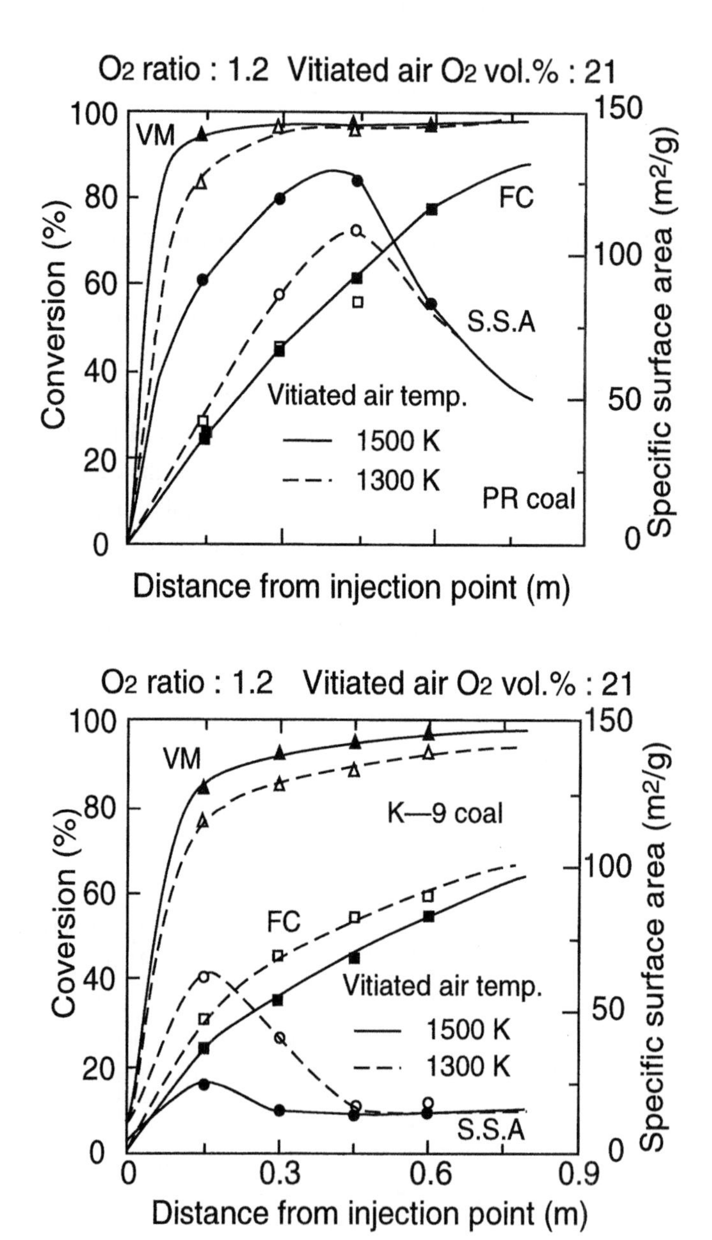

Fig.3-10 Influence of gas tempeature on volatile matter and fixed carbon conversions

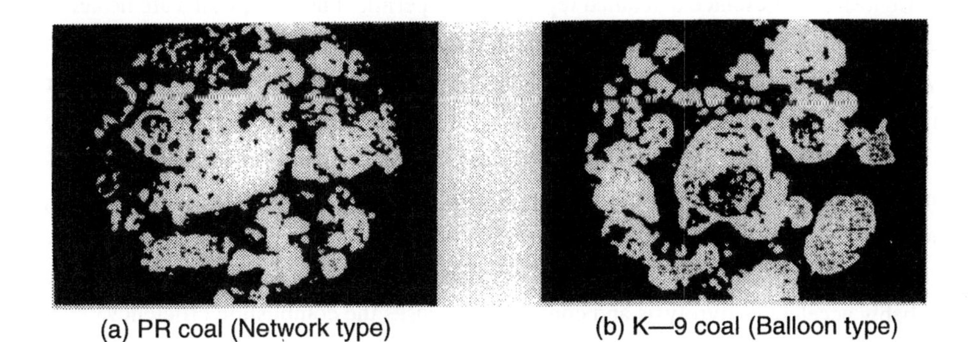

(a) PR coal (Network type) (b) K—9 coal (Balloon type)

Fig.3-11 Cross-sectional structure of burning particles

the case of K-9 coal, the rise in hot blast temperature increases the volatile matter emission rate, although not as much as in the case of PR coal, but reduces the conversion of fixed carbon. It also decreases the specific surface area. This is probably because the sample of K-9 coal contained more low melting point ash, and the molten ash clogs the pores in the particle surface.

The difference in the behavior of the specific surface area between the two kinds of coal may be attributable to the difference in char structure. The photographs in Fig. 3-11 show the structures of both coal particles after combustion. The char particles from PR coal show a network-type structure, and thereby tend to have a large surface area, while many of those from K-9 show a balloon-type structure. These observations suggest that char structure may be a factor that affects the ability to ignite PC, or ignitability, which is an index for the reactivity of PC. The flame structure around a PC particle may be illustrated as shown in Fig. 2-7.

The combustion process of PC varies depending upon the type of coal, or more specifically, the fuel ratio (fixed carbon/volatile matter). The ambient temperature generally affects the volatile matter emission rate, but its effect upon the reaction efficiency of fixed carbon may vary with the char structure and ash content. These differences affect the flame structure of PC and are therefore considered to be important when the behavior of coal particle groups is analyzed.

3.2.3 PC pyrolysis just after injection

Ueno et al.[4] presented a detailed report of an experiment using a plasma arc heater to determine the effects of coal type, ambient temperature and oxygen concentration on the pyrolysis of PC. They specifically investigated the pyrolysis of PC immediately after injection and measured changes in the composition of gas generated by pyrolysis. Figure 3-12 shows the results of the measurements and indicates that the generated gas consists mainly of CH_4, CO, and H_2 in the lower temperature zone, but that the emission of H_2 increases in the higher temperature zone. The generated gas burns with oxygen in the blast at the tuyere of a blast furnace, raising the temperatures of gas and particles. This probably accelerates pyrolysis and consequently raises the combustion efficiency.

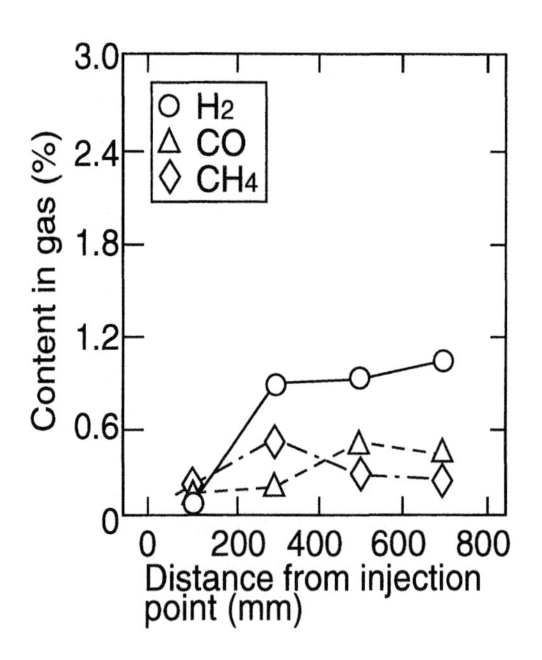

Fig.3-12 Generated gas in a stage of pyrolysis (blast temp.1250°C)

3.2.4 Effect of oxygen concentration

Figure 3-13 shows how the combustion behavior of PC varies when the oxygen concentration is increased widely from zero to 21% at a blast temperature of 1250°C. A proximate analysis of the coal showed that the volatile matter, fixed carbon and ash contents were 32.4%, 57.2% and 10.4%, respectively. The results of the test indicate that the combustion reaction between the gas generated by pyrolysis and oxygen raises the gas temperature and combustion efficiency. A rise in the ambient temperature may promote the rate of pyrolysis and consequently accelerate combustion in its initial stage. However, the effect of oxygen concentration on the ultimate combustion efficiency throughout the entire combustion process over the raceway is not necessarily large, as explained in 3.4.

3.2.5 Effect of additives to coal

It is reported that some additives like $CaCO_3$ or $KMnO_4$ to coal improve pulverized coal combustibitilty.[5)6)] Shen et al.[5)] confirmed that the burning time of coal can be shorten with the addition of $KMnO_4$. However, it is not clear that these behavior are

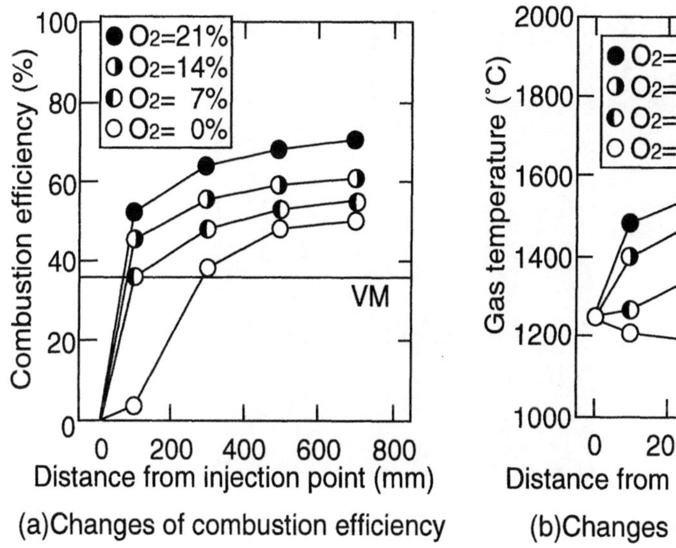

(a)Changes of combustion efficiency

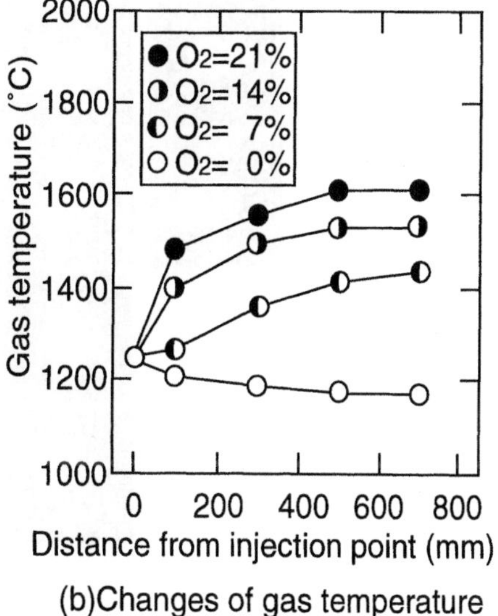

(b)Changes of gas temperature

Fig.3-13 Influence of oxygen concentration on combustion behavior

based on the oxidizing agent of additives or their catalytic effects.

3.3 Combustion mechanism of PC particle group in the blowpipe

3.3.1 Combustion of a single particle constituting a PC particle group

It was mentioned in the previous section that the movement of a particle group injected

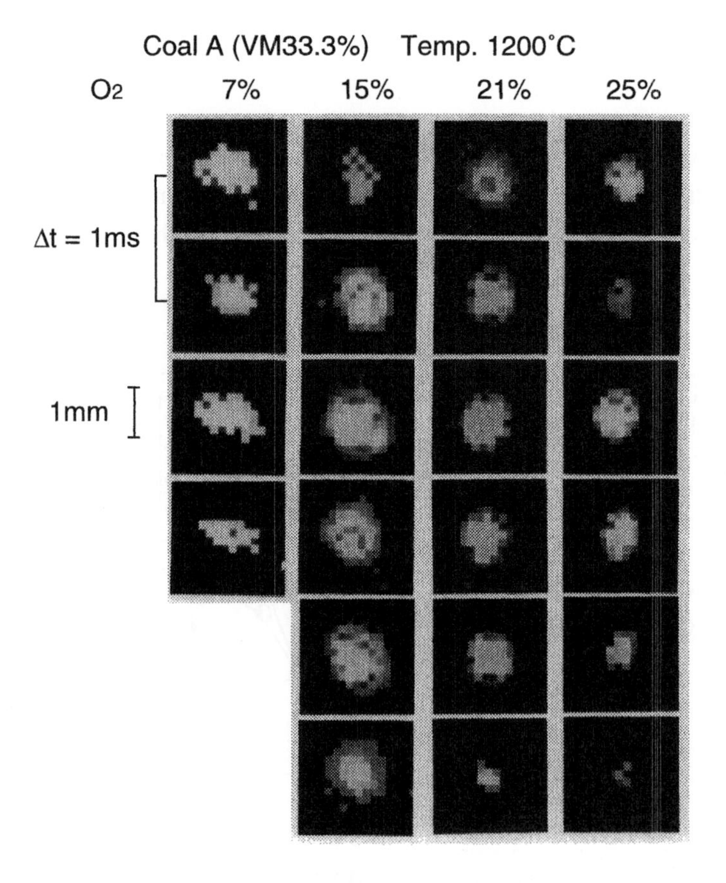

Fig.3-14 Combustion flame of a single coal particle

in a high-speed blast greatly affects its combustion characteristics. This section explains the combustion behavior of individual particles which constitute a particle group and collectively exhibit the combustion characteristics of the particle group.

Chapter 2 explains how a single particle burns. Direct observation of a single particle in a laminar-flow furnace has already been made with a view to understanding the behavior as a group.[7] Figure 3-14 shows a series of photographs of flames around a single particle immediately after being ignited. These photographs were taken by a high-speed camera and show changes of the flame patterns over time. They indicate how a single particle of the same coal containing volatile matter of 33.3% burns corresponding to changes in the ambient oxygen concentration. Almost instantly upon ignition, large flames of emitted volatile matter appear around the particle. The flames rapidly expand in size, and then shrink and disappear, being replaced by char combustion after a period of time as the

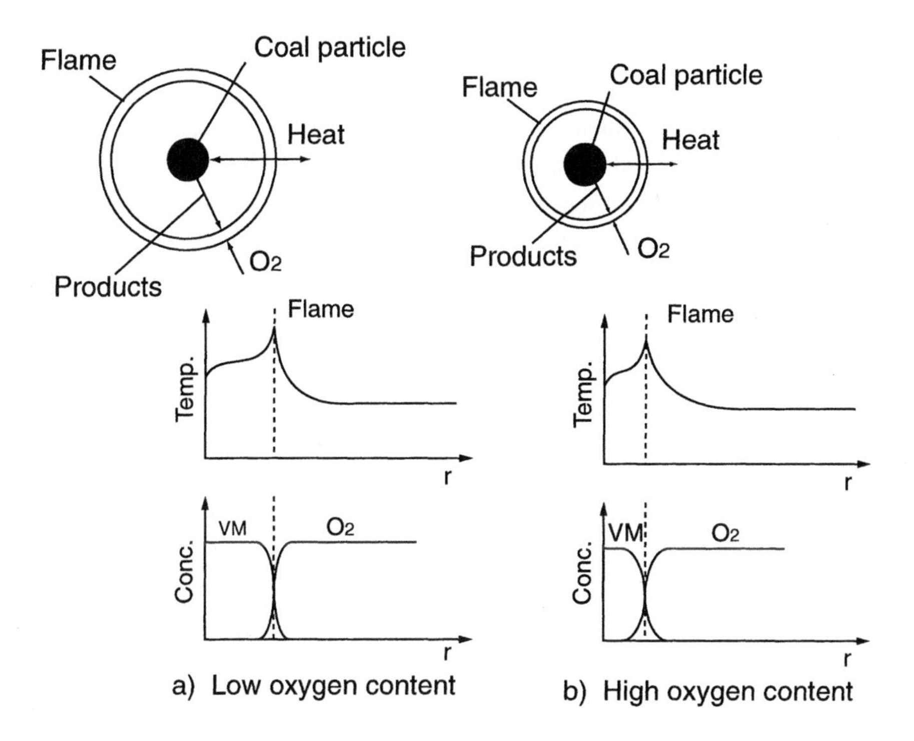

Fig.3-15 Combustion mechanism of a single coal particle

volatile matter burns out. By focusing on the expanded flames, it was noted that the combustion flames become smaller as oxygen concentration increases. The combustion efficiency of a particle at a distance of 300 mm is only 50% in an atmosphere containing 7% oxygen, but reaches 70% in an atmosphere containing 25% oxygen. The combustion mechanism of a single particle may be generally considered as indicated in Fig. 3-15. As shown in Fig. 3-15 (b), when oxygen concentration is high, the volatile matter emitted from the particle is consumed rapidly by combustion. The resultant flame is located in the immediate vicinity of the particle. The combustion heat is rapidly transferred to the particle, further promoting its pyrolysis. When the oxygen concentration is low, as represented by Fig. 3-15 (a), the flame expands in size, and the generated heat tends to dissipate into the surrounding area. It may be surmised by applying these findings to a particle group that, when the oxygen concentration is high, the individual flames tend to be compact, and the combustion of each particle is isolated. However, when the oxygen concentration is low, or the spatial concentration of particles is high, the flames tend to bulge and interfere with each other, which adversely affects the combustion rate.

3.3.2 PC group combustion

Figure 3-16 shows combustion patterns of localized particle groups.[8] When PC particles are sparsely dispersed in the space as shown on the left half of Fig. 3-16, or when the oxygen concentration is high, individual particles can burn rather independently. As the concentration of PC particles becomes more dense, the generation of combustion heat per unit volume of the space increases, leading to accelerated combustion. If the concentration of PC particles becomes even denser, and the stoichiometric oxygen ratio, which is an indicator of the oxygen concentration in combustion, becomes lower than 1.0, the flames tend to swell out. This was observed in the experiment using a laminar-flow furnace. Combustion of particles can then interfere with each other in the vicinity of the particle group, where the oxygen supply rate determines the combustion rate. Therefore, the combustion rate is expected to decrease, as shown in this figure.

The cross section of a particle group may be as shown in Fig. 3-17. When highly concentrated oxygen is supplied, the individual particles constituting the group behave as if they were independent of each other and burn vigorously as a single particle, as internal group combustion shown in Fig. 3-17 (a). When insufficient oxygen is supplied, or when the particles are too close to each other to supply sufficient oxygen within the given space, a single combined flame as external group combustion is formed. In this

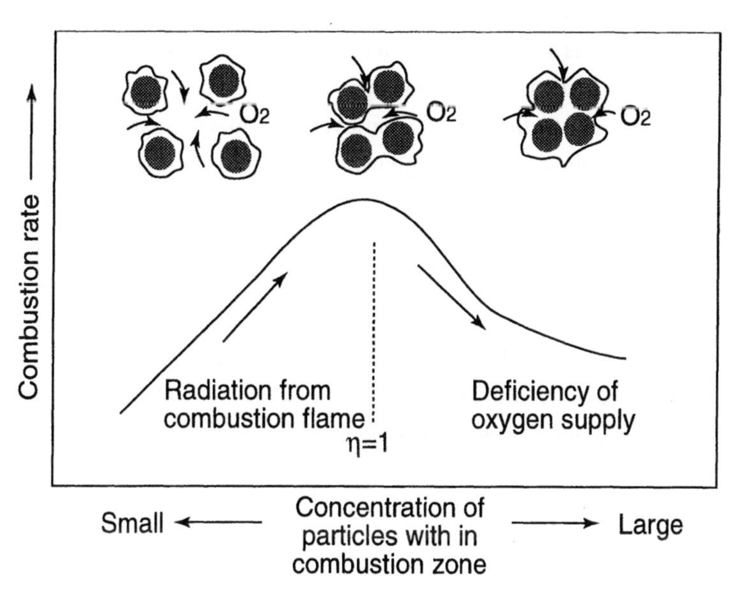

Fig.3-16 Relationship between coal particle concentration and combustion rate

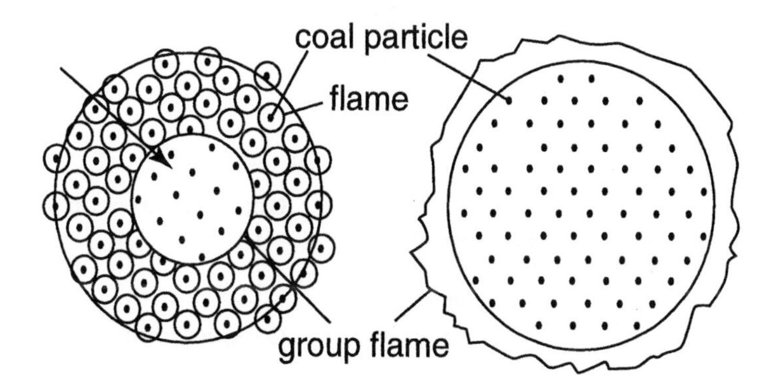

Fig.3-17 Group combustion mode of PC particles

case, the combustion rate tends to decrease. The injection of a large amount of PC (i.e., greater than 200 kg/thm) lowers the stoichiometric oxygen ratio to below 1.0. Control of the dispersion of particles is very important in assuring good combustion efficiency.

Figure 3-18 comprehensively shows the dispersion of PC particles resulting from the movement of particle groups in the blowpipe or tuyere of a blast furnace. The dispersion is indicated in terms of cross sectional distribution of the PC particle and oxygen concentration. Generally, as shown in Fig.3-18 (a), PC particle groups injected from a single lance form a confined stream, in which combustion rapidly proceeds and rapidly consumes oxygen. Oxygen around the stream remains unconsumed and is not effectively used for combustion because the high-speed blast prevents sufficient surrounding oxygen from being supplied into the particle stream. Under such conditions, combustion proceeds to a certain extent immediately after the PC particles are injected. However, combustion then tends to stagnate because of the insufficient availability of oxygen inside the PC

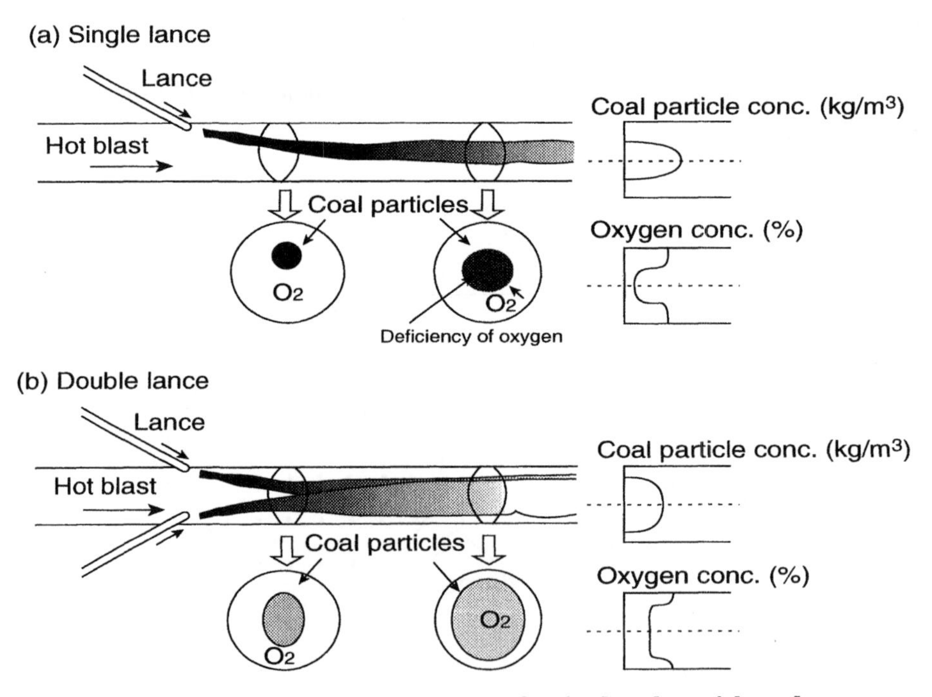

Fig.3-18 Influence of lance arrangement on pulverized coal particle and oxygen concentration

$$W_p = S(1 - \varepsilon)\rho_p u_p \quad \text{------------------------------------} \quad (1)$$

$$du_p / dL = (3/4)(C_d / D_p)(\rho_g / \rho_p)|u_g - u_p|(u_g - u_p)/u_p \quad \text{-------------} \quad (2)$$

W_p : pulverized coal injection rate (kg/s)
S : area including pulverized coal (m^2)
u_p : particle velocity (m/s)
u_g : gas velocity (m/s)
ε : void fraction in the area including pulverized coal (—)
ρ_p : particle density (kg/m^3)
ρ_g : gas density (kg/m^3)
C_d : drag coefficient (—)
D_p : particle diameter (m)
L : distance from injection point (m)

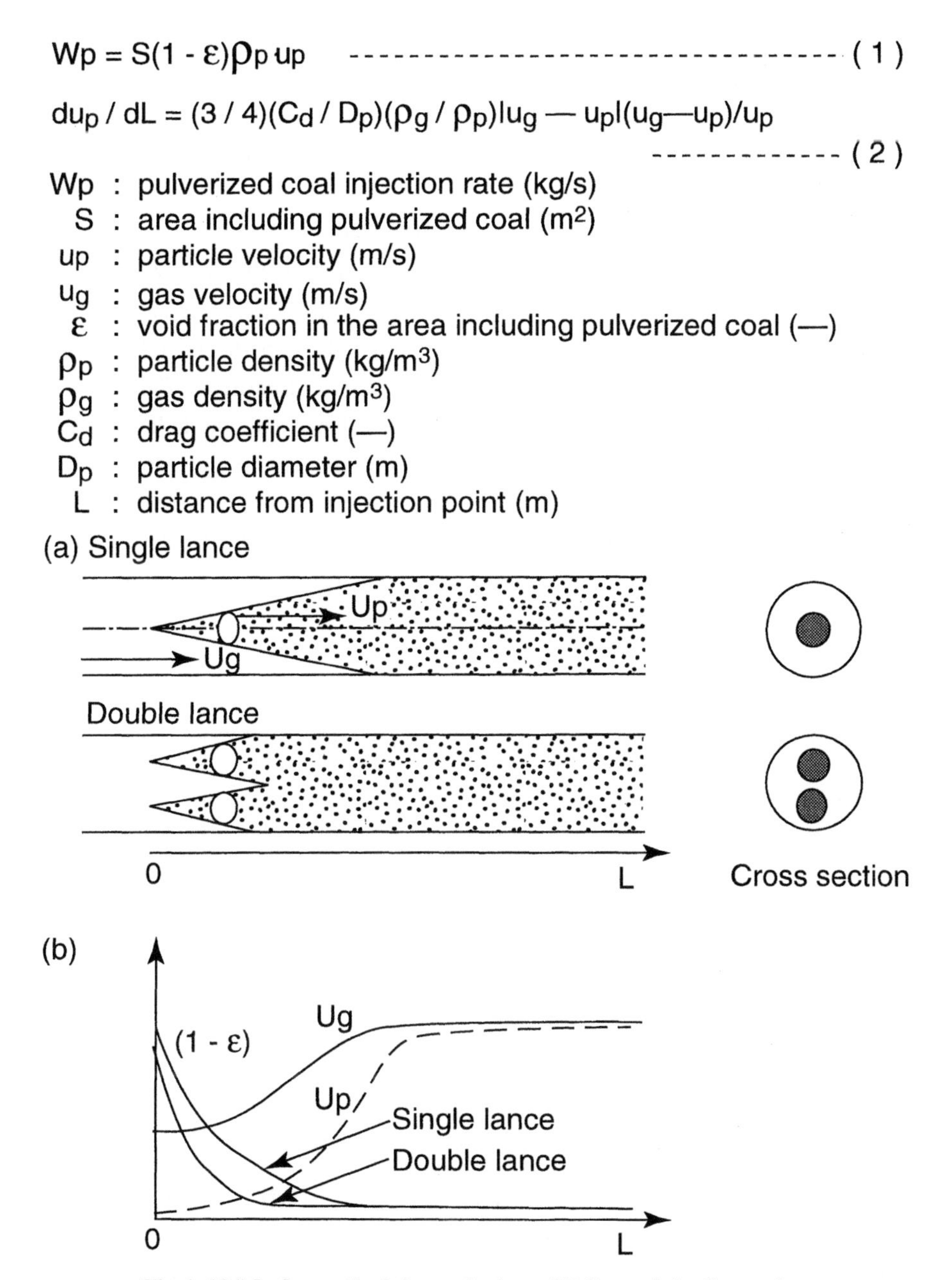

Fig.3-19 Mathematical formulation of PC particle dispersion

particle group, as shown in the right half of Fig. 3-16 or Fig. 3-17 (b). Therefore, control of the movement of PC particle groups is very important. With double lance, a more uniform cross sectional dispersion of particles is attainable compared to a single lance for the same amount of PC injection. The average particle-to-particle distance is longer, and the oxygen is more effectively utilized, resulting in a higher combustion rate.

3.3.3 Mathematical formulation of PC particle dispersion

Chapter 5 explains in detail how a mathematical model treats the movement of PC particle groups. This section explains how to establish a mathematical equation to determine the particle-to-particle distance, in recognition of the importance of the spatial particle concentration as a factor governing the combustion of particles. Figure 3-19 shows approaches to mathematical formulation for single lance and double lance.[1] As shown in this figure, the model assumes that PC particles disperse at a pre-determined angle from a given point for both cases. The PC particles disperse not only in the transverse direction, but also in the longitudinal direction, as they are accelerated by the blast. Figure 3-19 (b) shows the distribution of the void fraction at each cross section. Figure 3-20

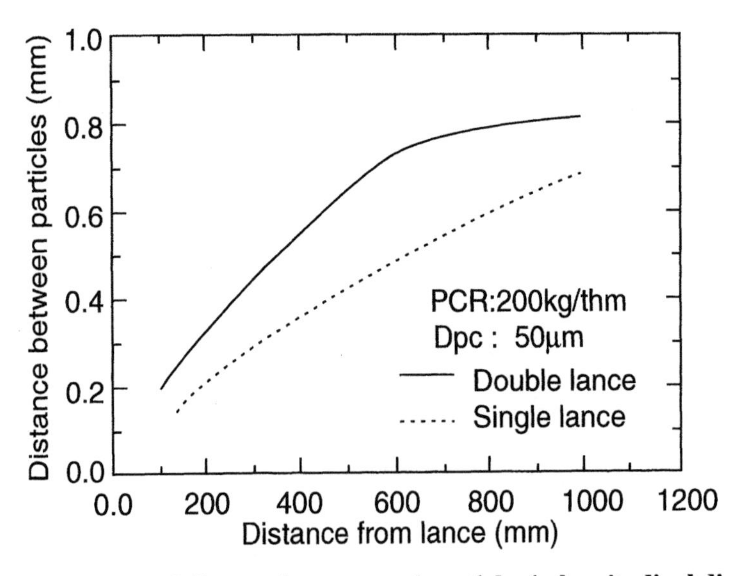

Fig.3-20 Change of distance between coal particles in longitudinal direction

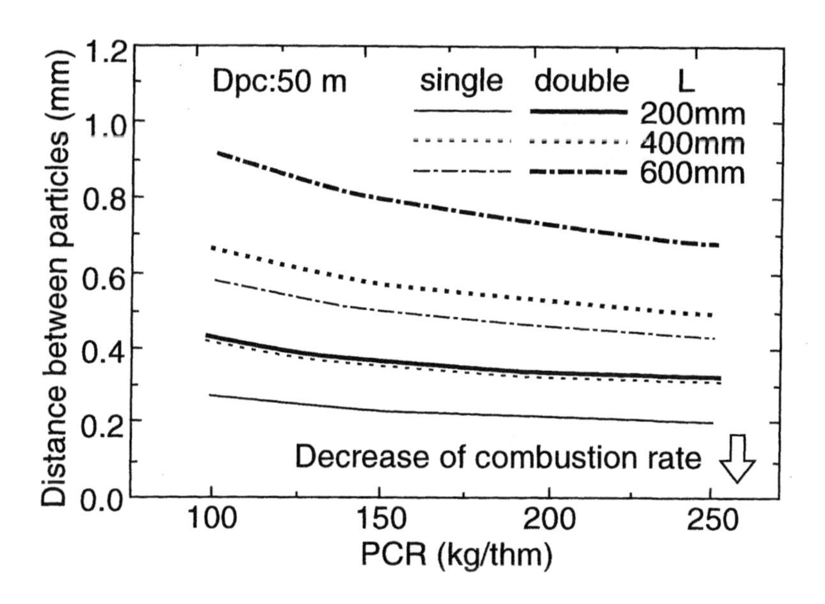

Fig.3-21 Influence of injection rate and lance arrangemant on distance between coal particles (L:distance from lance)

shows the calculated particle-to-particle distance as a function of distance from the lance.[1] This figure shows that the PC particles are dispersed both in the cross sectional and longitudinal directions by the high-speed blast. However, when injected by more than one lance, such as with double lance, they can disperse more widely and have a greater initial particle-to-particle distance at any given point. Figure 3-21 shows particle-to-particle distance as a function of the injection rate. The particle-to-particle distance decreases as the injection rate increases, forming combustion zones that are more concentrated. As explained previously, this will decrease the combustion rate. Here again, the use of more than one lance could increase the particle-to-particle distance, creating a more favorable condition for increasing the combustion rate.

3.4 Combustion of PC in the raceway

In the raceway, the combustion of PC competes with that of coke, and the combustion behavior of PC becomes more complicated. Experimental analyses have been done on the combustion characteristics in the raceway. In an experiment, the hot model shown in Fig. 3-1 was used to measure gas composition in the raceway. At the same time, the two-dimensional temperature distribution was measured using optical fibers.[9] The optical fibers used in this experiment were covered with metal tube and inserted into the raceway through a water-cooled probe. The moment the optical fibers are inserted, thermal radiation from the solid is detected and measured as a temperature. The solid temperature in the raceway is measured in this way. Further, a two-dimensional temperature distribution is measured by rotating the tip of the probe. Figure 3-22 shows the results of such

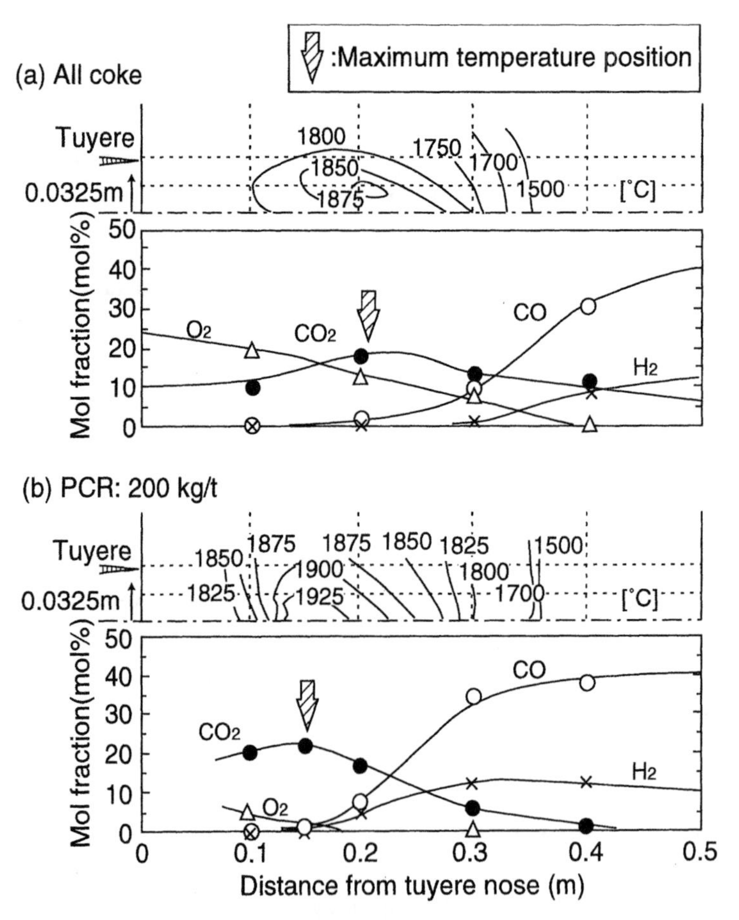

Fig.3-22 Temperature and gas distribution in raceway measured in hot model

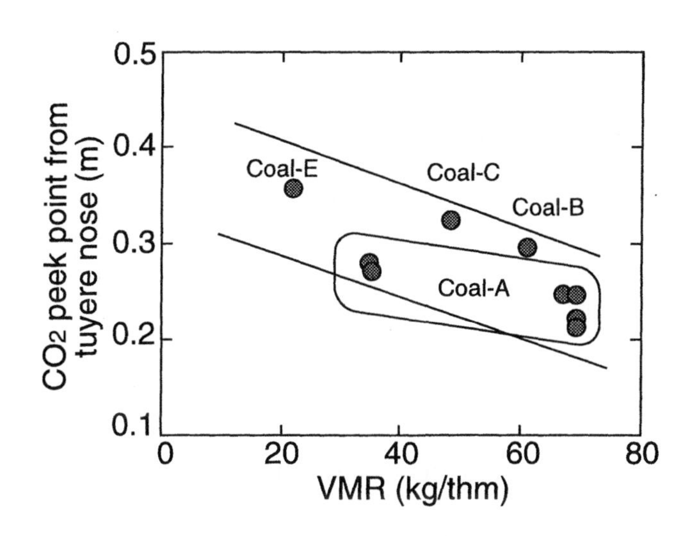

Fig.3-23 Influence of volatile matter on CO_2 peak point in raceway

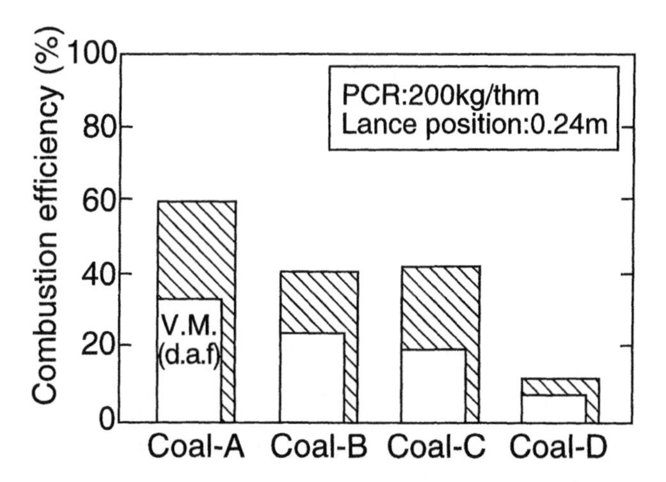

Fig.3-24 Influence of coal species on combustion degree

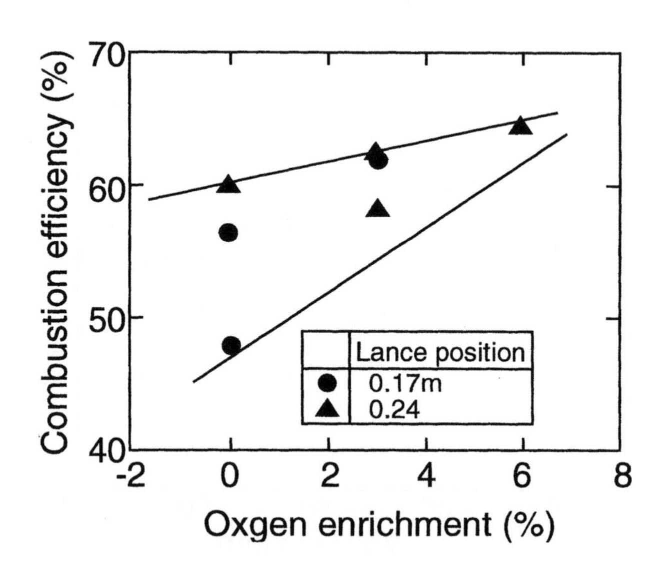

Fig.3-25 Influence of oxygen enrichment and lance position on combustion degree

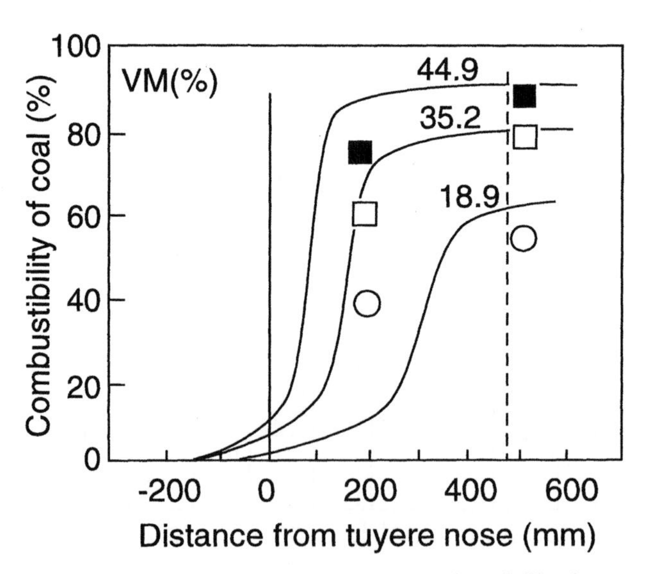

Fig.3-26 Influence of volatile matter content on combustibility in raceway

measurements. The highest temperature is situated approximately on the axis of the tuyere. Oxygen in the raceway is rapidly consumed as PC is injected, and the point of maximum CO_2 concentration coincides with that of the highest temperature. In the case of all-coke, the gas composition and temperature distribution are determined by the contact between hot coke and oxygen entering the raceway. When PC is injected, the volatile matter generated in the initial period of PC combustion burns very rapidly, and the combustion characteristics of the PC govern the gas composition and temperature distribution in the raceway. The relation between the location of combustion focus (i.e., the point where the concentration of CO_2 becomes highest) and the PC injection point was studied by widely varying the PC properties (volatile matter ranging from 24.3 to 34.9%) and the injection rate.[9)] Figure 3-23 shows the location of CO_2 peak position versus the input of volatile matter, VMR (VM (%) x Injection rate(PCR) (kg/thm) x 100). The figure indicates that the location of the CO_2 peak is greatly influenced by the input of volatile matter. For a given type of coal, as the PC injection rate increases, the consumption of oxygen increases, and the combustion focus moves closer to the tuyere. As the volatile matter content in PC increases, the combustion focus moves closer to the tuyere.

Figure 3-24 shows the relationship between the combustion efficiency and the type of coal [9)]. The combustion efficiency is obtained by the coke replacement rate. It is seen from the figure that coal types with higher volatile matter achieve a higher combustion efficiency. Figure 3-25 indicates the change in combustion efficiency with respect to oxygen enrichment and lance position. When the lance is located 0.24 m from the tuyere, the combustion efficiency increases gradually by oxygen enrichment. However, the increase is not significant: only 6% at 6% oxygen enrichment. This test did not produce a clear result as to the effect of the lance position. Considering that the starting point of combustion is determined by the lance position, the combustion efficiency achievable after passing through the raceway is probably affected by the lance position.

Iwanaga et al.[10)] conducted a similar combustion experiment using a coke-packed bed and measured the combustion efficiency of PC samples taken from the furnace. Figure 3-26 shows the result of the experiment. It shows that the volatile matter content of PC greatly affects the attainable combustion efficiency, as expected. However, the combustion efficiency attained was much greater than that obtained previously. The method of evaluating the combustion efficiency after passing through the raceway, including calculation methods, requires further study.

3.5 Conclusion

On the basis of recent studies, the mechanism of PC combustion in the blast furnace conditions has been clarified. However, measurements of the behavior of PC particle groups during combustion have not been sufficient to provide any direct relationship to the total combustion mechanism. A quantitative expression of the behavior of particle groups would be very helpful in developing a lance design that will improve combustibility. Research and development effort in this direction is greatly anticipated.

REFERENCES

1) T. Ariyama, M. Sato, Y. Yamakawa ,Y. Yamada and M. Suzuki: Tetsuto-to-Hagane, 80(1994), 288.
2) T. Ariyama, M. Sato and R. Murai: Research Group of Pulverized Coal Combustion in Blast Furnace, Rep-29(1995)
3) H. Kugizaki, K. Oshima, I.Naruse and K.Otake: CAMP-ISIJ,9(1996), 6.
4) I. Ueno, K. Yamaguchi and K. Tamura: ISIJ Int.,33(1993), 640.
5) F. Shen, T. Inada, K. Yamamoto, and Y. Iwanaga: Tetsuto-to-Hagane,80, (1994), 7.
6) K. H. Peters, M. Peters, B. Korthas, K. Mülheims and K. Kreibich: Proceedings of the 6th Int. Iron and Steel Congress, ISIJ, Nagoya, (1990), 498.
7) T. Ariyama, M. Sato, and Y. Yamakawa: Tetsuto-to-Hagane,81, (1995), 703.
8) T. Ariyama: Research Group of Pulverized Coal Combustion in Blast Furnace,Rep-16(1993)
9) T. Ariyama, M. Sato, R. Murai, K.Miyakawa, K.Nozawa and T.Kamijo: Tetsuto-to-Hagane,81(1995), 1114.
10) Y. Iwanaga and T.Inada :Research Group of Pulverized Coal Combustion in Blast Furnace,Rep-12(1993)

CHAPTER 4

PC Combustion in blast furnace

In order to achieve stable high rate PCI operation, securing of gas permeability and suppression of increase of heat loss are essential, and the combustibility is an important factor that affects these two.

While the PC injection rate increases in the blast furnace, the combustibility of pulverized coal lowers as the excess oxygen coefficient lowers, and increase of unburned pulverized coal is forecast, and the increased unburned pulverized coal flows from the raceway to the furnace, giving rise to concerns in that gas permeability and liquid permeability in the furnace may be impeded, the replacement ratio of PC to coke is lowered due to discharge of pulverized coal as dust from the furnace top, and the fuel rate may be increased. Consequently, a large number of investigations have been done on the combustibility in the raceway by the use of model combustion furnaces, mathematical models, etc.[1)] However, the actual condition has not yet been completely clarified because of the difficulty of sampling inside the blast furnace in operation.

Therefore, in order to be familiar with the combustion behavior in the raceway of the actual blast furnace, based on the measured values, discussion will be made on the effects of coal properties, lance construction, blast conditions, etc. on the combustion condition and combustibility of the combustion zone in front of tuyere.

4.1 Gas composition and temperature at the combustion zone

4.1.1 Distribution of gas composition

Figure 4-1 shows the distribution of gas composition in the raceway when pulverized coal is intensively injected in Kakogawa 1 BF.[2)] In all cases, the O_2 concentration rapidly decreases from the tuyere nose towards the furnace inside, and the CO_2 concentration rapidly increases, and it is indicated that combustion of pulverized coal takes place only in an extremely short distance in the raceway.

Based on Fig. 4-1, the reactions of pulverized coal in the raceway are roughly illustrated with different three reaction steps as Fig. 4-2. In first step, the volatile matters in coal are gasified. In next step, the carbon and the hydrogen gas are oxidized to CO_2

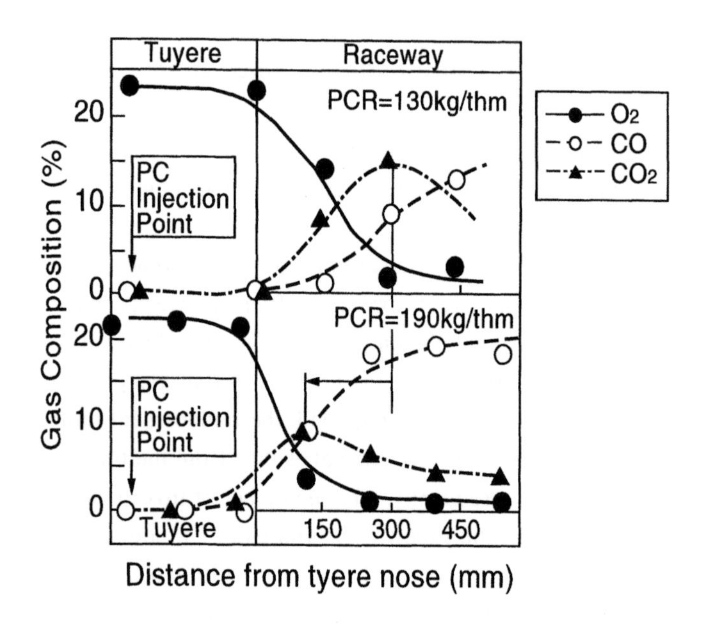

Fig. 4-1 Distribution of gas composition in raceway

and H_2O. Generated CO_2 and H_2O gas are reformed to CO and H_2 gas by carbon solution-loss reaction in last step. According this reaction steps, reaction rate of PC in the raceway is basically depending on the oxygen content in the hot air (Stoichiometric oxygen ratio).

In Fig. 4-1, when the relation to the PC injection rate is observed with the peak CO_2 concentration position designated as the combustion focus where combustion reaction takes place most actively, it is assumed that the combustion focus, that is, the highest temperature region, comes closer to the tuyere nose at a higher PC injection rate.

When effects of the coal properties on the combustion focus are observed on the basis of the results in a packed-bed type test combustion furnace (Fig. 4-3),[3] there is recognized an phenomenon in that injecting pulverized coal of high volatile matter content causes the combustion focus to come closer to the tuyere side, and the wall temperature rises. The combustion focus position is also varied by the effect of ash.

When the effect of coke properties is observed with the same test furnace, the ultimate combustion efficiency of pulverized coal is not varied by the difference of coke reactivity

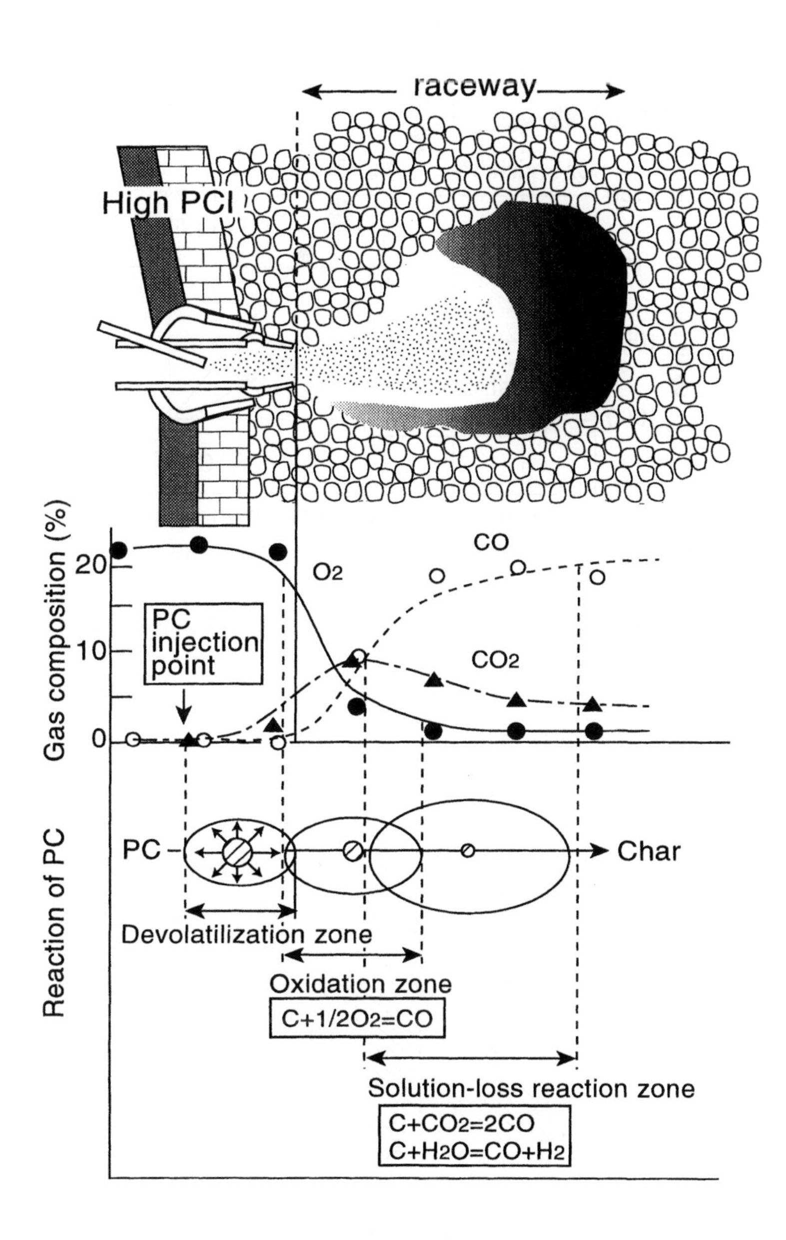

Fig. 4-2 Schematic drawing of PC reaction in raceway

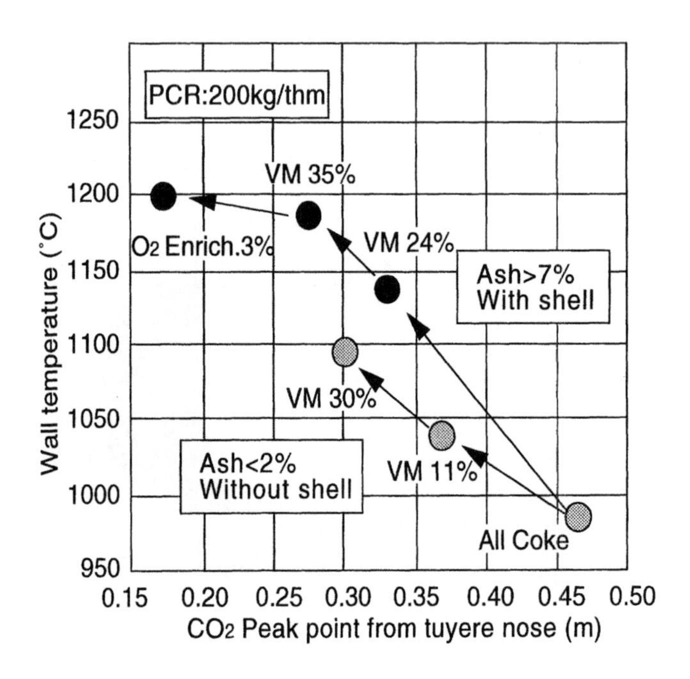

Fig. 4-3 Relationship between CO_2 peak point and wall temperature

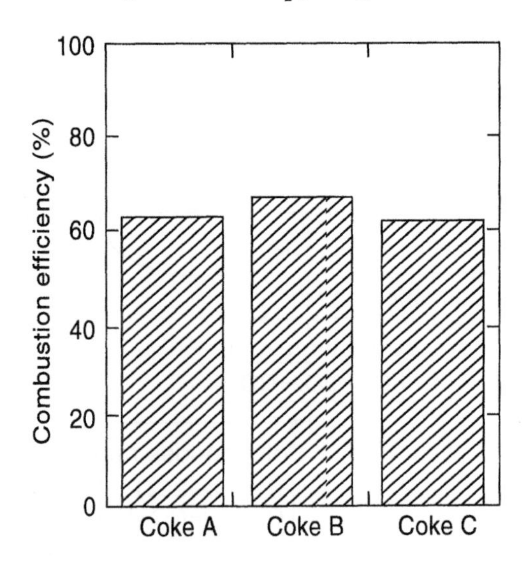

Fig. 4-4 Effect of coke property on combustion efficiency

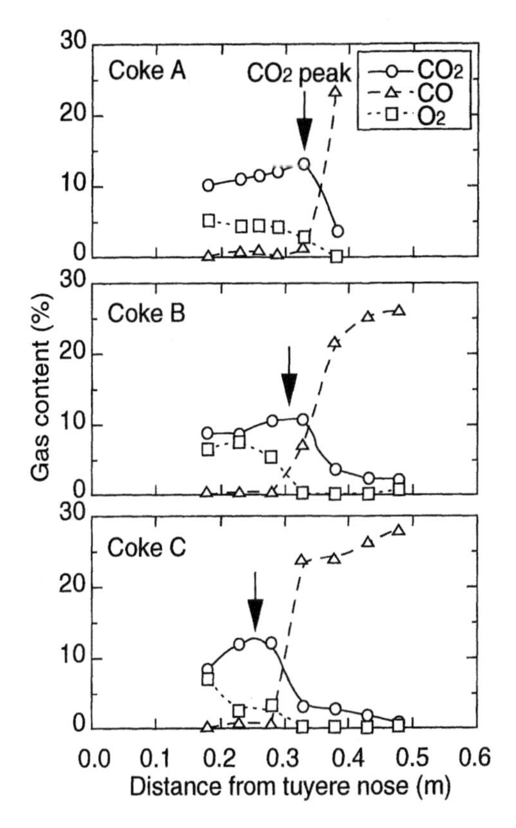

Fig.4-5 Distribution of gas content in raceway

Table 4-1 Coke properties

Coke No.	DI^{150}_{15} (-)	$I^{600}_{13.2}$ (-)	RI (-)	RSI (-)
A	85.2	88.8	23.0	70.5
B	84.5	85.4	35.6	53.5
C	81.9	83.6	40.2	43.2

(Fig. 4-4), but the gas composition distribution is varied, and since the combustion focus moves to the tuyere side as coke reactivity rises (Fig. 4-5, Table 4-1), it can be assumed that the combustion focus also depends on the combustion behavior of coke properties flowing into the raceway.

Here, combustion efficiency is generally calculated as follows.

Combustion efficiency (%)
=(Reduction rate of combustible portion in coal/combustible portion in coal) x 100

4.1.2 Temperature distribution

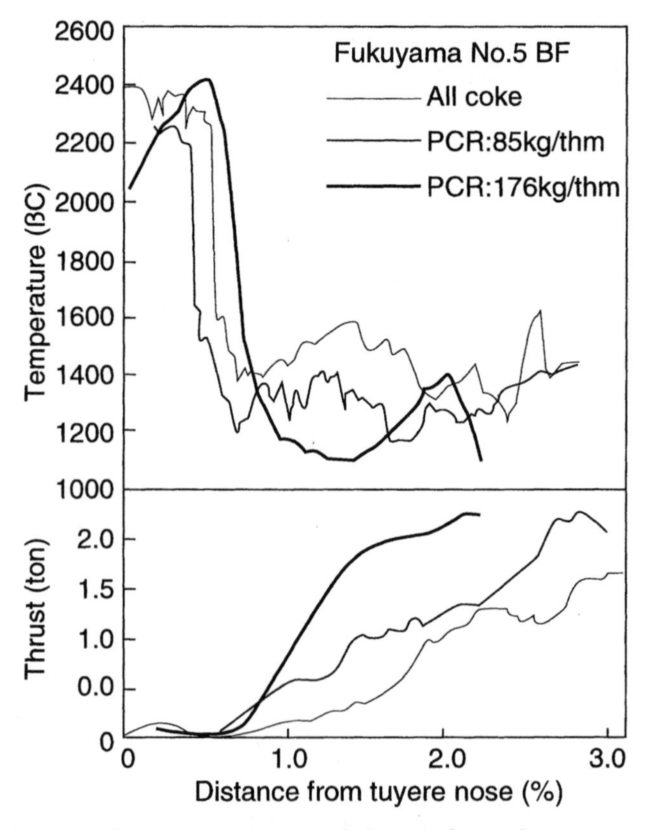

Fig.4-6 Change of gas temperature and thrust of sampler at tuyere level

Figure 4-6 shows temperature distribution in the raceway obtained when pulverized coal was injected from one tuyere at Fukuyama 5 BF.[4] While the temperature distribution is comparatively flat at the tuyere nose portion in all-coke operation, when pulverized coal is injected, the temperature distribution has a peak and temperature drop occurs rapidly at the depths of the raceway. In addition, the rapid rise of thrust of the tuyere sampler is simultaneously recognized, suggesting that a region with comparatively strong and low gas permeability is formed at the depths of the raceway.

When the estimated temperature of coke collected by deadman sampling at an injection rate of 150 kg/thm is compared in Muroran 2 BF with that at all-coke operation, the high-temperature zone moves to the exit side when pulverized coal is injected, and temperature temporarily lowers at the depths of the raceway (Fig. 4-7).[5] When the estimated temperature is further compared with that at high PC rate in Kobe 3 BF (Fig. 4.8), the temperature distribution in the vicinity of the tuyere changes dependently on the furnace condition,[6] and it is assumed that the raceway depth is shallow and its profile rather extends upwards when the furnace condition is unstable.

That is, the temperature distribution in the raceway seems to be varied not only due to

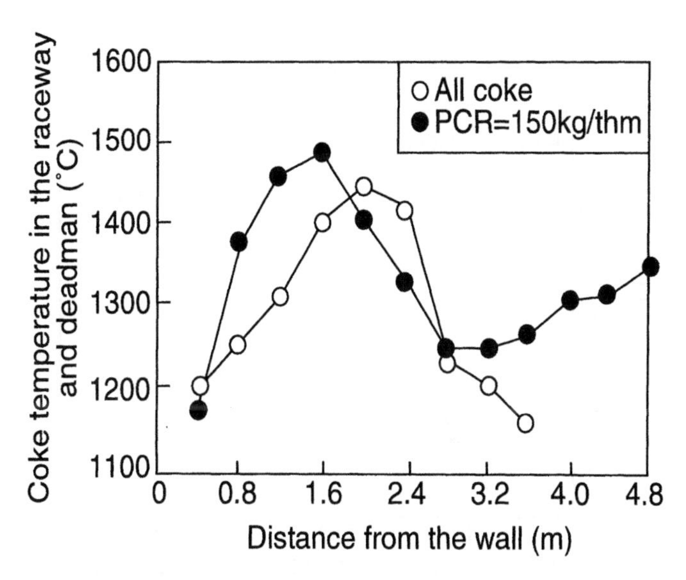

Fig.4-7 Radial distribution of estimated coke temperature at tuyere level (Muroran 2 BF)

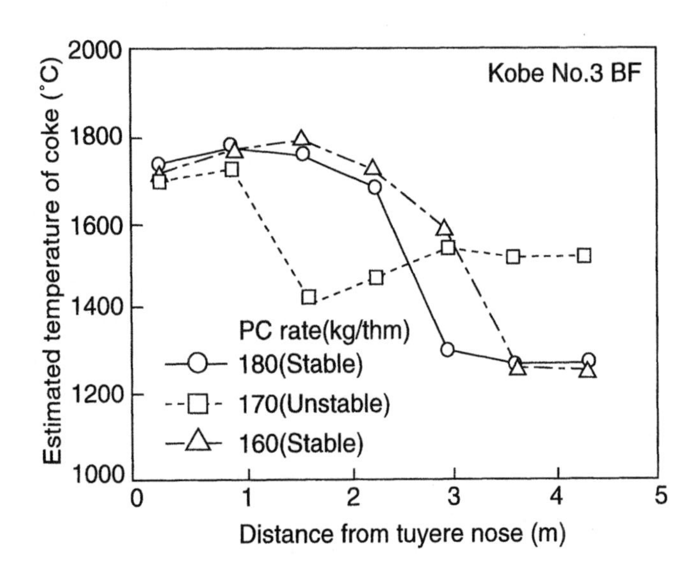

Fig.4-8 Radial distribution of estimated coke temperature at tuyere level (Kobe 3 BF)

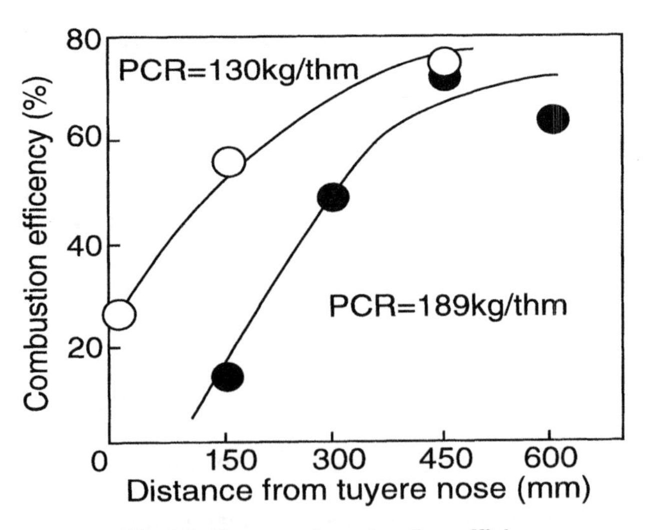

Fig.4-9 Change of combustion efficiency

the effect of combustibility of pulverized coal but also due to the effect of changes of gas flow from the raceway inside as a result of injection of pulverized coal.

4.2 Evaluation of PC combustion efficiency

4.2.1 Effect of PC injection rate

Figure 4-9 shows changes of combustion efficiency in the raceway, which is converted from the ash balance for the pulverized coal sampled from the vicinity of the tuyere in Kakogawa 1 BF in operation. As the injection rate of pulverized coal increases, the combustibility tends to decrease as a whole and the PC rate lowers to about 70% at 189 kg/thm. Assuming from this result, about 57 kg/thm unburned pulverized coal flows into the furnace. However, because the carbon content in the dust discharged from the furnace top indicates a nearly constant value (Fig. 4-10)[7] without depending on the pulverized coal injection rate, it is assumed that unburned pulverized coal is nearly completely consumed by reactions in the furnace.

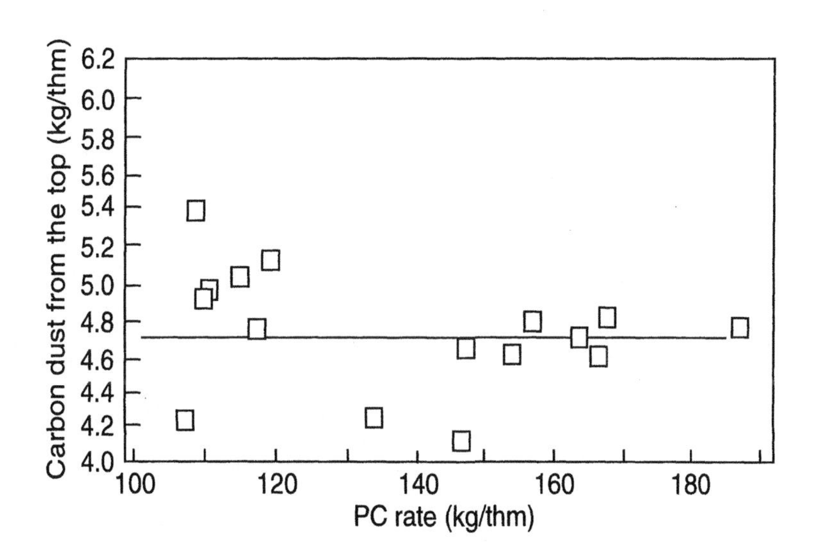

Fig.4-10 Relationship between carbon dust from the top and PC rate

4.2.2 Effect of coal properties

Figure 4-11 shows a sampling apparatus, Table 4-2 the injected pulverized coal properties, and Figure 4-12 the test results, when pulverized coal injection test was carried out at 200 kg/thm using one tuyere in Wakayama 5 BF.[8] Figure 4-12 indicates that the pulverized coal with high volatile matter contents provides better combustibility in the vicinity of tuyere nose and about 70% ultimate combustibility is achieved. However, as pulverized coal moves towards the depths of the raceway, the difference in combustibility between coal types tends to decrease.

With respect to the effect of coal properties, the test results in the packed bed type combustion furnace (Fig. 4-13)[3] indicate that the coal with low volatile matter content tends to provide a lower combustion efficiency but no difference is generated in the

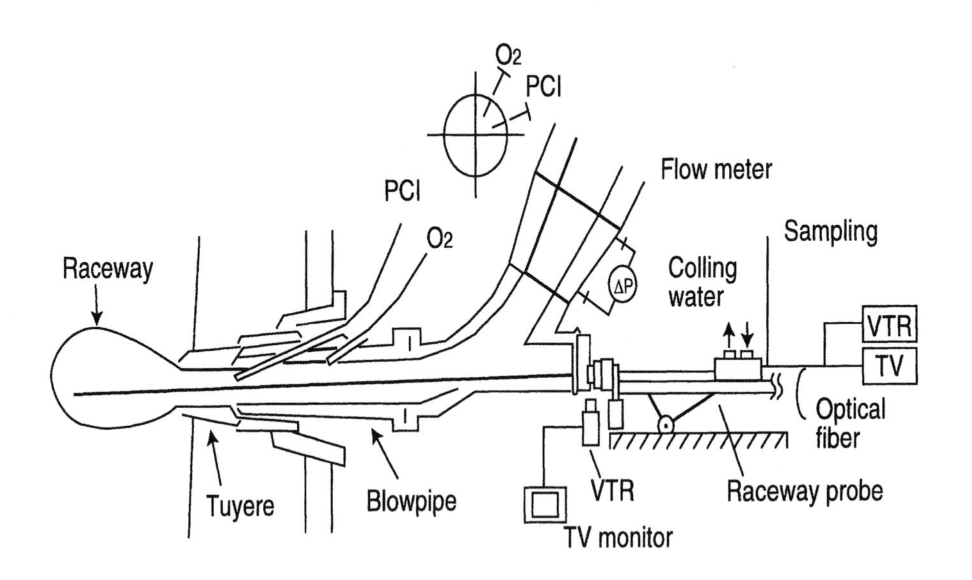

Fig.4-11 Schematic diagram of test equipment at Wakayama Works

Table 4-2 Properties of coal

Coal	Approxlmate analysis(%)			Calorific value (kcal/kg)
	Ash	VM	FC	
A	10.8	17.4	71.8	7760
B	9.3	35.4	55.3	7280
C	4.3	45.9	49.3	7180
D	3.6	5.9	90.5	

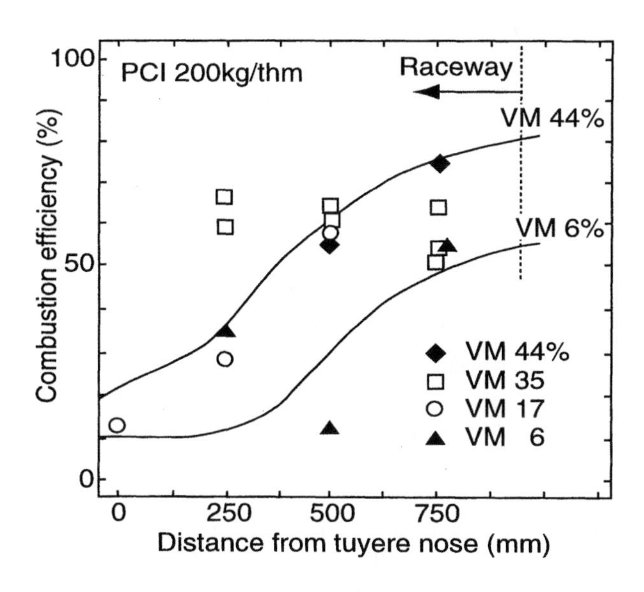

Fig.4-12 Effect of volatile matter on combustibility

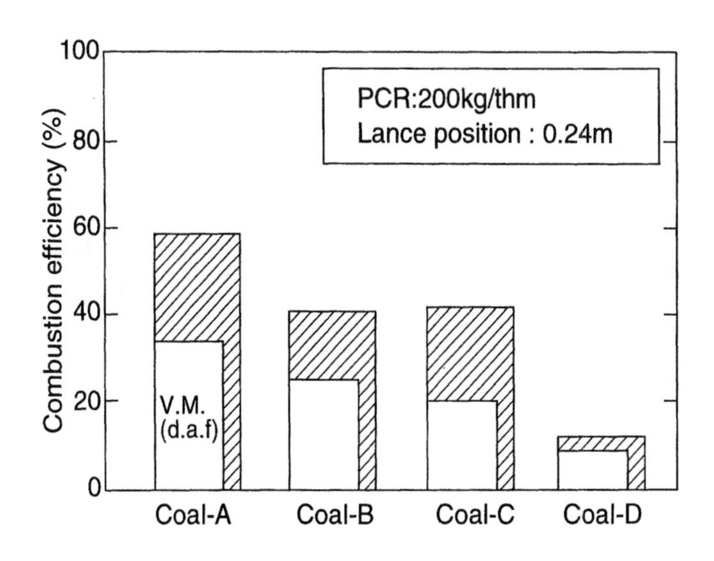

Fig.4-13 Effect of coal properties on combustion efficiency

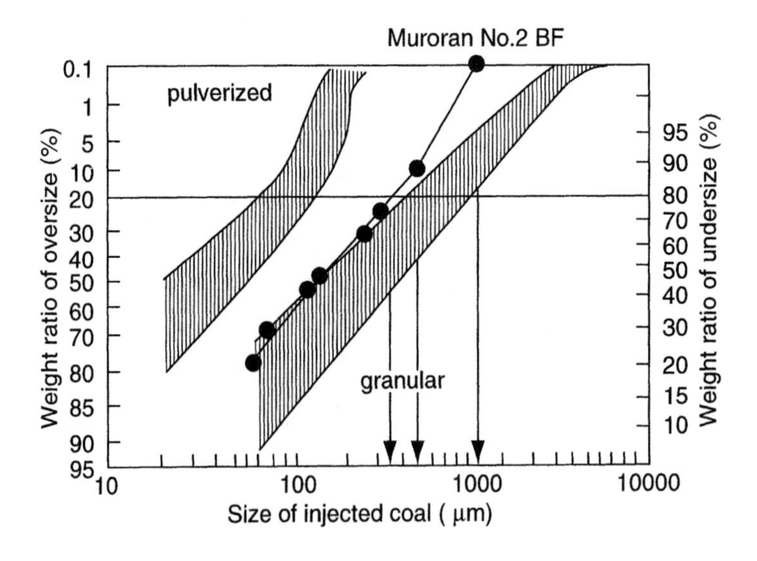

Fig.4-14 Size distribution of injected coal

ultimate combustion efficiency between coals B and C of similar fixed carbon rates. This can be assumed that because ignition and combustion of volatile matter take place quickly, the ultimate combustion efficiency shifts to be governed by reactions of fixed carbon rather than by reactions of the volatile matter as the volatile matter content lowers.

According to the coarse particle coal injection test at Muroran 2 BF,[5] when coarse particles of typical particle size of 350 μ m (Fig. 4-14) were injected, stable operation was achieved at PCR of 150 kg/thm, and change of pressure loss at the lower furnace portion and char discharge at the furnace top were neither recognized. In addition, the ratio of -3 mm coke powder in the deadman by tuyere coke sampling at blast furnace downtime scarcely varies from that at all-coke operation (Fig. 4-15). Consequently, it is assumed that lowering of combustibility which is foreseen as a result of coarsening of particles does not always have effect on the blast furnace condition.

4.2.3 Effect of lance construction

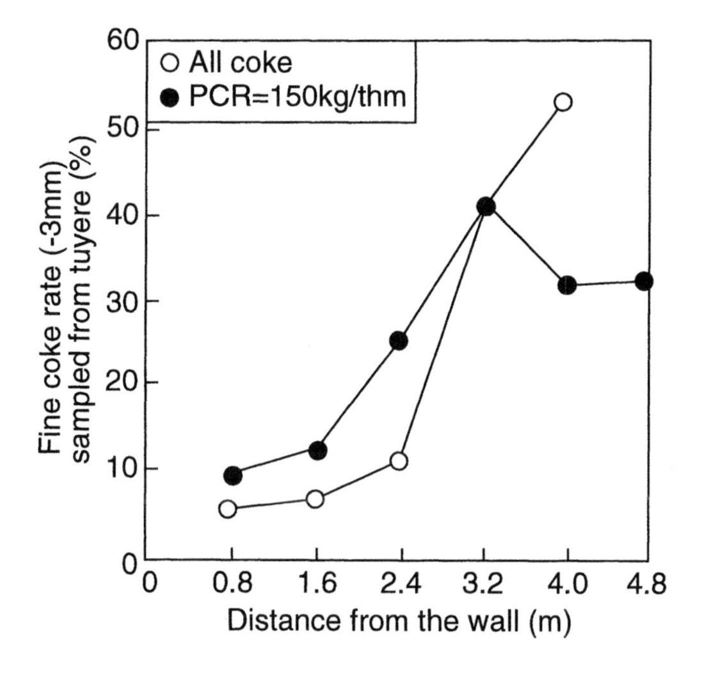

Fig.4-15 Fine coke distribution at tuyere level

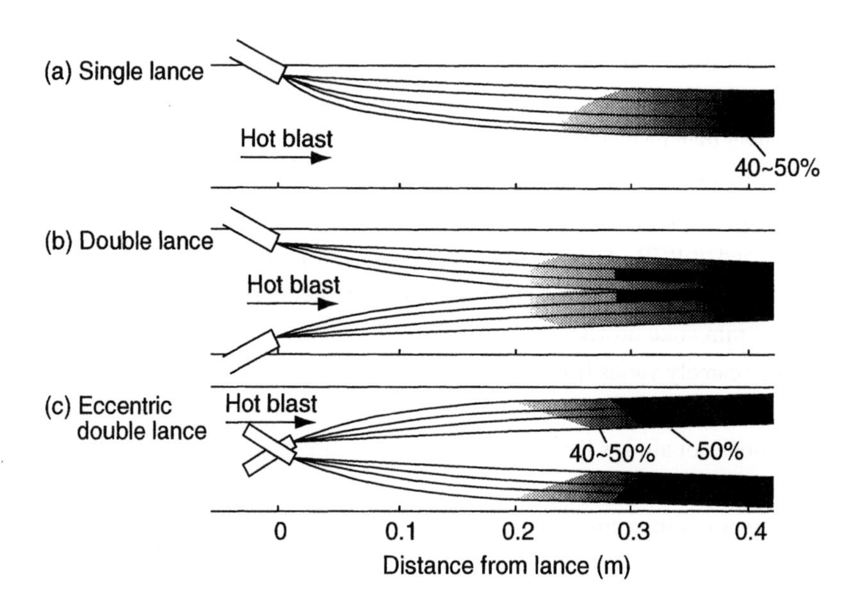

Fig.4-16 Effect of lance arrangement on the flow and combustion efficiency of PC

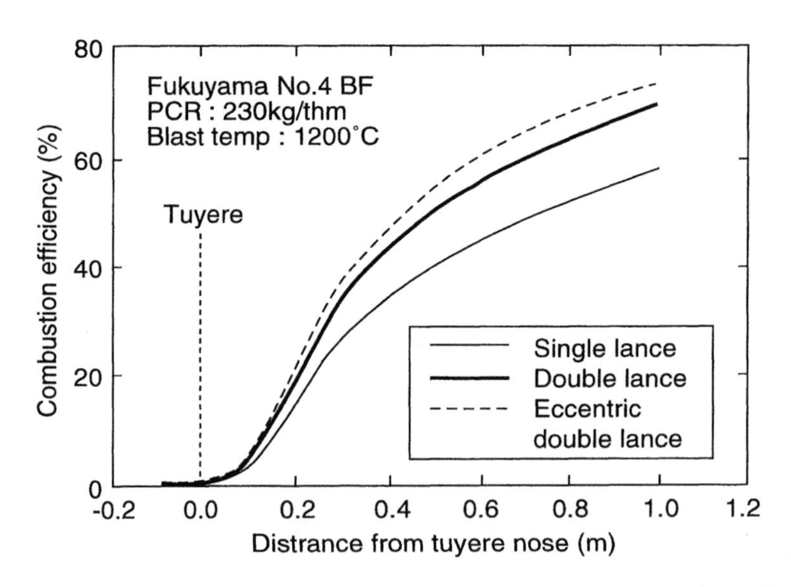

Fig.4-17 Effect of lance arrangement on combustion efficiency of PC (calculated)

When injection at PCR exceeding 200 kg/thm is assumed, the consumption capabilities of unburned pulverized coal in the blast furnace as a result of lowering of combustion efficiency cause problems. For a control means, optimization of the lance types and arrangements is important.

It must be mentioned that one of the factors of the successful 200 kg/thm level injection in Fukuyama 4 BF is attributed to an adoption of the eccentric double lance. This eccentric double lance has two lances arranged unsymmetrically so that collisions of pulverized coal flows can be prevented in order to strengthen dispersion of pulverized coal (Fig. 4-16), and computing the combustion efficiency at the pulverized coal injection rate of 230 kg/thm and under the conditions of Fukuyama 4 BF, the 74% combustion efficiency is achieved with the eccentric double lance as against 58% with a general single lance; more than 15% higher combustion efficiency can be expected (Fig. 4-17).[9] That is, it indicates that the improvement in dispersibility directly results in the improvement in combustion efficiency.

When an intensive PC injection experiment was carried out using one tuyere in Chiba 5 BF which was operating at low blast temperature (750-800°C), a lance thicker than

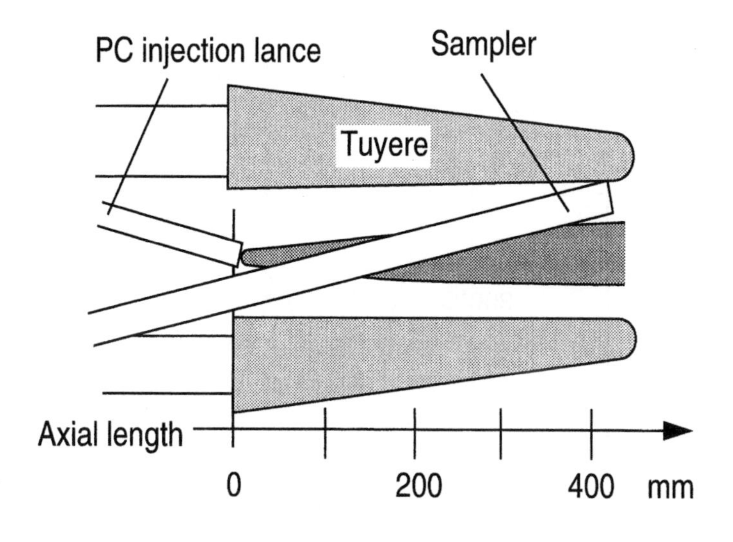

Fig.4-18 Schematic drawing of the sampler and PC lance in tuyere

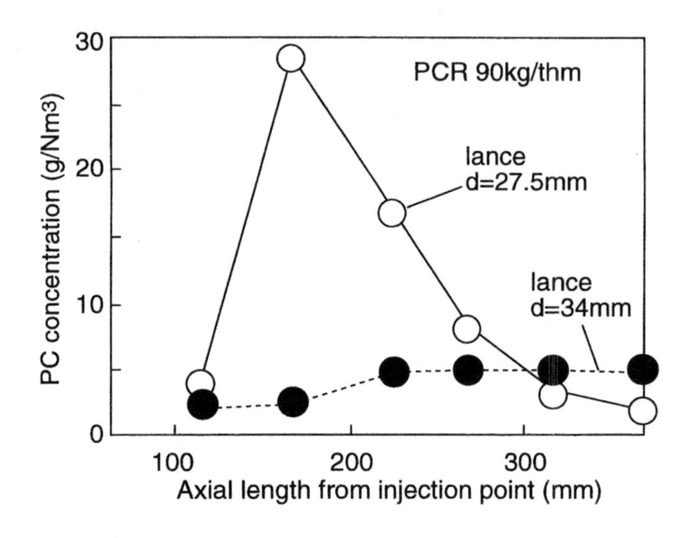

Fig.4-19 PC concentration in tuyere

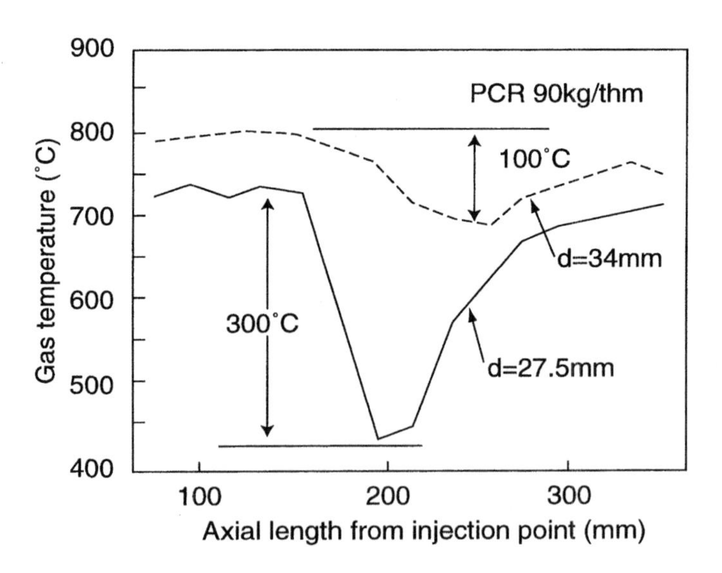

Fig.4-20 Gas temperature distribution in raceway

usual was adopted to improve the combustibility, and at the same time, sampling of pulverized coal was carried out (Fig. 4-18).[10)]

Figure 4-19 shows the distribution of pulverized coal concentration in the tuyere. In the case of the general lance (27.5 mm in outside diameter), pulverized coal is localized on the downstream side of the lance top end and is scarcely dispersed, but in the case of thick-wall lance (34 mm in outside diameter), the distribution of pulverized coal concentration becomes flat, indicating that the dispersibility is improved. Comparison of gas temperatures at the main stream of injected PC particles shown in Fig. 4-20 indicates that the gas temperature lowers by 300°C from the surrounding temperature with the general lance, but this is equivalent to the temperature drop associated with injection of pulverized coal, and since the gas composition in the raceway is nearly similar to that of blast, it is indicated that ignition of pulverized coal has not taken place at blast temperatures lower than 800°C.

On the other hand, with the thick-wall lance, the range of gas temperature drop decreased to 100°C, and the CO_2 concentration in gas also increased. That is, it is assumed

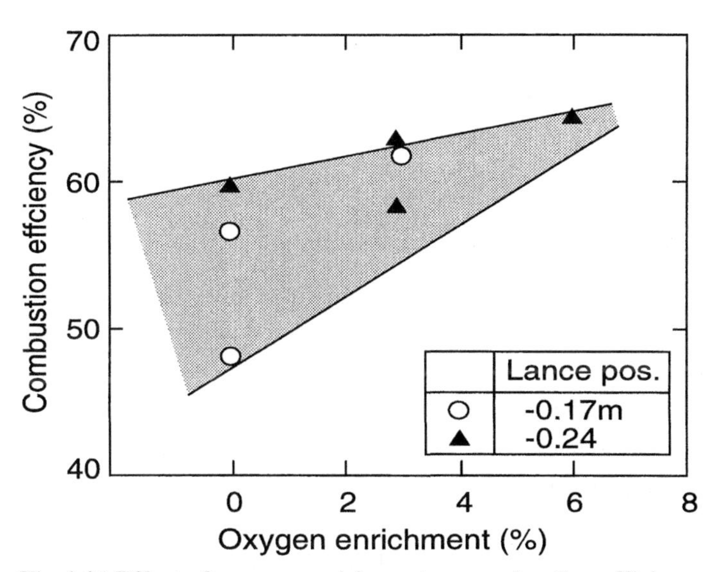

Fig.4-21 Effect of oxygen enrichment on combustion efficiency.

that turbulent energy caused by the lance itself increased downstream the thick-wall lance, and the dispersibility of pulverized coal was improved, resulting in combustion.

4.2.4 Effect of blast condition

Figure 4-21 shows the improvement effect of the combustion efficiency when oxygen is enriched in the test combustion furnace.[3] When the lance is located 0.24 m from the tuyere nose, the combustion efficiency gradually increases by the oxygen enrichment, but the combustion efficiency improves by about 5% at 6% oxygen enrichment ratio, and no conspicuous oxygen enrichment effect is observed in this range.

When the effect of oxygen enrichment is observed from the relation between the combustion focus and wall temperature shown in Figure 4-3, the increase of wall temperature is small even if the combustion focus comes greatly close to the tuyere side by oxygen enrichment. This results from the increased heat flux ratio at the upper tuyere portion because of oxygen enrichment. In addition, effects of lowering of the raceway depth associated with lowering of blast flow rate are also assumed. Consequently, oxygen enrichment would rather play a more important role as a gas flow control means in the furnace than in the combustion control means of pulverized coal.

Furthermore, the relationship between the gas flow rate in front of tuyere and stave cooler (Fig. 4-22) [2] suggests that heat loss can be lowered by combustibility control while securing the raceway depth and pulverized coal residence time through a proper gas flow rate.

Consequently, it is essential to be familiar with optimum conditions including reactions and physical behavior in front of tuyere, tuyere diameter, lance construction, etc. as control of blast condition.

4.3 Control of combustibility

In order to be familiar with combustion behavior of pulverized coal in the actual blast furnace, factors which have effects on the combustibility have been discussed, and the importance of combustion focus has been identified, as well as ultimate combustion efficiency. To achieve the increased pulverized coal injection rate and stable operation, it becomes more and more important to prevent an increase of heat loss of the blast furnace and to secure gas permeability in the furnace.

The combustibility is a factor that has effects on all the subjects mentioned above, and in particular, because it is assumed that the combustibility has effects on gas

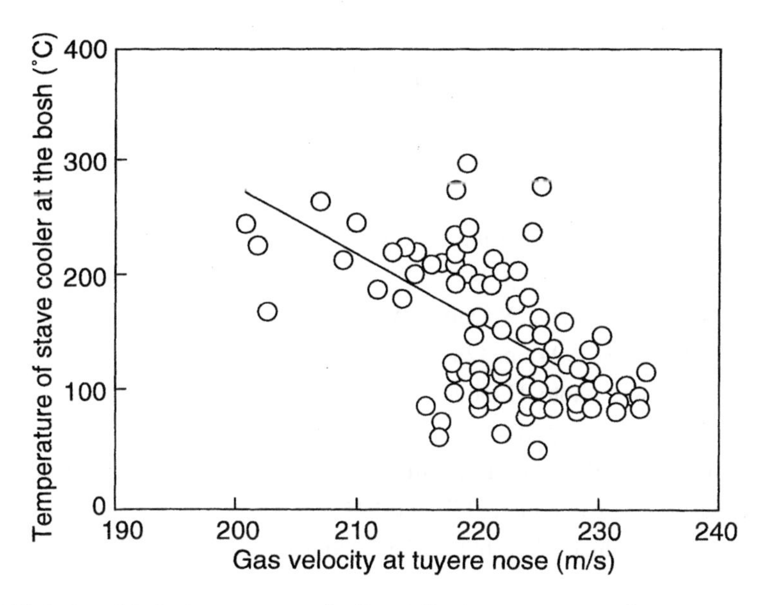

Fig.4-22 Relationship between gas velocity at the tuyere nose and temperature of stave cooler at bosh

permeability at the lower portion of blast furnace through generation of unburned pulverized coal, and in addition, because the combustion behavior in front of tuyere represented by the combustion focus, which is dependent on coal properties, blast condition, lance construction, and properties of coke that flows in, has effects on the in-furnace gas flow and profile of cohesive zone based on changes of the gas flow behavior around the raceway, it would become more and more important to develop proper combustibility control and its detection means.

REFERENCES

1) X.Xingguo, K.Nozawa, S.Sasahara, M.Shimizu and S.Inaba : Tetsu-to-Hagane, 78 (1992), p. 1230
2) S.Shibata, K.Nozawa, M.Shimizu, R.Ono, T.Okuda, R.Ito and K.Hanao : Materials and Processes, 6 (1993), p. 112
3) K.Miyagawa K.Nozawa, Y,Kamijo, M.Sato, Y.Yamakawa and T.Ariyama : Materials and Processes, 7 (1994), p. 128
4) R. Murai, S.Kishimoto, H.Inoue, H.Mitsufuji, M.Sakurai and H.Wakai : Proceeding of 1st ICSTI (1994), p. 272
5) K.Yamaguchi : Material of 4th Study Meeting of Blast Furnace Pulverized Coal Combustion, 54 Committee, JSPS (1994)
6) T.Yabata K.Hoshino, Y.Yoshida, S.Ishiwaki and S.Kitayama : Tetsu-to-Hagane, 78 (1992), T61
7) T.Kamijo, M.Shimizu, R.Ito and K.Hoshino : Materials and Processes, 9 (1996), p. 26
8) Iwanaga : Material of 2nd Study Meeting of Blast Furnace Pulverized Coal Combustion, 54 Committee, JSPS (1994)
9) M.Sato, Y.Yamakawa, T.Ariyama, M.Suzuki and T.Tuboi : Materials and Processes, 7 (1994), p. 958
10) N.Ishiwatari, T.Uchiyama, K.Takeda,H.Itaya, Y.Sakuma and H.Nishimura : Materials and Processes, 8 (1995), p. 325

CHAPTER 5

Modeling of pulverized coal combustion

A mathematical simulation model based on heat, mass and momentum balance and reaction kinetics is one of the most useful tools for design and optimization of pulverized coal combustion devices. For the estimation of pulverized coal combustion characteristics, the combustion phenomena should be taken apart once so that unit processes can be described by some mathematical equations or simple model component. Then, these equations should be solved simultaneously to express their entire burning behavior. Even though pulverized coal combustion is a quite complex phenomenon since a large number of chemical and physical processes occur within it, chemical engineering treatment is very useful to understand, separate and formulate unit processes.

This chapter first outlines combustion behavior of a single particle of pulverized coal. Next, formulation of unit processes related to burning of coal particles is shown. Then, some descriptions of the combustion fields and treatment of particle behavior within the fields are presented. Finally, several practical applications of mathematical simulation to the pulverized coal combustion process in the lower part of blast furnace are shown.

5.1 Combustion behavior of a single pulverized coal particle

In most practical combustion devices, pulverized coal particles are pneumatically carried with primary air and introduced into the furnace. Subsequently, this suspension of cold air and coal particles mixes with hot secondary (combustion) air and combustion takes place during the suspension flows through the furnace. The burning coal suspension flow in the practical furnace is regarded as a turbulent two-phase flow accompanied by simultaneous heat, mass and momentum transfer in several modes and a large number of chemical reactions. These phenomena strongly affect each other and interactions between particle and gas phases play important roles in the combustion fields. Therefore, the model should formulate not only behaviors of both the gas and particle phase but also interactions between two phases. The major model components constituting the total mathematical model are:

1. fluid mechanics

a) turbulence, b) particle dispersion

2. reaction of coal particle

a) devolatilization, b) char reaction

3. gaseous reaction

a) turbulent combustion

4. heat transfer

a) convection, b) radiation, c) reaction heat

5. others

a) pollutant formation, b) particle deformation, c) fragmentation, etc.

It is known that the combustion characteristics of pulverized coal in practical furnaces highly depends on coal type, thus the properties of coals should be reflected to the total simulation. The burning behavior of a single coal particle is one of the most important components of the mathematical model. The first section of this chapter starts with describing the burning behavior of a single coal particle.

The combustion process of pulverized coal is roughly divided into three steps as shown in Fig. 2-4. In the first step, 'pre-heat zone', pulverized coal particles are heated up by thermal energy supplied from their atmosphere. The following step is 'flame zone' where coal particles release volatile matter and its homogeneous reaction in gas phase forms a flame. The last is 'post-flame zone' in which heterogeneous reaction of char mainly occurs.

In the pre-heat zone, pulverized coal particles are heated up enough to begin combustion reactions. Coal particles are supplied to the furnace by pneumatic transportation, and are surrounded by cold gas just after they are injected into the furnace. Therefore, thermal radiation from the furnace wall, flame and burning particles is the predominant heat transfer process in this period. With the progress of mixing of coal suspension and hot combustion gas, the contribution of convective heat transfer to the pre-heating process increases. Moisture in the particles evaporates during the pre-heating process. Although no reaction occurs in the pre-heat zone, this step is still important among the combustion processes because it closely relates to ignition and flame stability.

Regarding to the blowpipe and raceway zone in the blast furnace, although pulverized coal particles are surrounded by a high-temperature atmosphere (blowpipe wall, coke particles in raceway and blast), the particles have a quite short burning time since they move with the high-velocity blast. Then pre-heating is a predominant process for the combustion efficiency, in other words, the generation of fine powder. On the contrary, too early ignition in the blowpipe results in the increase of blast gas volume and raises the blast pressure. Thus the pre-heating characteristics is an important factor to design pulverized coal injection operation and should be discussed in detail.

When coal particles are heated up to a high enough temperature, thermal decomposition of the carbon network in coal occurs and it releases volatile matter and tar into the gas phase (devolatilization). The released gaseous combustibles react homogeneously with oxygen in the surrounding gas and form a flame. This period of the combustion process is called `flame zone'. The rate of devolatilization is controlled by the rate of the thermal decomposition reaction (and mass transfer), while the homogeneous reaction rate in practical combustion devices is governed by turbulent mixing as well as the chemical reaction rate. The period of devolatilization depends on particle size, heating rate, and so on, and this process finishes within 0.1 s under typical pulverized coal combustion conditions. During this short period, coal particles undergo softening and resolidification and it results in drastic changes of particle structure, that is an important reaction parameter of the subsequent heterogeneous reaction.

The heterogeneous reaction of solid residue of devolatilization takes place after the homogeneous reaction of gaseous fuel (volatile matter combustion) and this reaction step corresponds to `post-flame zone'. This solid residue is called `char' and it contains mineral matter as well as carbonaceous components. The char combustion takes one to several seconds and this step possesses most part of burnout time for the coal particles. Therefore, it has a major effect on the combustion efficiency (burnt ratio) in limited residence time in combustion devices. Additionally, existence of solid residue is a significant difference from the combustion of gaseous or liquid fuel, because char particles contribute to radiative heat transfer as well as chemical reaction.

For simplicity, the latter two stages, flame zone and post-flame zone, are explained as sequential and separate processes. However, it is reported that volatile matter release and heterogeneous reaction of the particles can proceed in parallel, due to jet-like volatile matter release [1-5] and entrainment of oxidizer gas to the particle surface caused by the volatile jet. The concurrent progress of devolatilization and heterogeneous reaction of char needs to be taken into account to estimate reaction efficiency of pulverized coal precisely.

In addition, it is reported that this volatile jet has spun particles and twisted these trajectories. Thus it is expected that the volatile jet affects transportation of heat, mass and momentum around the particles. The volatile jet forms at random and has yet to be discussed in detail. However, the effect of the jet-like volatile release is an important issue to be clarified for the precise simulation of single particle burning behavior.

Ignition of pulverized coal, which is the transition step from the pre-heat zone to the flame zone, is usually treated as formation of initial flame. The general definition of ignition is that the combustion reaction becomes sustainable, and this process should be discussed based on the particle heat balance including reaction heat. Coal is a partially pyrolytic material, and there are two fundamental modes of ignition classified by the reaction, which supplies reaction heat to the particle. One is homogeneous reaction of volatile matter in the gas phase and the other is heterogeneous reaction at the particle surface.

Different theories are applied to each ignition process [6)] and the ignition temperatures in both modes are correlated with particle diameter, gas composition, and so on. The steady state analysis revealed that the same kind of coal showed the transition of ignition mechanism with changes in the combustion condition.[6)] The generation rate of reaction heat closely relates to the coal properties such as volatile content, volatile composition, kinetics of devolatilization and particle reactivity, thus these factors have to be included in the ignition theory. In addition, in the actual combustion situation coal particles have a wide variety of chemical and physical properties, and the particles in the cloud have interactions among themselves. No exact formulation of these effects on the ignition process has been reported so far, and the precise definition of pulverized coal ignition in the practical combustor has yet to be given.

Most recent pulverized coal combustion models do not have exact definitions of ignition since they solve the heat balance equation and other conservation and rate equations, and are able to estimate location or time of reaction commencement. However the ignition criterion should be a useful index to determine the location of lance tip (nozzle tip) in the blast furnace blowpipe.

5.2 Formulation of unit processes

5.2.1 Devolatilization

When coal particles are heated up beyond about 700 K, gaseous and liquid products are released from the particles. The gaseous product contains the hydrocarbons of low molecular weight, carbon monoxide, hydrogen, etc. and the liquid product consists of polynuclear aromatic hydrocarbons. This process is called `devolatilization' and resulted from decomposition of bridges and bonds in coal molecules. The coal molecule comprises many kinds of polynuclear aromatic unit, side-chain, functional group, etc., and has a complicated and huge structure. Moreover even a single type of coal includes a wide variety of molecule structures. In the practical combustion analysis, it is not realistic to simulate all decomposing behavior of coal molecule based on the actual molecular structure. Therefore, a simplified model of the devolatilization is required for practical applications.

An industrial standard defines the volatile matter content of coal as a weight loss when the coal specimen is heated up to and retained at a specified temperature and duration (for example, 900±20°C and 7 min in the Japanese Industrial Standard [7)]). The definitions in most standards are determined to evaluate coal properties for cokemaking, thus temperatures specified are usually much lower than particle temperatures attained in the actual combustion condition. Many efforts were made to examine the effect of temperature, heating rate, pressure, particle size, etc. on devolatilization behavior. They reported that both amount and rate of devolatilization increased with the temperature. For example, Kobayashi, et al.[8)] reported that volatile matter released from a bituminous coal (VM: 40.7 wt%) was 40-60 % larger than its proximate analysis value when it was heated up to 1510 K. Hence, the devolatilization model should describe such an increase in volatile release as well as releasing rate.

The most primitive mathematical model for devolatilization is an overall first-order reaction model.[9)]

$$\text{(Raw coal)} \xrightarrow{K_V} \text{(Volatile)} + \text{(Residue)} \tag{5.1}$$

This model postulates that the devolatilization rate is proportional to the amount of volatile matter remaining in the coal:

$$\frac{dw_V}{dt} = k_V\,(w_V^* - w_V) \tag{5.2}$$

where w_V is the released amount of volatile matter and w_V^* is the total volatile matter yield usually given from the proximate analysis. The total volatile yield, however, increases with temperature as mentioned above. One solution for this problem is introducing the parameter, so called Q-factor [10)], which adjusts the value of w_V^* according to the expected temperature range. Even though this parameter is introduced, the value of total volatile yield needs to be specified prior to the simulating calculation. The volatile yield depends on the thermal history of the burning particle and the other burning conditions that are usually parts of model analysis. Thus the problem of total amount of volatile matter still remains.

Anthony, et al.[11)] postulated that the decomposition of raw coal (unreacted organic matrix in coal particle) consists of an infinite series of parallel first order irreversible reactions and volatile matter and solid residue are generated through the intermediate product of these parallel reactions:

$$\text{(Raw coal)} \xrightarrow{\sum k_i} \text{(Metaplast)} \begin{cases} \xrightarrow{K_v} \text{(Volatile)} \\ \xrightarrow{K_r} \text{(Residus)} \end{cases} \tag{5.3}$$

They assumed continuous Gaussian distribution of activation energies for the first step reactions and gave a reaction rate constant for pyrolysis of raw coal as:

$$k_{vp} = -\frac{d}{dt}\ln\{\int_{-\infty}^{\infty} \exp[-\int_0^t A\exp(-\frac{E}{RT})dt']\, f(E)dE\} \tag{5.4}$$

$$f(E) = \frac{1}{(2\pi)^{1/2}\sigma}\exp[-\frac{(E-E_0)^2}{2\sigma^2}] \tag{5.5}$$

The rate constant of the second step includes the overall mass transfer coefficient which accounts for the effect of pressure. These treatments allow this model to evaluate the effects of pressure, particle size and temperature history. However, this model needs to determine the total amount of volatile matter, so that the same problem as the overall first-order reaction model remains in this model.

Kobayashi et al.[8)] described devolatilization reaction by only a pair of competing, first-order, irreversible reactions:

$$\text{(Raw coal)} \begin{cases} \xrightarrow{K_{v1}} \alpha_1\text{(Volatile)}+(1-\alpha_1)\text{(Residue)} \\ \xrightarrow{K_{v2}} \alpha_2\text{(Volatile)}+(1-\alpha_2)\text{(Residue)} \end{cases} \tag{5.6}$$

Releasing rate of volatile matter is expressed by the following equation in this model.

$$\frac{dw_v}{dt} = (\alpha_1 k_{v1} + \alpha_2 k_{v2})w_{RC} \tag{5.7}$$

where, α_i, k_{vi} and w_{RC} are the stoichiometric coefficient, reaction rate constant and mass of unreacted raw coal in a coal particle, respectively. The rate constants, k_{v1} and k_{v2}, are given by Arrhenius type equations and path 2 always has a higher activation energy than path 1. This combination of activation energies makes path 2 activate at a higher temperature range. Burgess et al.[12] tried to reflect coal properties to this two-competing reaction model and proposed a determination of stoichiometric coefficients α_1 based on the proximate analysis of coals instead of constant value.[8,13] They used volatile matter content in proximate analysis as α_1, and the smaller values of $2\times\alpha_1$ or unity as α_2. In the low temperature condition, the devolatilization proceeds mainly through path 1, which results in volatile matter release close to the proximate value. On the other hand, more volatile matter is generated through path 2 in high temperature range because the value of α_2 is larger than the proximate volatile matter content. This model can express an increase of volatile matter in the high temperature range without specifying the total volatile yield, although it uses a fairly simple representation.

These models assume that 'raw coal' is only a reactive material in the coal particle and calculate the generation rate of 'volatile matter'. However, some analyses require releasing rates of a particular chemical species as a result of simulation. Such rates highly depend on chemical composition and molecular structure of coal. One class of devolatilization models to reflect chemical nature on pyrolysis behavior is the `network model'. Network models describe coal molecules as a network or chains of aromatic nuclei connected by various bridges, and functional groups are attached to their periphery. Rates of dissociation of aromatic units, elimination of peripheral group, char formation (cross-linking of the clusters), and the other reaction steps are statistically determined based on the specified molecular structure. The network models are further classified by means of the description of coal structure. Three typical models of them are the functional group, depolymerization, vaporization and cross-linking (FG-DVC) model [14], the FLASHCHAIN model [15] and the chemical percolation devolatilization (CPD) model. [16]

These models have successfully reflected the chemical features of coals on devolatilization behavior and can be applied to various burning conditions (temperature, heating rate, pressure, etc.). However, these models need more calculation steps compared with first-order reaction models, so that it is preferable to simplify the calculation procedure for applying these simple models to practical combustion analysis.

5.2.2 Char combustion

The solid residue of coal pyrolysis, char, usually has porous structure and shows various textures such as network, balloon, skeleton, etc. dependent on the maceral contents of parent coal and the condition of devolatilization. The reaction rate of such porous medium strongly relates to structural factors (particle diameter, porosity, specific surface area, pore diameter, etc.) as well as its intrinsic reactivity and diffusion of reactants to the particle surface. The rate-limiting step in a heterogeneous reaction of porous particle depends on the temperature range. A typical variation of overall reaction rate constant with temperature is shown in Fig. 5-1. In low temperature range, chemical reaction is the rate limiting step (zone I). The reaction rate in middle temperature range is controlled by both chemical reaction and pore diffusion (zone II). Diffusion of reactants and products in the boundary layer limits the reaction rate at high temperatures (zone III). In the actual combustion field, reactions of char particles occur under various temperatures and gaseous atmospheres. Therefore, it is necessary to apply treatment of chemical reaction engineering to estimate reaction rate of char particles.

Under the assumptions of first order reaction and the combined rate limiting step of chemical reaction and bulk diffusion, reaction rate of a char particle is expressed as:

$$R_c = k_{ov} A_p m_{ox} = \frac{A_p m_{ox}}{1/k_f + 1/k_c} \tag{5.8}$$

where k_c and k_f are respectively the chemical reaction constant and mass transfer

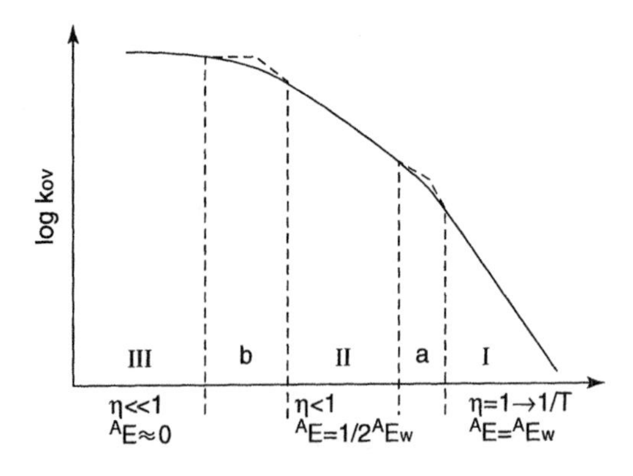

Fig.5-1 Variations of overall reaction rate constants with temperature

coefficient, and A_p is the external surface area of the particle. The chemical reaction constant k_c is given as the external-area basis and is expressed by using an effectiveness factor and intrinsic chemical reaction rate constant as:

$$k_c = \eta_i k_s A_s d_p \rho_p / 6 \tag{5.9}$$

$$\eta_i = \frac{1}{\phi_e}\left[\frac{1}{\tanh(3\phi_e)} - \frac{1}{3\phi_e}\right], \quad \phi_e = \frac{d_p}{6}\sqrt{\frac{k_s A_s \rho_p}{D_e}} \tag{5.10}$$

The value of the effectiveness factor η_i approaches unity when ϕ_e becomes less than 0.1, so that the reaction occurs uniformly within the particle. In contrast, reaction takes place only near the surface portion of the particle when ϕ_e is larger than 5.

In the raceway region of the blast furnace, char and fine coke particles co-exist and react competitively. These competitive reactions relate to the generation of fines in the lower part of the blast furnace. Although these two carbonaceous materials have the same reaction mechanism, their reactivity can be different from each other because of the differences of physical and chemical structures. These properties of char particles depend on the devolatilization conditions as well as parent coal properties. Therefore it is important to clarify the relationship between the char reactivity and coal pyrolysis condition for evaluating the char reaction rate precisely.

5.2.3 Heat and momentum exchange between particle and gas phase

Particle dispersion is an indispensable phenomenon for estimating the multi-dimensional combustion analyses because it is an essential factor to form temperature and concentration fields in a combustor. The particle dispersion is determined by the trajectories of all particles within the gas flow field. The momentum exchange between gas and particles plays an important role in particle motion in the combustion gas stream. Equation of motion of a particle, in viscous fluid under low particle Reynolds number ($Re_p = d_p \mid u - v \mid \rho_f / \mu_f$) conditions, is given by the following equation.

$$\frac{\pi d_p^3 \rho_p}{6}\frac{d\vec{v}}{dt} = 3\pi\mu_f d_p(\vec{u} - \vec{v}) + \frac{\pi d_p^3 \rho_f}{6}\frac{d\vec{u}}{dt} + \frac{1}{2}\frac{\pi d_p^3 \rho_f}{6}\left(\frac{d\vec{u}}{dt} - \frac{d\vec{v}}{dt}\right)$$
$$+ \frac{3}{2} d_p^2 \sqrt{\pi \rho_f \mu_f}\int_{t0}^{t} \frac{(d\vec{u}/dt' - d\vec{v}/dt')}{\sqrt{t - t'}} dt' + \vec{F} \tag{5.11}$$

Under the ordinary conditions of pulverized coal combustion, second through fourth term on right hand side can be omitted because $\rho_p >> \rho_f$ and $\mu_f << 1$ are satisfied.

Then, momentum changes of the particle are expressed by drag and external forces. Although the drag force in Equation (5.11) is described by the Stokes's law, which is a consequence of an analytical solution, it can only be applied to the low Reynolds number condition. The generalized form of the drag force for the spherical particle is given by the following:

$$R = C_D \frac{\pi d_p^2}{4} \frac{\rho_f}{2} |\vec{u} - \vec{v}|(\vec{u} - \vec{v}) \tag{5.12}$$

where C_D is the drag coefficient and is usually expressed as a function of the Reynolds number, for example[17]:

$$C_D \begin{cases} \dfrac{24}{Re_p}[1+0.1315 Re_p^{(0.82-0.55w)}] & \text{for } 0.01 < Re_p < 20 \\ \dfrac{24}{Re_p}(1+0.1935 Re_p^{0.6305}) & \text{for } 20 < Re_p < 737 \\ 0.44 & \text{for } 737 < Re_p \end{cases} \tag{5.13}$$

$$w = \log Re$$

The equation of motion of a spherical solid particle in the gas stream is described as follows.

$$\frac{d\vec{v}}{dt} = \frac{3\rho_f C_D}{4\rho_p d_p} |\vec{u} - \vec{v}|(\vec{u} - \vec{v}) + \frac{\vec{F}}{w_p} \tag{5.14}$$

Temperature difference between fluid and a particle causes heat exchange between them. Heat transfer rate from the fluid to the particle expressed as:

$$Q_{f\text{-}p} = A_p h_c (T_f - T_p) \tag{5.15}$$

where coefficient of proportionality h_c is the heat transfer coefficient. It has been measured under various conditions and was summarized as an equation in respect to the Nusselt number ($Nu = h_c d_p / \lambda_f$). Under the stagnant condition, thermal energy transmission occurs due to only heat conduction only and was theoretically derived for the spherical particle. For forced convection around a sphere, the Nusselt number is expressed as a function of the Reynolds number and the Prandtl number.

$$Nu = 2.0 + CRe^{1/2} Pr^{1/3} \tag{5.16}$$

This equation with the coefficient of 0.6 is known as the Ranz-Marshall equation [18] and is widely used in practical applications. For the condition that natural convection is predominant, Nusselt number is given by the function of the Grashof number and the Prandtl number:

$$Nu = 2.0 + 0.43(Gr \bullet Pr)^{1/4} \tag{5.17}$$

5.2.4 Radiative heat transfer

Radiative heat transfer, concerned with coal and char particles, is one of the most remarkable features of pulverized coal combustion compared with gaseous and liquid fuel combustion. Thermal radiation is electromagnetic radiation emitted by a body as a result of its temperature. Total energy emitted from black body, that is an ideal thermal radiator, is given by the Stefan-Boltzmann law:

$$q_b = \int_0^\infty q_{b\lambda} d\lambda = \sigma T^4 \tag{5.18}$$

The emissive power is proportional to the absolute temperature to the fourth power. The constant σ is called the Stefan-Boltzmann constant with a value of 5.669×10^{-8} W/ m^2K^4. An actual surface always emits less emissive power than does a black surface having the same temperature. The ratio of emissive power from the actual surface to one from the black surface is defined as the emissivity.

When radiant energy impinges on a material surface, it is partly reflected, absorbed and transmitted. The fractions of these energies are reflectivity, absorptivity and transitivity, respectively. The sum of these three values must be unity and the absorptivity takes same magnitude as the emissivity (Kirchhoff's identity).

Some species in combustion gas, such as carbon dioxide, water vapor and various hydrocarbons, emit and absorb radiant energy. They usually take part in radiative transfer by certain narrow wavelength bands. The decrease in monochromatic radiative intensity in absorbing media is expressed as following equation by assuming the absorption in absorbing media is proportional to the distance which radiation traverses and radiative intensity at that point (Beer's law).

$$I = I_o \exp(-K_a x) \tag{5.19}$$

In the pulverized coal combustor, coal particles and soot are scattering media as well as radiative emitter and absorber. Thus absorption, emission and scattering should be taken into account for the radiant energy balance. The variation of radiative intensity during radiation travels through a particle cloud with thickness of dl is expressed by the following integral-differential equation.

$$dI / dl = -(K_a + K_s)I + K_a I_b + (K_s / 4\pi)\int_{4\pi} I(\Omega)P(\theta,\Omega)d\Omega \tag{5.20}$$

The first term on the right hand side represents extinction due to absorption and scattering in different directions by the cloud. The second term is the emissive power from the cloud. The third terms is the emissive power from the incident radiation scattered from different direction into the direction considered.

The characteristics of radiative heat transfer within a combustor are given by solving this equation. The radiative intensity, however, is a function of direction and position, so this equation needs to be solved over all solid angle to get the radiation field. In addition, the integral term includes multiple scattering. Therefore, the analytical solution of this integral-differential equation is obtained only for very simple conditions and the numerical solutions require enormous computation power. From such background, a number of models of radiative heat transfer have been proposed.

The radiative heat transfer model taking the most straightforward treatment is the zone method. [19] This method divides the computational domain into a number of volume and surface zones. Then the radiative heat fluxes exchanged between the two zones are calculated with respect to all combinations. The geometrical configuration between each zone pair is taken into account in the radiative exchange coefficient. For the estimation accuracy, each divided zone should be small enough that its temperature and properties can be regarded as being uniform within it. This method requires a large number of zones for the analysis of a large and complicated furnace. Thus, few applications of the zone method to practical pulverized coal combustor have been made, although it generally gives fairly accurate results.

The Monte Carlo method [20] also divides calculation domain into finite control volumes. This method emits many bundles of radiation from each volume to random directions and follows each bundle that can be absorbed, scattered and reflected until its extinction. Although this method has good flexibility on furnace geometry and gives accurate results, it needs to follow statistically a large number of bundles of radiation. Thus the method requires a large capacity for computation to perform precise analysis.

The radiative heat ray method [21] was proposed to reduce the computational load of the Monte Carlo simulation. It calculates attenuation of many radiative heat rays, due to absorption and scattering, similar to the Monte Carlo method. In contrast to random projection of radiation bundles in the Monte Carlo simulation, the paths of the heat rays are specified prior to the calculation in the radiative heat ray method. It is reported that this method successfully improves computational performance in combination with the radiative energy absorption distribution (READ).

The flux method simplifies the angular variation of radiant intensities and describes radiation transfer by the balance of radiative heat flux in several discrete directions. The angular simplification is also applied to the scattering, then integral-differential equation of radiative transfer reduced into ordinary differential equations. The directions of radiative flux are usually set along coordinates of flow analysis performed simultaneously. These treatments allow the reduced radiative transfer equations to be solved by the same scheme as flow and heat transfer calculations. For this reason, the flux model is often used as a radiation model of combustion analyses. However, it is pointed out that the accuracy of the results depends on the angular configuration between directions of peak intensity of actual radiation and specified flux. [22]

5.2.5 Formulation of single coal particle burning

The preceding sections described the unit processes relating to the burning of single coal particle and showed the formulation of them. This section gives the mathematical expressions of single coal burning by using some of these model components.

The following assumptions are used to formulate burning behavior:

1) no fragmentation of particles,

2) no evaporation of ash and,

3) uniform distribution of composition and temperature in a particle.

According to the first two assumptions, a weight change of a coal particle is obtained from moisture evaporation, volatile matter release and heterogeneous reaction of char.

$$\frac{dw_p}{dt} = -\dot{w}_w - \dot{w}_v - R_c \tag{5.21}$$

where $\dot{w}_w$, $\dot{w}_v$ and R_c are rates of evaporation, volatile release and char reaction, respectively.

The two competing reaction model of devolatilization (in section 5.2.1) gives mass variations of raw coal, released volatile and char as:

$$\frac{dw_{RC}}{dt} = -(k_{v1} + k_{v2})\, w_{RC} \tag{5.22}$$

$$\frac{dw_v}{dt} = \dot{w}_v = (\alpha_{v1}k_{v1} + \alpha_{v2}k_{v2})\, w_{RC} \tag{5.23}$$

$$\frac{dw_c}{dt} = -\frac{dw_{RC}}{dt} - \frac{dw_v}{dt} - R_c \tag{5.24}$$

Moisture evaporation rate is expressed as the following equation by assuming convective heat supply from the gas phase as the rate-limiting step of this process.

$$\frac{dw_w}{dt} = \dot{w}_w = \frac{A_p h_c (T_g - T_p)}{L_w} \quad \text{for } T_g > T_p \geq T_b \text{ and } w_w > 0 \tag{5.25}$$

The burning rate of char was given by Equation (5.8) as shown in the section 5.2.2. Thus the mass variations of coal composition (moisture, raw coal, char and released volatile) and particle itself can be given by integrating these differential equations simultaneously. In these equations, the reaction rate constants and the moisture evaporation rate are expressed as functions of the particle temperature. Hence, the particle temperature usually needs to be calculated at the same time with other variables.

The heat balance equation of a coal particle is

$$w_p Cp_p \frac{dT_p}{dt} = A_p h_c (T_g - T_p) + \xi_v Q_v + \xi_c Q_c + \frac{\kappa_p}{n_p}(I_{tot} - I_p) \tag{5.26}$$
$$- (Cp_p T_p) \frac{dw_p}{dt} - L_v \dot{w}_v - L_w \dot{w}_w$$

Additionally, the motion and trajectory of the coal particle in the gas stream is calculated from Equation (5.14). Therefore, the burning behavior of a single coal particle in arbitrary fields of gas flow, temperature, concentration and radiative flux can be calculated.

5.3 Coupling up unit processes and combustion field

In the previous section, the burning behavior of a single coal particle has been described as a set of differential equations. To solve these equations, the distributions of gas velocity, gas temperature, gaseous species and radiative flux need to be given. In the actual combustion field, the burning behavior of coal particles and gaseous fields relate to each other. Therefore interactions between the gas phase and coal particles is necessarily take into account in the calculation of a gaseous field. In addition, flow regime in practical combustor is turbulent and it requires appropriate treatments in simulation. This section first describes the gas flow field and gaseous homogeneous reaction rate, then shows two different treatments of particle behavior in the gas stream.

5.3.1 Turbulent model

The flow regime in practical combustion devices is usually turbulent. The turbulent flow is characterized by random and three-dimensional motions of fluid in addition to the mean motion. This random motion provides much higher transport rates of mass, momentum and heat compared with the ones in the laminar flow. The motion of viscous fluids is described by the Navier-Stokes equation and continuity equation.

$$\frac{\partial \rho_g}{\partial t} + \frac{\partial}{\partial x_j}(\rho_g u_j) = 0 \tag{5.27}$$

$$\frac{\partial}{\partial t}(\rho_g u_i) + \frac{\partial}{\partial x_j}(\rho_g u_j u_i) = - \frac{\partial p}{\partial x_i} + \frac{\partial}{\partial x_j}(\mu_g \frac{\partial u_i}{\partial x_j}) + S_{ui} \quad (5.28)$$

The turbulent flow characteristics, including the above-mentioned random motions, are given by a complete and time-dependent solution of these equations. However, the numerical solution of these equations under the turbulent conditions requires a huge computer capacity and speed, since the time step and grid spacing should be fine enough to resolve micro structure of the turbulence flow and its transient variation. Fortunately, we usually need the time mean value of process variables rather than instantaneous values as the results of process simulation, and such analyses can be performed with much less computer capacity and computational time than the direct solution requires. In contrast to this advantage, turbulent (diffusive) transport rates in the turbulent flow field are unable to be given by only the mean motion of fluid and the laminar fluid properties. Thus, it is necessary to introduce some methods which reflect effects of random motion of turbulent fine structure on the mean fluid behavior. Equation of motion that describes the mean fluid motion is given by introducing the concept of time averaging:

$$\bar{\phi} = \frac{1}{T}\int_t^{t+T} \phi dt, \quad \phi = \bar{\phi} + \phi', \quad \bar{\phi}' = 0 \quad (5.29)$$

where averaging time should be sufficiently longer than turbulent time scale while it should be sufficiently short to resolve transient variation of mean flow. This averaging procedure generates a large number of extra terms involving the fluctuating components. Some of them can be reduced by using the assumption of constant physical properties; but some may not. The remaining terms having the form of $\rho \overline{u_i' u_j'}$ are called Reynolds stress. As a consequence of the generation of these new unknown variables, the closure problem of the equation set arises because the number of variables to be solved is more than that of conservation equations. Therefore, these cross-correlation terms should be approximated or modeled by using the mean dependent variables or properties. The procedure of closure is the heart of turbulent modeling.

The turbulence model, which is most widely used for combustion analysis, is the k–ε model. [23] This model adds two conservation equations, namely ones of turbulent kinetic energy and its dissipation rate, to governing equations to be solved. The Boussinesq hypothesis [24] and isotropic turbulence are employed in k–ε model. The Boussinesq gradient-diffusion hypothesis expresses the Reynolds stress as the product of gradient of mean velocity and turbulent viscosity.

$$-\overline{u_i' u_j'} = \nu_t\left(\frac{\partial \overline{u_i}}{\partial x_j} + \frac{\partial \overline{u_j}}{\partial x_i}\right) - \frac{2}{3} k\delta_{ij} \tag{5.30}$$

The turbulent viscosity, ν_t, is given by the Prandtl-Kolmogorov relationship:

$$\nu_t = C_\mu k^2 / \varepsilon \tag{5.31}$$

The local value of turbulent viscosity is calculated, based on the distributions of k and ε which are obtained by solving their transport equations. The Reynolds stress terms are given by the Boussinesq hypothesis and then mean flow field can be calculated.

The k–ε model is widely used in practical combustion simulation as a result of its simplicity and computational efficiency. However, the assumptions used in this model, namely Boussinesq hypothesis and isotropic turbulent viscosity, causes a limitation of its applicability. It is reported that the prediction accuracy of the k–ε model decreases in application to strongly swirling flows and wall jets [25,26] although several refinements [25,27,28] have been proposed for such applications. Thus the applicability of the turbulence model to the flow field of interest should be examined well prior to the simulating calculation.

5.3.2 Homogeneous gaseous combustion in turbulent flow field

Mass transport rate in turbulent flow field is much higher than that in the laminar field because the chemical components are carried by the turbulent random motions and its rate is faster than the molecular diffusion rate. The turbulent random motions stretch the reaction interfaces in the diffusion flame. Consequently, turbulent fields can provide higher combustion rates even with the same furnace volume. In other words, the turbulent motion closely relates to the supply and mixing of reactants in the turbulent combustion process, thus its effect should be taken into account in the estimation of overall reaction rates.

The flow fields calculated by the above-mentioned turbulence model are time-mean fields and the reaction rates substituted into the transport equations of chemical species should be also time-mean value. However, time-mean reaction rates in the turbulent fields are usually different from the reaction rate that is calculated based on the time-mean variables:

$$\bar{r}_A = \overline{m_A m_B \rho^2 A\exp(-E/RT)} \neq \bar{m}_A \bar{m}_B \bar{\rho}^2 A\exp(-E/R\bar{T}) \quad (5.32)$$

Under some conditions, two values can be equal, such as the case in which characteristic time of chemical reaction is much longer than that of turbulent mixing. Therefore it is necessary to introduce a model which describes time-mean reaction rates based on the time-mean variables and turbulent properties.

One modeling concept considers that the chemical reactions take place when molecular scale mixing of reactants occurs. It relates the reaction rate to the turbulent decay rate. The reaction models following this concept are 'eddy break up model' [29] and 'eddy dissipation model'. [30] Both models assume that time scale of combustion reaction is much shorter than that of turbulence. Therefore, the chemical reaction rate can be treated as infinity. The combustion rates in these models are expressed as follows:

Eddy break up model:

$$\dot{w}_{fu} = C_{ebu}\rho_g \frac{\varepsilon}{k} g_f^{1/2} \quad (5.33)$$

Eddy dissipation model:

$$\dot{w}_{fu} = C_{ed}\rho_g \frac{\varepsilon}{k} \min(m_{fu}, \frac{m_{ox}}{\alpha}) \quad (5.34)$$

where g_f in the eddy-break-up rate equation is the square of the mixture fraction fluctuation and is obtained by solving its transport equation. [31] In contrast to the eddy break up model, the reaction rate of the eddy dissipation model is related to the time-mean fraction of reactants. By this simplicity, the eddy dissipation model has been widely used in various combustion simulations. [32-40] However, this rate is based on the time scale of large eddy (k/ε) and does not follow the basic modeling concept precisely. Magnussen, et al. [40] proposed a refinement of the model based on the similarity between the energy transfer from macro-scale to fine structures and the mass transfer from surrounding fluid to the fine structure. In this refinement, the reaction rate is related to the characteristic velocity and characteristic length of Kolmogorov micro-scale structure and is expressed as:

$$\dot{w}_{fu} = 23.6\left(\frac{\nu \cdot \varepsilon}{k^2}\right)^{1/4} \frac{\varepsilon}{k} \chi \min(m_{fu}, \frac{m_{ox}}{\alpha}) \tag{5.35}$$

Another modeling concept also postulates instantaneous chemical reactions. This assumption denies fuel and oxidizer to exist at the same location and time, thus they appear intermittently. The time mean fractions of these reacting species can be determined by the probability density distribution, and reaction rate can be determined based on the variations of mean fractions. The models of this concept presume the form of probability density functions (PDF). Next, the conservation equations of mixture fraction or reactedness and its fluctuation are solved. The characteristic values of the PDFs are calculated from the local value of these variables, then mean fraction of reacting species are obtained. However, the forms of PDFs depend on flame structures and flow conditions, and are given for only a few conditions.

5.3.3 Particle motion in turbulent combustion fields

As mentioned above, the dispersion of pulverized coal particles closely concerns the formation of the combustion field in the furnace. Even though the coal particles have the same burning history, different particle dispersion results in different combustion fields. Therefore, the dispersion of pulverized coal particles should be estimated correctly for precise combustion simulations. Two major methods to describe particle motion in flow field are the `Lagrangian method' and the `Eulerian method'.

(1) Lagrangian method

The Lagrangian method tracks behavior of each particle, such as velocity, temperature, mass, composition, reaction rates, while the particle moves through combustion field. It is practically impossible to trace all coal particles in the furnace. Thus the particles are usually classified into several groups by their properties (diameter, coal type, etc.), and the equations of change described in section 5.2.5 are integrated for the representative particles.

In the Lagrangian method, particle behaviors are calculated along the discrete particle trajectories that represent classified particle groups. In contrast to this, the gaseous combustion field is described by the transport equations of the continuum phase. The Particle-Source-In cell (PSIC) model [41)] couples particle behavior to gaseous fields described by these two different schemes. The PSIC model records the changes in particle velocities, temperatures, weight and composition while the particle passes through the computational cell on its trajectory. These changes are translated into the interactions between particle and gas phases and substituted into the gas-phase governing equations as source terms.

The actual computation procedure of this technique starts with the calculation of the non-particle laden flow fields. Particle trajectories and changes in momentum, mass, enthalpy and composition of particles along their trajectories are calculated based on this gaseous field. The variations of these variables within each computational cell are recorded as particle source terms. New gaseous-fields are calculated by using these particle source terms. The particle source terms are renewed based on the new gaseous combustion field. The computations of particle behavior and gaseous fields are alternatively repeated until variation of the gaseous fields with the iteration becomes negligible. This iterating procedure allows the two-way interaction between the continuum and dispersed phases.

The turbulent motions of fluid have great affects on particle dispersion as well as transport of gaseous species, thus it should be taken into consideration for the particle trajectory calculation. For the Lagrangian approach, turbulent fluctuations are put into the fluid velocity in the particle-phase equation of motion. The turbulent fluctuations are stochastically generated and its intensity and interval of renewal are determined based on the turbulent properties. One method gives the fluctuation intensity following the probability density function having normal distribution with the standard deviation of $(2k/3)^{1/2}$. One turbulent eddy affects motion of a particle during the particle's stay within the eddy. The residence time of the particle in a turbulent eddy is considered to be equivalent to the time periods taken by the particle passing through the eddy or the eddy-dissipation time scale (eddy lifetime).

$$t_{tr} = l_e / |\vec{u} - \vec{v}| \tag{5.36}$$

$$t_e = l_e / (2k/3)^{1/2} \tag{5.37}$$

In the integration of the particle-phase equation of motion, the fluid-phase velocity fluctuation is regenerated at the shorter time of these two.

(2) Eulerian method

The Eulerian approach takes the balance of particle mass (or number density), momentum or other process variables with respect to a fixed space and describes particle behavior by conservation equations similar to the gaseous ones.

The steady-state equation of motion of non-reacting and deformable particle phase is expressed as:

$$\rho_p \frac{\partial \phi_p v_j}{\partial x_j} = 0 \tag{5.38}$$

$$\rho_p v_j \frac{\partial \phi_p v_i}{\partial x_j} = F_{pi} \tag{5.39}$$

where ϕ_p is the volumetric fraction of the particle phase, F_{pi} is the source term accounting the momentum exchange with the gas phase. The effect of the gaseous turbulent motions on particle dispersion can be evaluated by taking a similar method in the gas-phase turbulence modeling. The gradient-diffusion hypothesis is also applied to the cross correlation terms that appear in the time-mean equations of the particle phase.

$$-\overline{\phi'_p v'_i} = D_p(\partial \bar{\phi}_p \bar{\alpha} / \partial x_i) \tag{5.40}$$

$$-\overline{v'_i v'_j} = \nu_p(\partial \bar{v}_i / \partial x_j + \partial \bar{v}_j / \partial x_i) \tag{5.41}$$

Several methods to evaluate the turbulent kinetic viscosity of the particle phase have been proposed. Chen et al. [42] gave the following expression:

$$v_p = \frac{v_t}{1 + t^*/t_e} \tag{5.42}$$

where t^* is the relaxation time of turbulence.

$$t^* = d_p^2 \rho_p / (18 \mu_g) \tag{5.43}$$

Lagrangian approach is a straightforward method to express particle behavior within combustion fields, and it can easily reflect the particle properties to the particle-burning behavior. However, this method has to track a large number of particle trajectories and tends to require a long calculation time. In the Eulerian approach, the same solution scheme to the gas-phase equations can be applied to the particle-phase equations, thus it makes the simulation program simple. For the analysis of particles having broad or varying properties, classifications of the particles are introduced similar to the Lagrangian approach. However, each class of the particles needs the sets of governing equations and it results in an increase of complexity of solution. Regarding the particle reactions and the burning of coal particles, the Eulerian approach evaluates the reaction rates of particle clouds in the space of interest instead of tracking reactions of each particle. One of the methods to describe combustion behavior of fuel suspension is group combustion theory [43)], however, this theory has not yet be well established for the pulverized coal combustion.

5.4 Mathematical modeling of pulverized coal combustion

The modeling of unit processes of pulverized coal and the formulation in the combustion field were generally described in several previous sections. This section explains some practical applications of mathematical simulation to the pulverized coal combustion in a blast furnace.

For combustion modeling of pulverized coal injected to the blast furnace tuyere, one-dimensional models were developed by He et al. [44,45)] and Jamaluddin et al. [46,47)] in the 1980s. After the successful evaluation of combustion characteristics in the blowpipe and tuyere with the one-dimensional model, two or three-dimensional mathematical models were further developed in the 1990s. [48,49)]

5.4.1 One-dimensional mathematical model

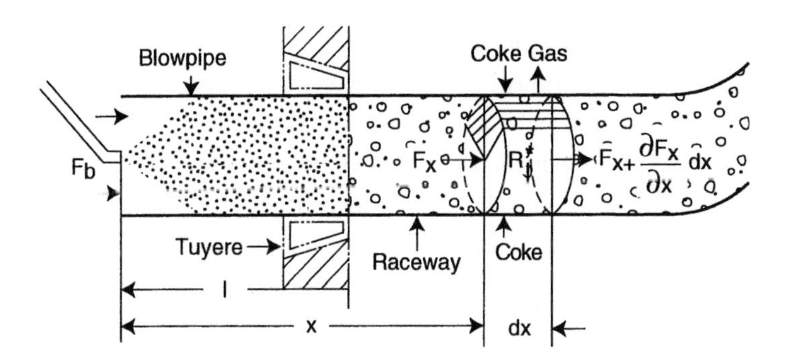

Fig.5-2 Schematic view of combustion zone with the injection of pulverized coal

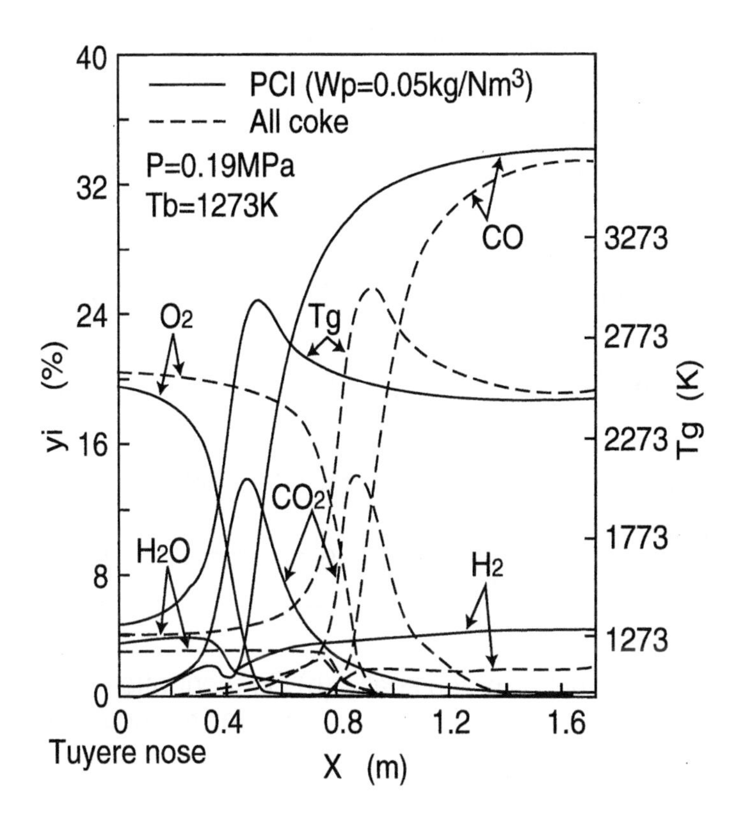

Fig.5-3 Comparison of the distribution of process variables in PCI and all coke operation

Features common to many one-dimensional models for the raceway zone are that the raceway is assumed as a loosely packed bed with high porosity through which the hot blast flows as a cylindrical and non-spreading jet. Gas leaves from the main flow zone through the roof of the cylindrical path. Competitive combustion of coke and pulverized coal is considered in the one-dimensional model by He et al. [45] as shown in Fig. 5-2. The single overall first order reaction model is employed for devolatilization. The combustion processes of injected pulverized coal along the tuyere axis are calculated in the model with additional heterogeneous reactions of O_2, CO_2 and H_2O with char and coke particles.

Computational results of the gas composition and temperature distribution in front of the tuyere are given in Fig. 5-3 for both with and without pulverized coal injection. Peak positions of CO_2 and temperature and depletion of oxygen content is moved toward the tuyere by the pulverized coal injection. The combustion of pulverized coal in the tuyere and raceway generates excess heat for an early rise of gas temperature and enhances all relevant reactions such as coke and coal combustion. A slightly lower peak value of the gas temperature is obtained in the pulverized coal injection due to the decomposition

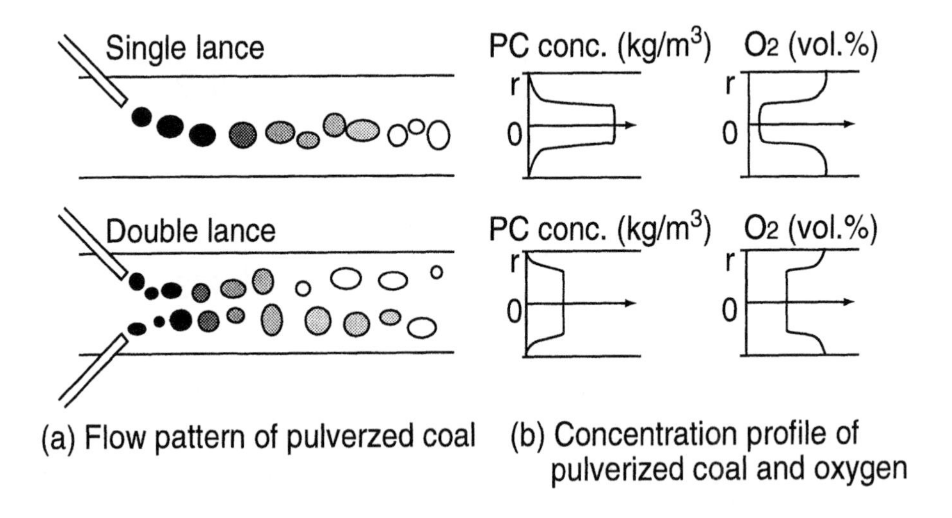

Fig.5-4 Flow pattern of pulverized coal particles and distribution of oxygen concentration in a blow pipe

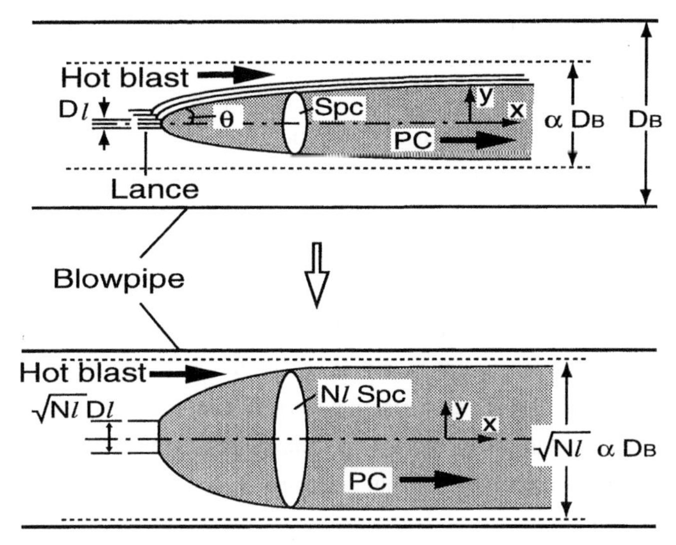

N l : Number of lance (-)
Spc : Cross sectional area of one PC flow (m^2)

Fig.5-5 Modeling concept of pulverized coal flows in blow pipe

heat of coal.

Sato et al. [50)] introduced a mixing rate of pulverized coal and hot blast air in the blowpipe as shown in Fig. 5-4. According to the observation by Ariyama et al. [51)] that the mixing rate of pulverized coal and the surrounding air are relatively small and the pulverized coal combustion remains within the main coal flow, the coal dispersion zone is presumed in the blowpipe as in Fig. 5-5. Combustion calculation with a one-dimensional model is carried out within the dispersion zone. Parameters for coal flow dispersion are chosen to describe various injection conditions such as the number of injection lance and lance geometry. The single overall first order reaction model is used for devolatilization and the mixed control model of diffusion and chemical reaction is chosen for char combustion.

The outline of the simulation result with the one-dimensional model is explained as follows. The combustion efficiency of pulverized coal in the raceway is improved by the enhanced dispersion of coal in the blowpipe which results in the sufficient heating of coal particles before the competitive combustion in the raceway and provides better mixing with the surrounding oxygen. The enhanced dispersion is achieved typically by the double lance arrangement.

Xiao et al. [52] has developed a mathematical model of the raceway zone with pulverized coal and fine ore injection. The two competing reaction model of devolatilization and the mixed control model of diffusion and chemical reaction for char combustion are used in the model. The reaction rate of char is assumed as a function of the fractional burnout. The model considers reactions of fine iron ore such as thermal decomposition, heterogeneous reduction and smelting reduction of iron oxide.

Figure 5-6 demonstrates typical examples of simulation results, the effect of oxygen enrichment and blast temperature on reaction degree of iron ore and coal at 1.5 m from the tuyere nose. An increase in the blast temperature gives a higher reaction degree while the oxygen enrichment improves the final reaction degree of iron ore and coal.

5.4.2 Two-dimensional mathematical model

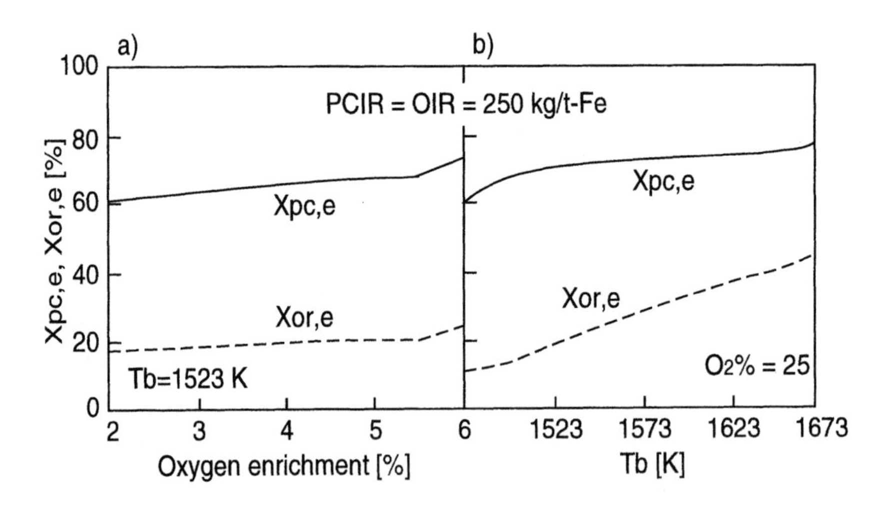

Fig.5-6 Effect of oxygen enrichement and blast temperature on reaction degree at 1.5m from tuyere nose
a) Effect of oxygen enrichment b)Effect of blast temprature

A two-dimensional mathematical model of pulverized coal combustion, which consists of the pulverized coal combustion model in the blowpipe and competing combustion model in a coke packed bed, has been developed by Nogami et al. [48] to clarify the flow, heat transfer and reactions of pulverized coal in the tuyere and raceway.

Computations of pulverized coal combustion in the blowpipe are carried out with the axi-symmetrical two-dimensional model in a cylindrical computational region fitted at the center of the tuyere. The two-equation turbulence model, $k-\varepsilon$ model, was used to calculate turbulent features of the gas phase. The Lagrangian method for particle phase is applied to calculate the turbulent dispersion of coal flow, heat transfer and reactions.

On the other hand, the competing combustion model in a coke packed bed considers

Table 5-1 Reaction parameters for single coal particle combustion

Devolatilization rate $k=A_i \exp(E_i/T_p)$			
Devolatilization path	α_{vi}	A_i [s^{-1}]	E_i [K]
1	VM/100	3.7×10^5	18000
2	2x α_{vi}	1.46×10^{13}	30200
Char combustion rate $k=A_c \exp(E_c/T_p)$			
	A_c [kg/m^2 s atm]		Ec
Tp < 1273K	87100		18000
Tp > 1273K	$-4.9 + 3.85 \times 10^{-3}\ T_p$		0

VM : Volatile matter content (wt% - d. a. f.)

Table 5-2 Kinetic coefficients for heterogeneous reaction

Reaction	Coefficient
$C + 1/2\ O_2 = CO$	
$C + O_2 = CO_2$	$7260 \exp(-18000/T_m) R\ T_g$
$C + CO_2 = 2CO$	$8.31 \times 10^9 (\rho / A_b) \exp(-30200 / T_m)$
$C + H_2O = CO + H_2$	$13.4\ (\rho / A_b)\ T_m \exp(-17300 / T_m)$

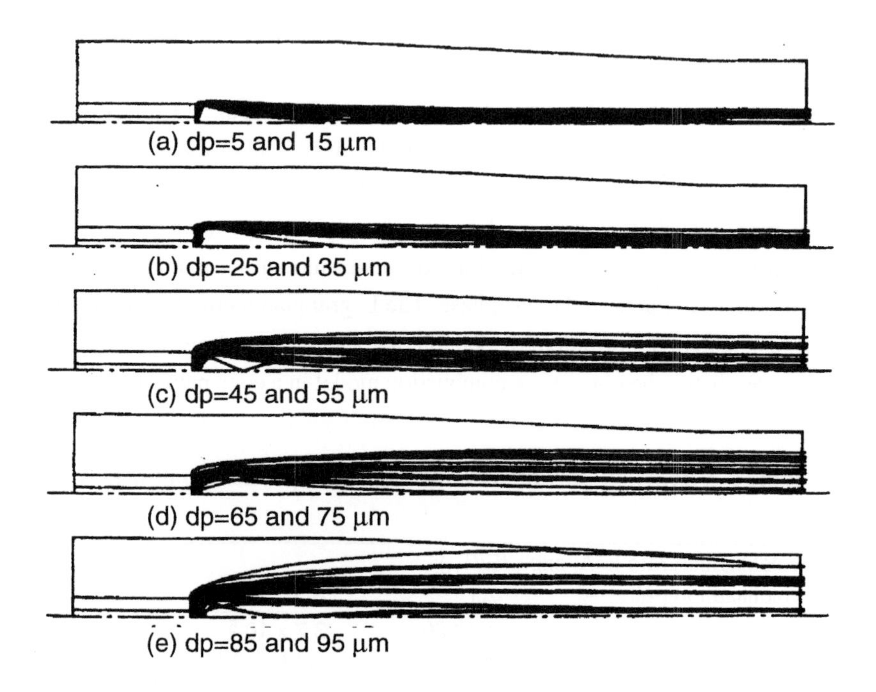

Fig.5-7 Calculated particle trajectories of low-VM coal

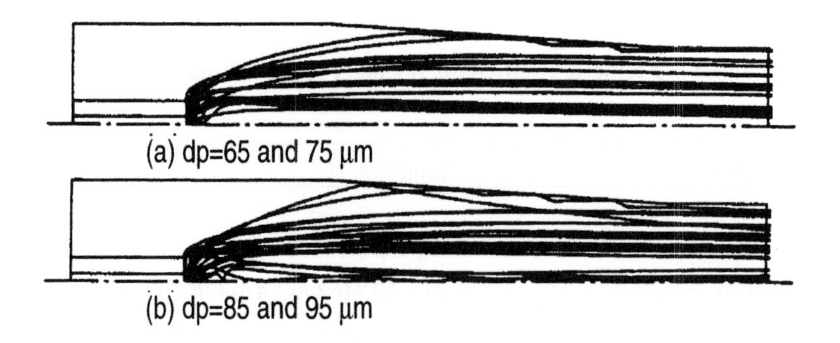

Fig.5-8 Calculated particle trajectories of high-VM coal

all relevant transport phenomena such as flow and combustion of both pulverized coal particles and coke particles and gas flow. The computational region is a two-dimensional $x-y$ plane set on the tuyere and furnace axis. The output of the pulverized coal combustion model in the blowpipe is used as in-flow conditions of the coke packed bed model. Coke particles in the raceway are assumed as a quasi-fluid and the two-phase flow of coke and gas phase was calculated in the model.

The two competing reaction model of the devolatilization and the mixed control model of diffusion and chemical reaction for char combustion are used in this model. The model used the heterogeneous reactions of O_2, CO_2 and H_2O and those kinetic parameters shown in Tables 5-1 and 5-2.

Some examples of particle trajectories obtained with the mathematical model are shown Figs. 5-7 and 5-8. The dispersion of coal flow in the blowpipe is relatively small for all particle sizes at the low-volatile-matter coal injection. Most particles flow within the diameter of injection lances. This is caused by the small radial diffusion of fine particles

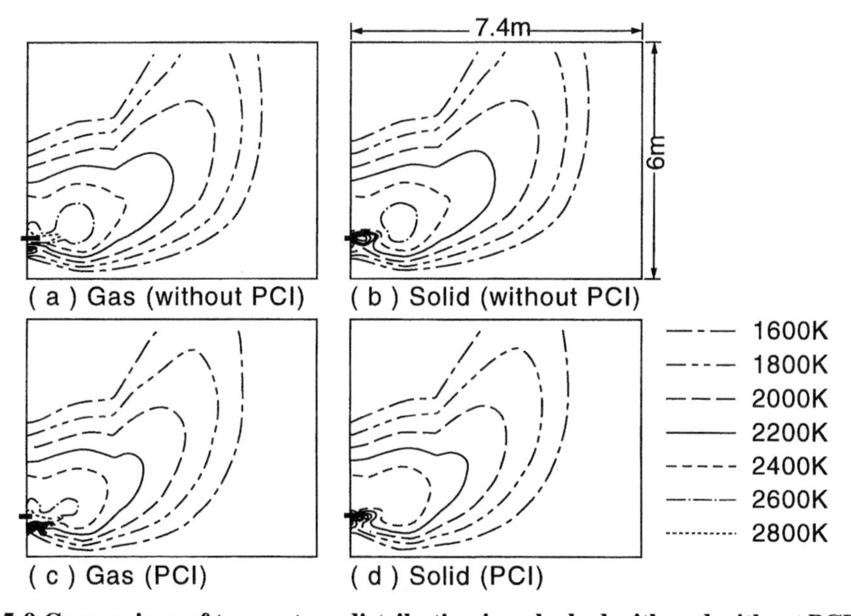

Fig.5-9 Comparison of temperture distribution in coke bed with and without PCI

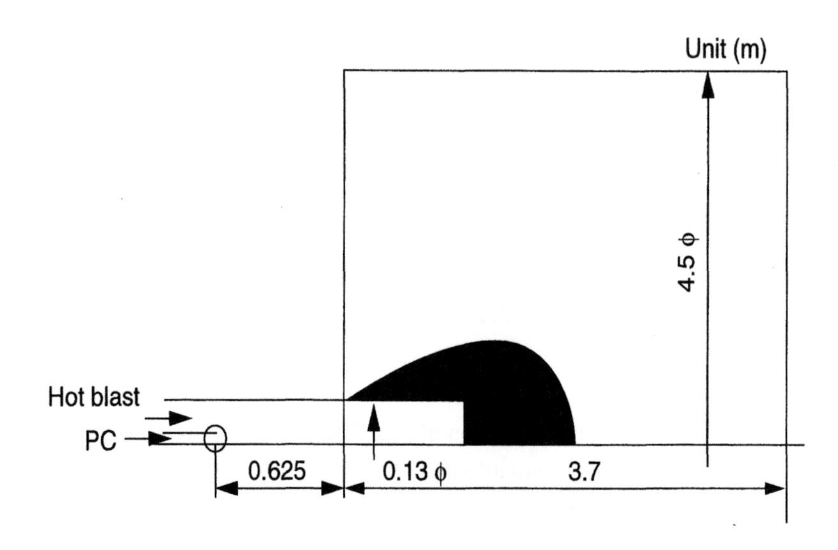

Fig.5-10 Simulation conditions of the coke and PC combustion

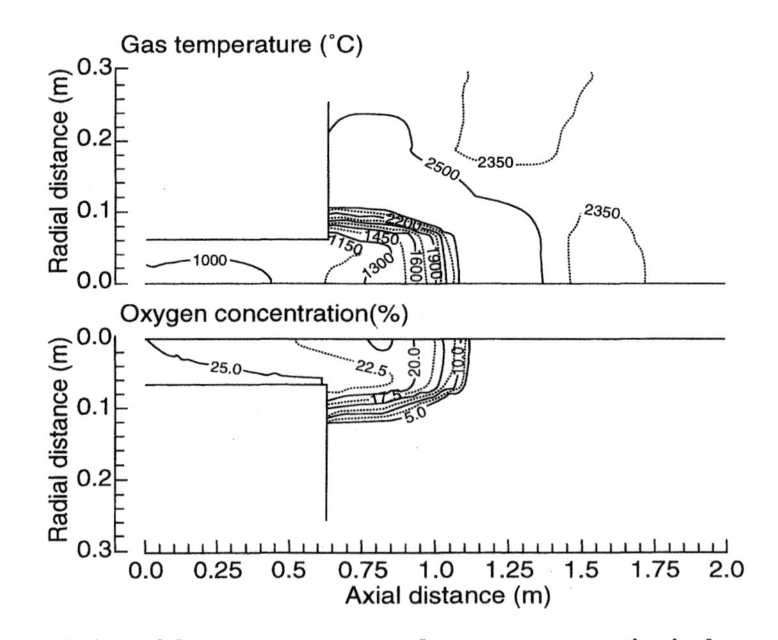

Fig.5-11 Predictions of the gas temperature and oxygen concentration in the raceway

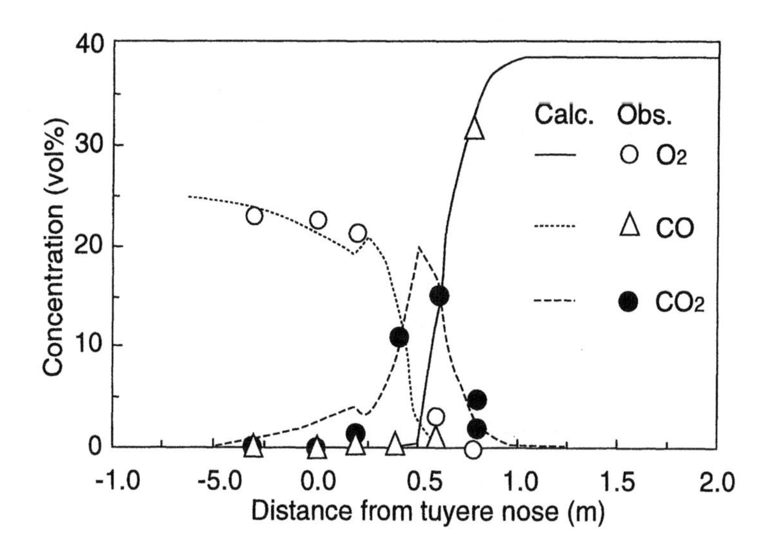

Fig.5-12 Predictions of axial profiles of gas compositions in the raceway

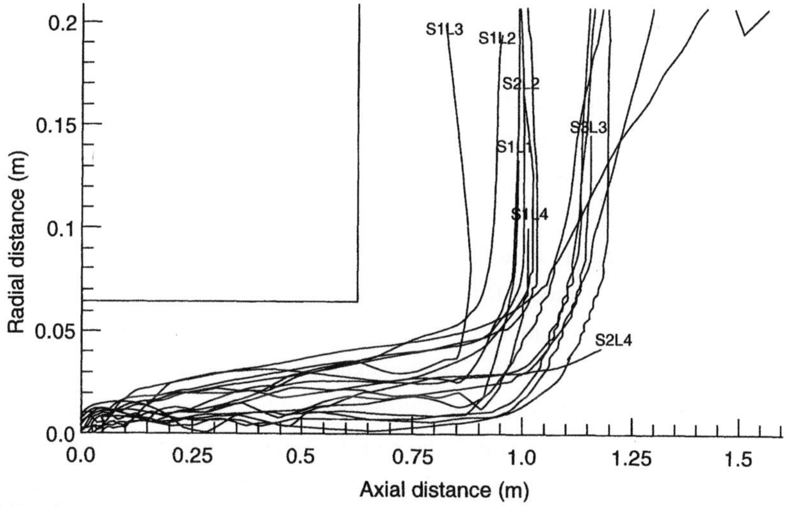

Fig.5-13 Typical coal particle trajectories predicted by the stochastic model for the pulverized coal injection.
The symbols, SiLj denote the size group i starting location j for the coal particle trajectories

with a diameter less than 30 μm in the blowpipe where the gas velocity is extremely high, 188 m/s. On the other hand, the injection of high-volatile-matter coal results in rapid devolatilization downstream the injection lance tip, and then large radial dispersion of coal. Some particles reach the inner wall at the exit of the tuyere. The simulation illustrates that particle dispersion in the blowpipe is strongly affected by the turbulent fluctuation, particularly by the intensive turbulence induced by the injection lance. The dispersion behavior is affected by the volatile matter content and particle size as well as lance geometry and configuration.

The temperature distribution around the raceway obtained with the coke packed bed model is given in Fig. 5-9. The hot blast introduced from the tuyere reacts with coke and coal particles in the raceway and shows a rapid temperature increase. After the peak temperature in the vacant space of the raceway, the gas temperature gradually decreases by the endothermic reaction of CO_2 and carbon in the coke particle. The temperature change with pulverized coal injection shows faster heating and lower peak temperature than those without pulverized coal injection. Although those changes with pulverized coal injection are in good agreement with the calculation by the one-dimensional model, more detailed and clear insight of the combustion processes of pulverized coal and coke in the raceway can be drawn with the two-dimensional mathematical model.

Takeda and Lockwood [49] has developed an integrated two-dimensional mathematical model of pulverized coal combustion in the blowpipe and raceway. As shown in Fig. 5-10, the axis of an axi-symmetrical computational area is fitted at the center of the blowpipe. As a single computational area covers the blowpipe, tuyere and raceway, the effect of injection design can be directly evaluated with the final combustion degree at the outer boundary of the raceway.

Similar to the Nogami's model, Takeda and Lockwood [49] have used the two-equation turbulence model, $k-\varepsilon$ model for gas turbulence. The Lagrangian method for particle phase with stochastic fluctuation is applied to calculate turbulent dispersion of coal flow, heat transfer and reactions. In the coke packed bed, they have developed an original turbulence model, $k-l_m$ model for gas turbulence, which describes randomness of gas and coal particles flowing through the bed. Particle trajectories and reactions were calculated with the Lagrangian method using a turbulence kinetic energy from the $k-l_m$ model.

Typical computational results of temperature and oxygen content distribution are shown in Fig. 5-11. Only a small dispersion and combustion of injected coal flow are observed in the blowpipe as indicated with a minor temperature increase and oxygen depletion. After the coal flow enters the raceway, rapid changes of gas temperature and oxygen concentration are found with the radiation heat transfer from the surrounding coke particles and resultant pulverized coal combustion. Calculated gas composition along the tuyere axis agrees with the measured gas composition as shown in Fig. 5-12.

Typical examples of particle trajectories in the blowpipe and raceway are shown in Fig. 5-13. Injected coal particles are dispersed by the stochastic fluctuation of turbulence. The dispersion in the blowpipe and tuyere is relatively small and the coal flow does not reach the tuyere wall even at the tuyere nose. On the other hand, rapid dispersion is found in the raceway due to the intense turbulence generated by the coal combustion and interaction with the packed bed. As demonstrated in Fig. 5-13, modifications of burner design and injection conditions can be directly accessed with this integrated model.

An existing two-dimensional mathematical model has already included all basic features for the development of an advanced three-dimensional model of pulverized coal combustion. The three-dimensional model is not so difficult to develop with the present level of computational fluid dynamics.

Although Clixby et al. [53)] reported three-dimensional analysis of gas flow in the raceway region, no three-dimensional model of pulverized coal combustion has developed yet. Practical problems on the three-dimensional model are insufficient three-dimensional measured data in the raceway and mathematical modeling of coke movement in the raceway and pulverized coal combustion in a packed bed.

5.4.3 Future subjects on a mathematical model

Recent developments of a mathematical model of pulverized coal combustion and their basic applications are explained in this section. These models provide clear understandings on the combustion process of pulverized coal in the blowpipe and raceway. Practical and industrial applications have recently spread and will be explained in a later chapter such as quantitative analysis of combustion efficiency for various coal characteristics and injection conditions.

Pulverized coal combustion proceeds in a relatively short period and in the blowpipe and raceway, which are very small regions in a modern large blast furnace. However, the combustion process affects overall performances of a blast furnace through temperature, gas composition and packing structure changes. Requirements for the quantitative evaluation of the combustion process are increasing further for better blast furnace performance.

The following are future subjects for the development of a mathematical model of pulverized coal combustion:

1) Three-dimensional mathematical model validated against the three dimensional measurement in the raceway and surrounding coke packed bed,
2) Dispersion and reaction of pulverized coal in a vacant space,
3) Modeling the turbulent features of a packed bed and behavior of pulverized coal in a packed bed,
4) Modeling the coke movement in the raceway and reaction and degradation of coke particles in the raceway,
5) Flow and reactions of unburned pulverized coal in a blast furnace.

REFERENCES

1) I. W. Smith: 19th Symp. (Int'l) Combust., (1982) , p. 1045
2) M. Sadakata: Nensho Kenkyu, 72 (1986), 18
3) T. Saotome, H. Murata, M. Saito, M. Sadakata and T. Sagai: 26th Proceedings of the Japanese Symposium on Combustion , (1998), p. 218
4) S. W. Kang, A. F. Sarofim and J. M. Beer: 22nd Symp. (Int'l) Combust., (1988), p.145
5) Y. Yamamoto, T. Ohe and K. Ohtake: Nippon Kikai Gakkai Ronbunshu, B60 (1994),p. 649
6) K. Annamalai and P. Durbetaki: Combust. Flame, 29 (1977), p. 193
7) JIS M 8812 "Proximate analysis of coal and coke" (1984)
8) H. Kobayashi, J. B. Howard and A. F. Salofim: 16th Symp. (Intn'l) Combust., (1977), p. 411
9) D. W. van Krevelen, C. van Heerden and F. J. Huntgens: Fuel, 30 (1951), p. 253
10) S. Badzioch and G. G. Hawksley: Ind. Eng. Chem. Proc. Des. Dev., 9 (1970), p. 521
11) D. B. Anthony, J. B. Howard, H. C. Hottel and H. M. Meissner: 15th Symp. (Intn'l) Combust., (1975), p.1303
12) J. M. Burgess, A. S. Jamaluddin, M. J. McCarthy, J. G. Mathieson, S. Nomura, J.

S. Truelove and T. F. Wall: Proc. Joint Symp. ISIJ and AIMM, Tokyo, Japan, (1983), p.129

13) S. K. Ubhayakar, D. B. Steckler, C. W.von Rosenberg and R. E. Gannon: 16th Symp. (Intn'l) Combust., (1977), p. 427

14) P. R. Solomon, D. G. Humblcm, R. M. Carangelo, M. A. Serio and G. V. Despande: Energy Fuels, 2 (1988), p. 405

15) S. Niksa and A. Kerstein: Energy Fuels, 5 (1991), p. 647

16) T. H. Fletcher, A. R. Kerstein, R. J. Pugmire, M. S. Solem and D. M. Grant: Energy Fuels, 6 (1992), p.414

17) R. M. Clift, J. R. Grace and M. E. Weber: "Bubbles, Drops and Particles", Achademic Press, New York, (1978), p.97

18) W. E. Ranz and W. R. Marshall: Chem. Eng. Prog., 48 (1952), p. 141

19) H. C. Hottel and E. C. Cohen: AIChE J., 4 (1958), p.3

20) H. Taniguchi and M. Hunatsu: Nippon Kikai Gakkai Ronbunshu, 36 (1970), p.610

21) H. Hayasaka, K. Kudo, H. Taniguchi, I. Nakamachi, T. Omori and T. Katayama: Nippon Kikai Gakkai Ronbunshu, B52 (1986), p. 1731

22) A. F. Sarofim: 21st Symp. (Intn'l) on Combust., (1986), p. 1

23) W. P. Jones and B. E. Laundar: Int. J. Heat Mass Trans., 15 (1972), p. 302

24) J. Boussinesq: Mem. Acad. Sci., 23 (1877), p. 1

25) T. Okamoto and T. takagi: Nippon Kikai Gakkai Ronbunshu, B53 (1987), p. 3338

26) T. Kobayashi: "Numerical Fluid Dynamics (in Japanese)", ed. by M. Hobara and H. Daiguji, Univ. of Tokyo press, Tokyo, (1992), p. 237

27) T. Kobayashi and M. Yoda: Nippon Kikai Gakkai Ronbunshu, B52 (1986), p. 3230

28) B. E. Launder, C. H. Priddin and B. I. Sharma: J. Fluid Eng., 99 (1977), p. 231

29) D. B. Spalding: 13th Symp. (Intn'l) Combust., (1971) , p. 649

30) B. F. Magnussen and B. H. Hjertager: 16th Symp. (Intn'l) Combust., (1977) , p. 719

31) D. B. Spalding: Chem. Eng. Sci., 26 (1971), p. 95

32) A. S. Novick, G. A. Miles and D. G. Lilley: J. Energy, 3 (1979), p. 95

33) A. D. Gosman, F. C. Lockwood and A. P. Salooja: 17th Symp. (Intn'l) Combust., (1979), p.747

34) F. C. Lockwood, A. P. Salooja and S. A. Syed: Combust. Flame, 38 (1980), p.1
35) D. G. Khalil, P. Hutchinson and J. H. Whitelaw: 18th Symp. (Intn'l) Combust., (1981) , p. 1927
36) T. Takagi and S. Kotoh: Nippon Kikai Gakkai Ronbunshu, B48 (1982), p. 2609
37)Y. Onuma, N. Morinaga, Y. Morinaga and K. Furushima: Nippon Kikai Gakkai Ronbunshu, B53 (1987), p. 3423
38) T. Takanohashi, S. Tanno, H. Aoki, T. Amagasa and S. Otani: Kagaku Kogaku Ronbunshu, 14 (1998), p. 272
39) H. Aoki, S. Tanno, T. Miura and S. Ohnishi: Nippon Kikai Gakkai Ronbunshu, B57 (1991), p. 2152
40) B. F. Magnussen, B. H. Hjertager, J. B. Olsen and D. Bhaduri: 17th Symp. (Intn'l) Combust., (1979) , p. 1383
41) C. T. Crowe, M. P. Sharma and D. E. Stock: Trans. ASME J. Fluid Eng., 99 (1077), p. 325
42) C. P. Chen: Ph. D. Thesis, Michigan State Univ. (1983)
43) H. H. Chiu, H. Y. Kim and E. J. Croke: 19th Symp. (Intn'l) Combust., (1982) , p. 971
44) M. Kuwabara, Y. Hsieh, K. Osobo and I. Muchi: Aust. Inst. Min. Metall. Symp. Ser. No. 26, (1981) , p. 7-1
45) J. He, M. Kuwabara and I. Muchi: Tetsu-to-Hagane, 72 (1086), p.35
46) A. S. Jamaluddin, T. F. Wall and J. S. Truelove: Ironmaking Steelmaking, 13 (1986), p. 91
47) A. S. Jamaluddin, T. F. Wall and J. S. Truelove: 21st Symp. (Intn'l) Combust., (1986) , p. 575
48) H. Nogami, T. Miura and T. Furukawa, Tetsu-to-Hagane, 78 (1992), p. 1222
49) K. Takeda and F. C. Lockwood: ISIJ Int., 37 (1997), p. 432
50) M. Sato, R. Murai and T. Ariyama: Tetsu-to-Hagane, 82 (1996), p. 731
51) T. Ariyama, M. Sato, Y. Yamakawa, Y. Yamada and M. Suzuki: Tetsu-to-Hagane, 80 (1994), p.288
52) X. Xiao, K. Nozawa, S. Sasahara, M. Shimizu and S. Inaba: Tetsu-to-Hagane, 78 (1992), p. 1230
53) G. Clixby and R. R. Willmers: Second European Ironmaking Congress, (1991), p.434.

NOMENCLATURE

A : area (m^2), pre-exponential factor
C : constant
C_D : drag coefficient
Cp : specific heat ($J \cdot kg^{-1}K^{-1}$)
C_μ : coefficient
D_e : effective diffusivity ($m^2 \cdot s$)
d : diameter (m)
E : activation energy ($J \cdot mol^{-1}$)
F : force (N)
g_f : square of fluctuation
h_c : heat transfer coefficient ($W \cdot m^{-2}K^{-1}$)
I : intensity of radiation
K : radiation coefficient
k_i : rate constant, turbulent kinetic energy (m^2s^{-2})
k_f : mass transfer coefficient ($m \cdot s^{-1}$)
L : latent heat ($J \cdot kg^{-1}$)
l_e : eddy length scale (m)
m : mass or mole fraction (-)
n_p : particle number density
Q : heat
q : emissive power ($W \cdot m^{-2}$)
R : gas constant ($J \cdot mol^{-1}K^{-1}$)
r : reaction rate ($kg \cdot s^{-1}$) or ($kg \cdot m^{-3}s^{-1}$)
S : source term
x : coordinates (m)
T : temperature (K)
t : time (s)
t^* : relaxation time of turbulence (s)
t_e : eddy life time (s)
u : gas velocity ($m \cdot s^{-1}$)
v : particle velocity ($m \cdot s^{-1}$)

w : weight (kg)
$\dot{w}$: reaction rate (kg • s^{-1}) or (kg • $m^{-3}s^{-1}$)

Greek symbols

α : stoichiometric coefficient
δ_{ij} : Kronecker delta
η : effectiveness factor
ε : dissipation rate of turbulent kinetic energy (m^2s^{-3})
λ : thermal conductivity (W • $m^{-1}K^{-1}$)
μ : viscosity (Pa • s)
ν : kinetic viscosity (m^2s)
ϕ : general variables
ϕ_e : Thiele modulus
ϕ_p : volumetric fraction of particle phase
Ω : angle
θ : angle
ρ : density (kg • m^{-3})
σ : standard deviation
ξ : fraction

Dimensionless numbers

Gr : Grashof number
Nu : Nusselt number
Pr : Prandtl number
Re : Reynolds number

subscript

a : absorption
b : black body, bulk
c : char
ebu : eddy break up
ed : eddy dissipation

f : fluid
fu : fuel
g : gas
ov : overall
ox : oxidizer
p : particle
RC : raw coal
r : residue
s : surface, scattering
v : volatile
w : water

CHAPTER 6

Advanced injection lances for high rate PCI

Recently, it is becoming more important to increase pulverized coal (PC) injection rate in excess of 200 kg/thm so as to reduce the cost of pig iron, substitute metallurgical coal to steam coal and prolong coke oven life. From these backgrounds, active research and development efforts are underway on PC combustion to improve combustibility. To develop PC lance with high combustibility, combustion mechanism of PC in the actual blast furnace condition must be clarified. As already described in Chapter 3, the behavior of PC particle groups injected from lances is greatly affected by the high-speed blast. The consumption rate of oxygen in the gas, which intimately affects combustion efficiency, is in turn affected by the behavior of PC particle groups. All these factors should be considered to design PC lance with favorable combustibility under the given combustion conditions.

6.1 Key factor to design injection lance with high combustibility

Figure 6-1 schematically shows the fundamentals for improving combustibility. One way is to control the PC flow to increase the dispersion of the particles in the cross

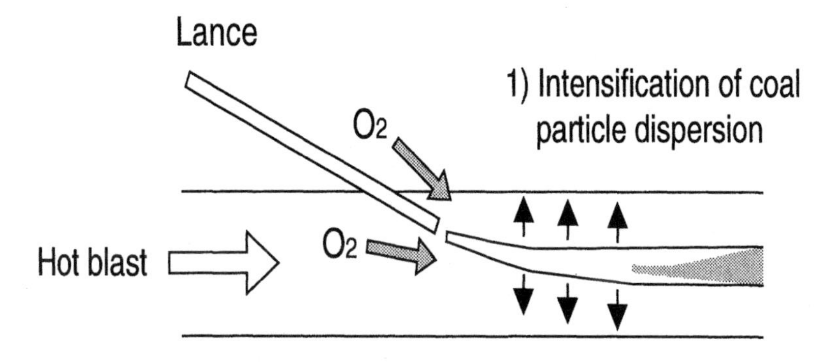

Fig.6-1 Key factors to design injection lance

sectional direction in the blow-pipe to most effectively utilize oxygen in this area. However, as the PC particles which are accelerated by the high-speed hot blast have a large momentum, it is not easy to control such PC particle flows in the cross sectional direction. An intensified dispersion of PC particles in such a condition inevitably generates pressure drop of gas flow, so this measure has certain constraints. It requires specific consideration to determine the suitable structure and arrangement of PC injection lance.

A second means would be to directly introduce oxygen into the particle flow. The contact between PC particles and oxygen can be improved if oxygen is locally enriched at some positions in the mainstream of the PC particle flow.

The principles for developing injection lance with high combustibility should be based on these above considerations described above. Additionally, from the viewpoint of developing practical PC injection lance applicable to actual blast furnace operation, other factors such as the deposition of ash from PC in the tuyere, erosion of the tuyere and the durability of the injection lance must be taken into consideration.

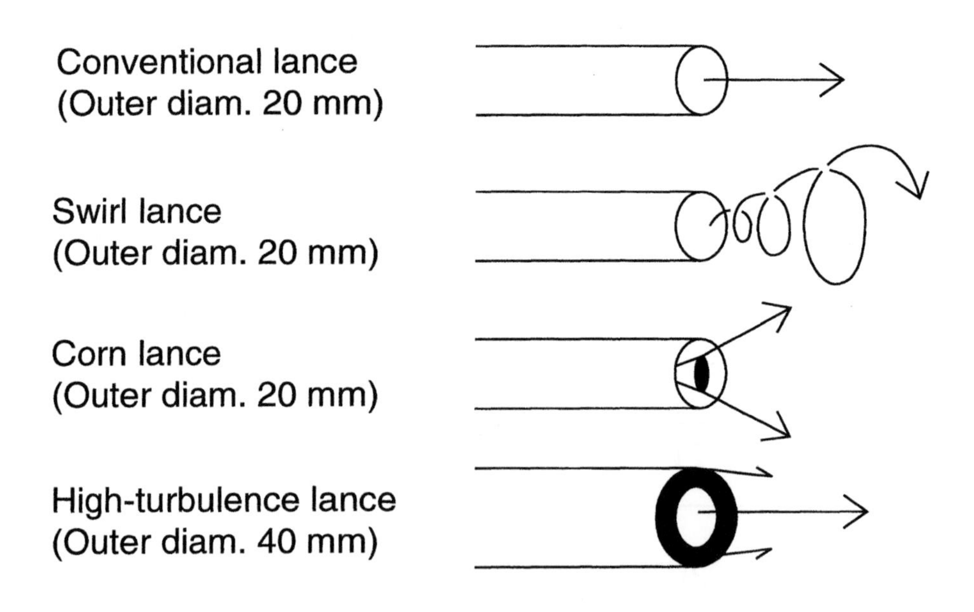

Fig.6-2 Lance geometry modifications for large dispersion of pulverized coal particles

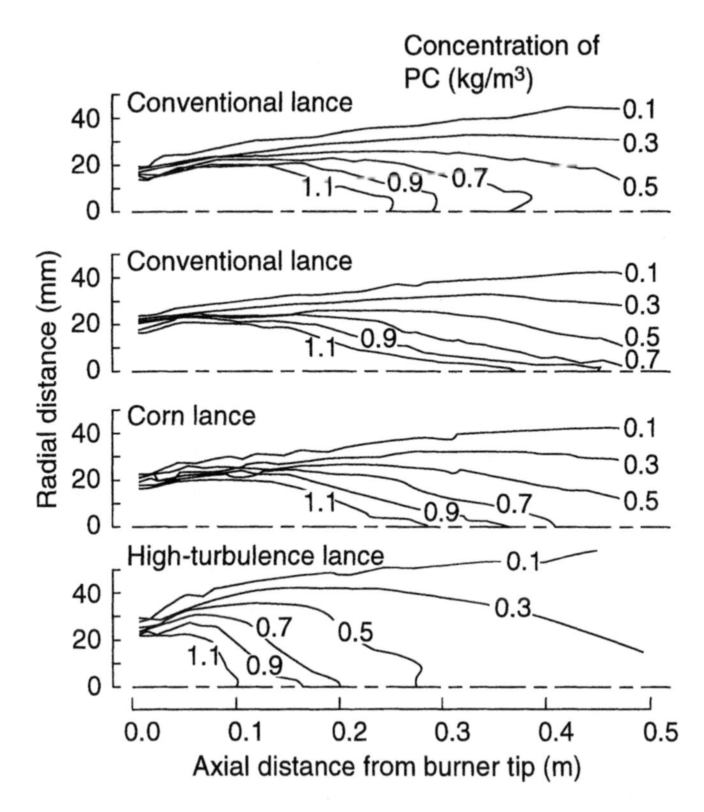

Fig.6-3 Effect of lance geometry on pulverized coal dispersion

Table 6-1 Effect of lance geometry on combustion degree

Lance type	Combustion efficiency (%)
Conventional lance	86.8
Swirl lance	87.2
Corn lance	87.1
High-turbulence lance	90.5

6.2 Advanced injection lance

6.2.1 High turbulence type lance

Takeda et al. investigated the effect of the four different structures of lances on combustibility as shown in Fig. 6-2 -- conventional lance, swirl lance, corn lance, and high-turbulence lance -- in order to increase the PC dispersion. They investigated the relationship between the lance structure and combustion efficiency using a mathematical model. They studied the states of dispersion of PC particles just beyond the lance, as shown in Fig. 6-3, and the combustion efficiency, as shown in Table 6-1. They consequently came to the conclusion that a thick-wall injection lance that can increase turbulence immediately downstream of the tip is effective for enhancing combustibility. The effect of the swirl lance is not clearly recognized. They confirmed these results by a small combustion furnace. The measurement results are shown in Figs. 6-4 and 6-5. They concluded that increasing the lance diameter increases the rate of ignited particles and improves the combustibility, as shown in Fig. 6-4. However, if the wall thickness is increased too much at a given diameter, the injection velocity becomes too large for particles to disperse, which adversely affects combustibility. They carried out the experiments with a high-turbulence, thick-wall lance in a coke-packed combustion furnace and obtained a nearly 10% increase in combustion efficiency over a conventional lance, as shown in Fig. 6-5.

6.2.2 Eccentric double lance

NKK proposed the eccentric double lance to achieve a better dispersion of PC particles over the cross section.[2] Double lance are more effective than a single lance in dispersing PC particles. According to the their experiments, the eccentric double lance shows more remarkable effect than double lance in obtaining a more uniform dispersion of PC particles over the cross section by preventing the two streams of PC particles from collision. Figure 6-6 compares PC particle streams from different lances using a two-dimensional mathematical model and the measured combustion efficiency obtained by a hot model experiment. The result of the mathematical model study indicates that eccentric double lance produces a more uniform dispersion of PC particles across the cross section of the blowpipe. An experiment using a hot model resulted in a 25% increase in the combustion efficiency over a conventional single lance at a position 600 mm downstream of the

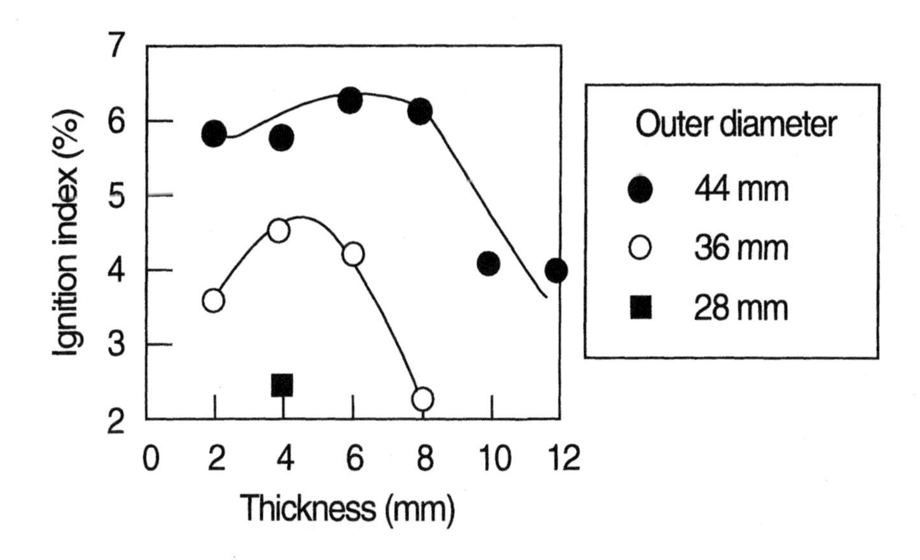

Fig.6-4 Effect of lance wall thickness on pulverized coal ignition

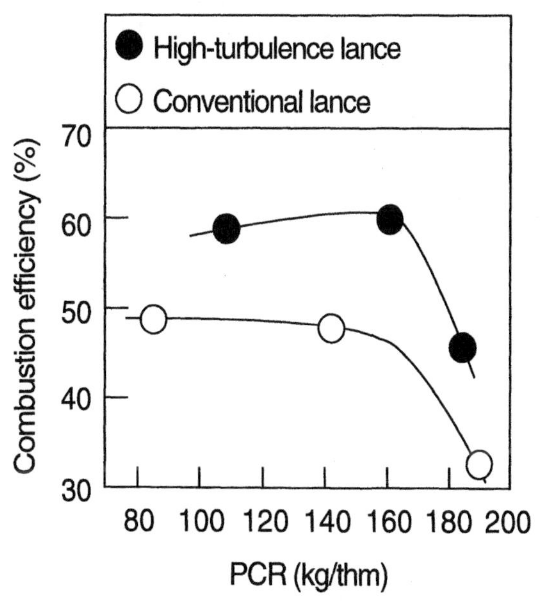

Fig.6-5 Combustion efficiency of PC with conventional and high-turbulace lance

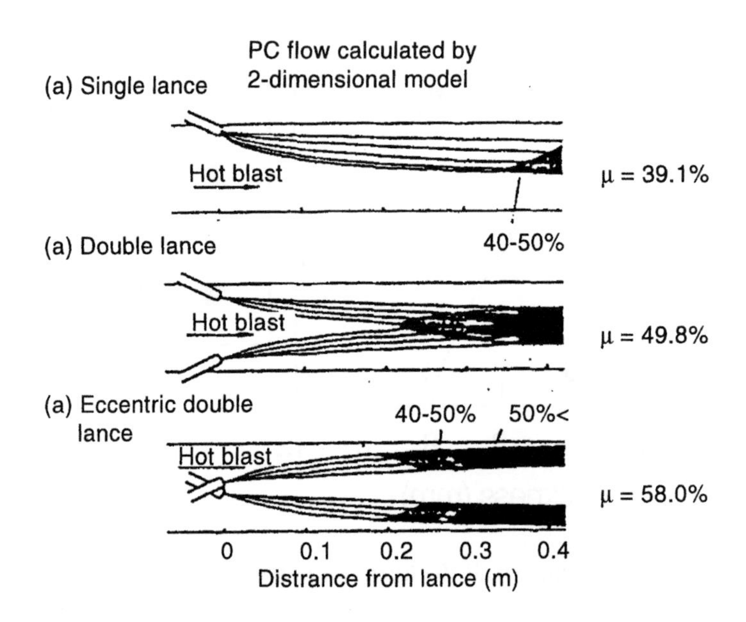

Fig.6-6 Influence of lance type on combustion efficiency by mathematical model (η : combustion degree measured by hot model)

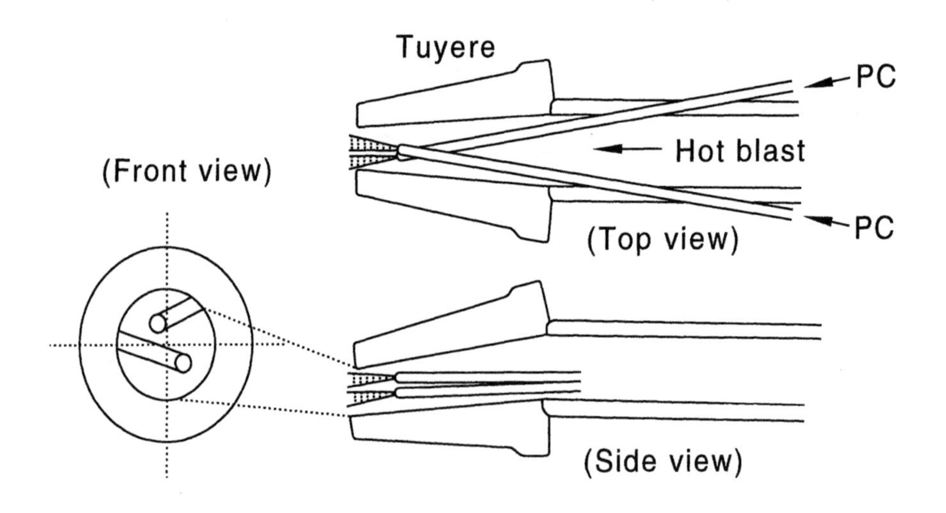

Fig.6-7 Eccentric double lance applied in Fukuyama No.4 BF

lance. Figure 6-7 shows the arrangement of the eccentric double lance for the Fukuyama 4 BF, which recorded a monthly-average injection rate of 218 kg/thm in October 1994.[3] Figure 6-8 gives the combustion efficiency for injected PC particles estimated by the mathematical model simulating the real conditions of a blast furnace. The eccentric double lance are estimated to attain 10% increase in combustion efficiency over the conventional single lance at a position just beyond the raceway.

6.2.3 Oxy-coal lance

Injecting cold oxygen directly into the PC particle flow by using a double-tube lance is actually being employed in some blast furnaces. SSAB in Sweden uses swirl-type oxy-coal lances that swirls the oxygen flow.[4] Thyssen studied the effect of the amount of cold oxygen injected on the gas temperature and gas composition by actual

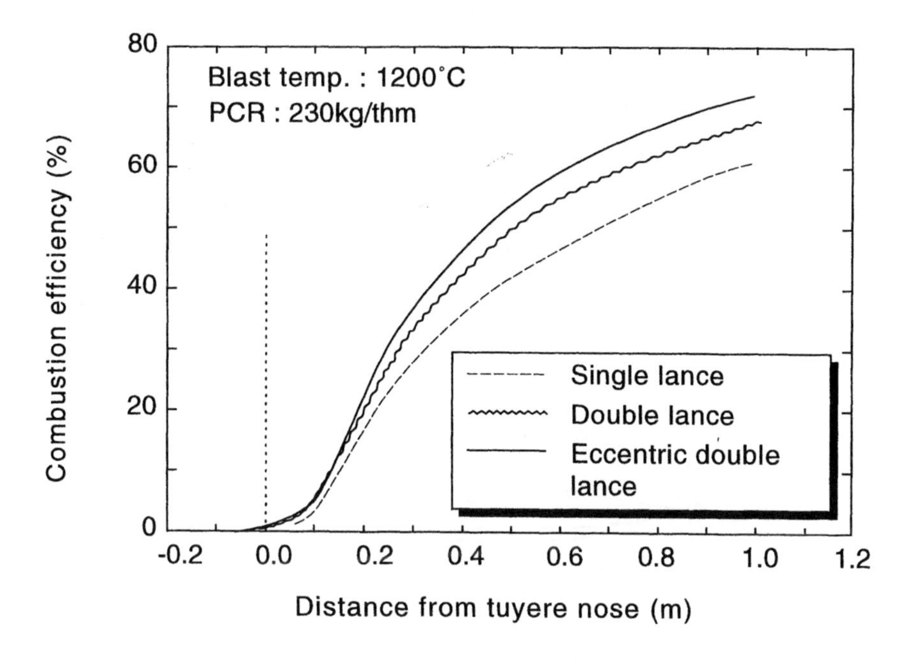

Fig.6-8 Combustion efficiency calculated by mathematical model with actual conditions of BF

measurements.[5] They inserted a probe for measuring temperature and gas composition from the opposite side of a straight oxy-coal lance to clarify the influence of the cold oxygen. It was concluded that oxy-coal lance greatly increases the combustibility.

NKK is conducting an experiment to confirm the performance of a newly developed oxy-coal lance,[6] as shown in Fig. 6-9. The lance introduces cold oxygen around the PC flow. In addition, cold air is introduced to cover the combined flow to prolong the life of the lance tip.

Nippon Steel is also conducting an experiment comparing the single lance, double lance and oxy-coal lance by means of a basic combustion testing apparatus.[7] The inside and outside diameters of the single and double lance are 9.8 and 13.8 mm, respectively. The outside diameter of the oxy-coal lance is 27.6 mm. The outer pipe of the oxy-coal

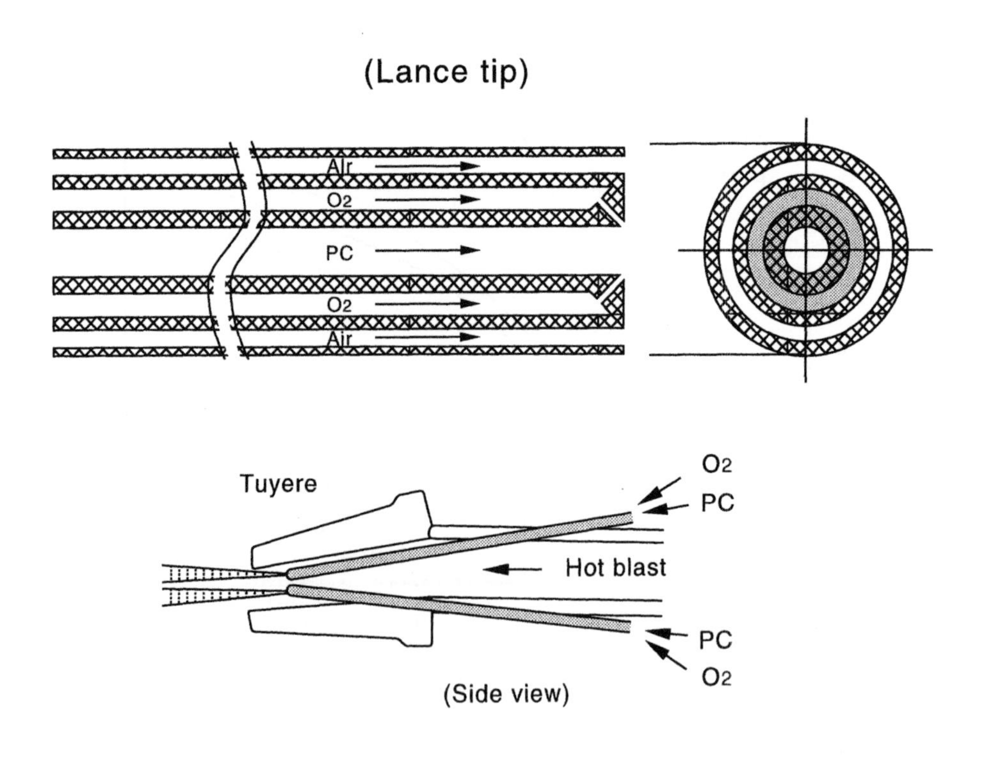

Fig.6-9 Oxy-coal lance in NKK

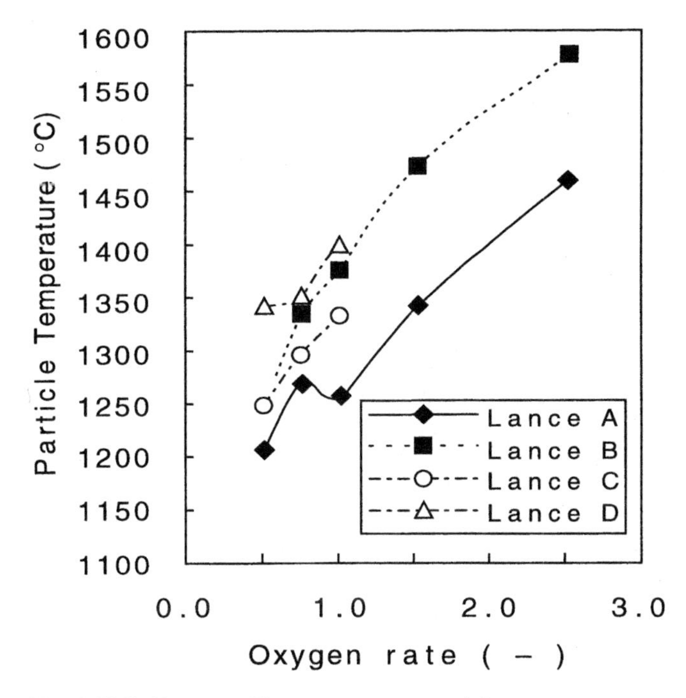

Fig.6-10 Influence of lance type on particle temperature (A : single lance, B : double lance, C,D : oxy-coal lance)

lance injects a mixture of oxygen and air. Figure 6-10 shows the results of the experiment in terms of particle temperatures for each lance. The oxy-coal lance D achieved high particle temperatures, even at a low air rate (oxygen concentration of 50% in the outer pipe), indicating improved combustibility.

6.3 Conclusion

Lances with high combustibility represent an essential technology for achieving a high PC injection rate. The injection lance must be designed under constraints imposed by the condition of a actual blast furnace. Recently, as the mechanism of PC combustion becomes clear, new types of injection lances are being developed and tried in actual production. To attain intensified PC injection, more appropriate injection lance must be pursued.

REFERENCES

1) T. Uchiyama, N. Ishizawa, K. Takeda and H. Itaya: Ironmaking Conference Proceedings, (1996), 67.
2) K. Takeda, N. Ishiwata and T. Uchiyama: Research Group of pulverized Coal Combustion in Blast furnace, Rep-35(1995)
3) A. Maki,A. Sakai,N. Takagaki,K. Mori,T. Ariyama,M. Sato and R. Murai: ISIJ Int.,36(1996), 650.
4) J. W. Wikström, B. E. Sköld and K. Kärsud: Ironmaking Conference Proceedings,(1996), 61.
5) H. W. Gudenau, M. Peters and M. Joksch: Stahl u. Eisen, 114(1994),81.
6) R. Murai, M. Sato and T. Ariyama: CAMP-ISIJ 10(1997),121.
7) S. Matsuzaki, M. Ichida, T. Sugiyama, K. Yamaguchi, T.Deno and S. Matsunaga: La Revue de Metallurgie-CIT, (1998), 398.

CHAPTER 7
Phenomena in blast furnace with high rate PCI

7.1 Change in burden distribution

7.1.1 Relationship between burden distribution and gas permeability

The ore/coke (O/C) of burden in blast furnace (BF) would increase with the increase of pulverized coal injection (PCI) due to the relative increase of the ore than the coke. In accordance with the high O/C, the gas permeability resistance increases and the burden distribution and the gas flow are varied. These variations of burdens and gas flow distributions could be affected not only by the O/C but also by the charging methods, which are conducted by the coke base or the ore base. These charging methods have major effects on the burdens and the gas flow distributions, and quite different behaviors could be resulted from the charging methods. Hence, the control of burden distribution is so important, and only one factor can be controlled by BF engineers. The basic procedure of operation is to maintain the center flow and to inhibit the heat loss from the body of blast furnace. For the optimum and effective control, it is necessary to understand the relationship between the characteristics of the gas permeability resistance and the variation of burden distribution according to the increasing O/C.[3)]

Figure 7-1 shows the variation of the distribution of gas permeability resistance, K value in two cases which are the ore base and the coke base operations. Generally the increases of O/C is carried out with the amount of charging ore (coke base operation) or with the amount of charging coke (ore base operation). Assuming the same inclined angle of the layers of ore and coke, the ratio of layer thickness of ore to coke, lo/lc from the middle region to the periphery (near wall) increases in the coke base operation. On the other hand, the lo/lc in the middle region largely increases in the ore base operation (These positions were shown by arrows in the figure). However, the distribution of K value does not always show the same tendency against the distribution of lo/lc. The K value in the center and the periphery increase in the coke base operation. On the other hand, the K value from the middle region to the periphery increases in the case of the ore base operation, and the variation in the center is small. Since the distribution of gas flow in the blast furnace is controlled by the relative pressure drop in the radial direction, the normalized gas permeability resistances with the Kc value at the center, K/Kc are also shown in Fig. 7-1. The variation of K/Kc in the radial direction shows a flat distribution in the coke base operation. From this feature of K/Kc variation, the distribution of gas flow tends to develop in the region from the periphery to the middle and is significantly weaken in the center region, when the O/C increased by the increase of the ore charged by the coke base operation. While the gas flow in the middle region strongly developed, when the amount of charged coke decreased (ore base operation).

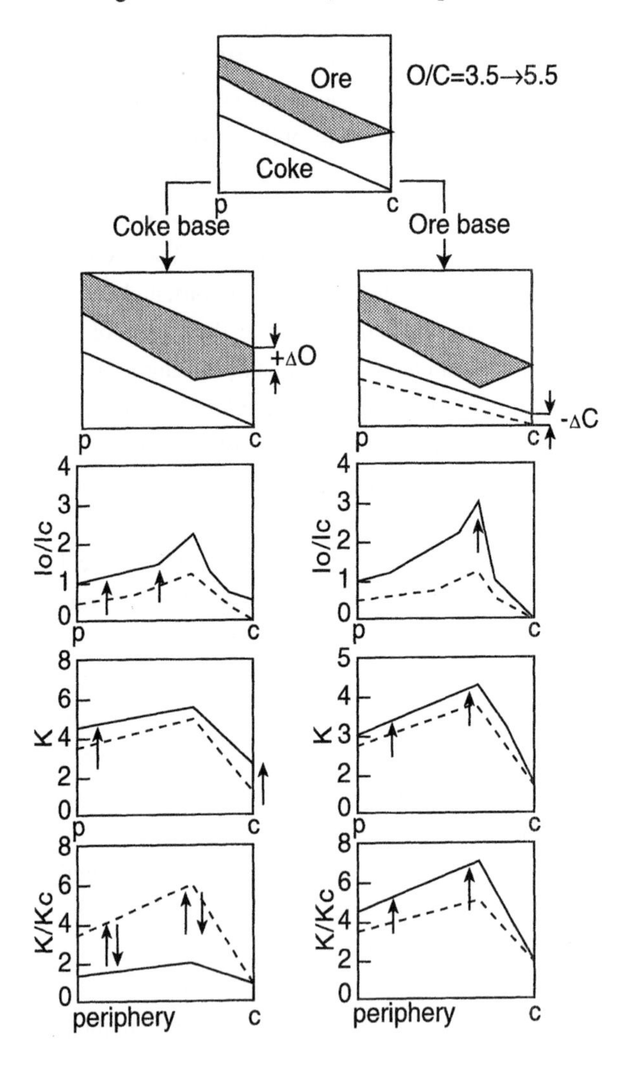

Fig.7-1 Changes of burden distribution and K value with increase of O/C

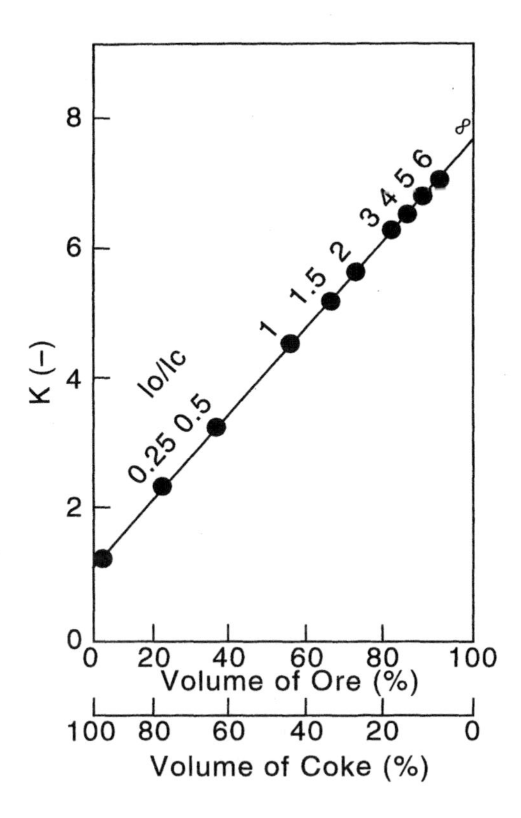

Fig.7-2 K value of layered packed bed

This relationship between the ratio of layer thickness (lo/lc) and the gas permeability resistance (K value) can be understood by Fig. 7-2. The total gas permeability resistance of overall layers has a linear relationship against the volume ratio of the each particle in the packed column, in which two kinds of particle have different gas permeability resistance and are packed layer by layer. The values of lo/lc are shown on the line in Fig. 7-2 and are not proportional to the K value. The variation of K value is large in the small lo/lc (< 1.0) and significantly small in the region more than 2 of lo/lc. That is, the distribution of gas flow might show a large variation in the low region of lo/lc.

Although a assumption was made in Fig. 7-1 as the inclined angles of ore and coke layers were not changed for convenience, the actual inclined angles would be changed by the amount of burdens charged and those charging method. In the two batch dumping system generally adopted in the blast furnace operation, the inclined angle of coke layer increases with the increasing ore (coke base operation). This feature acts to intensify the variation of the gas flow distribution mentioned above. Therefore, the ore base operation would be more preferable than the coke base operation in view point of the maintaining the stable central gas flow. However in the case of ore base operation, it would be important to control the optimum thickness of coke slit in the cohesive zone for maintaining the gas permeability.

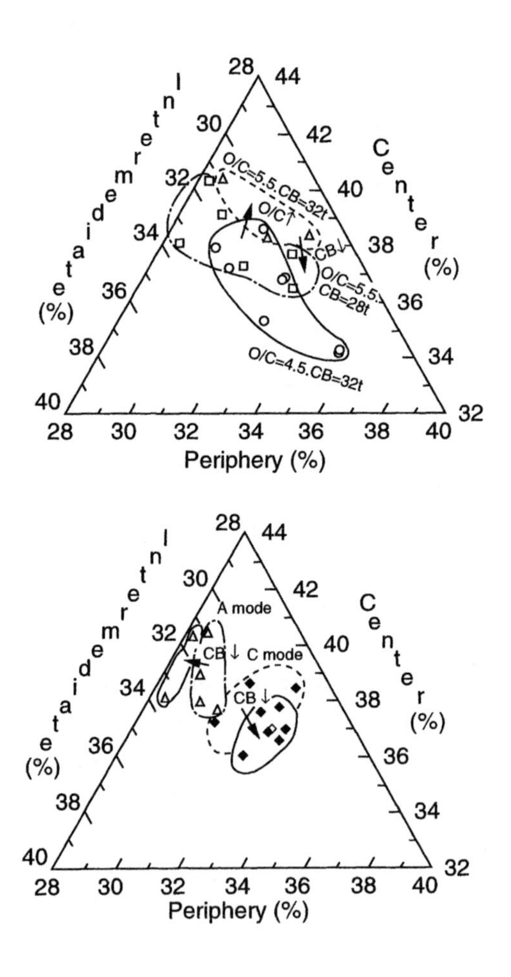

Fig.7-3 Distribution of gas flow

7.1.2 Burden distribution changes with increased O/C

Since the characteristics of collapse of ore and coke layers and segregation of particle size would be changed with the increase of O/C, these variation might show the different tendency in whether the coke base operation or the ore base. Then the shapes of layers of coke and ore were measured using the electrical thickness meter in 1/3 model of actual blast furnace, in which the similar particle sizes of coke and sinter in actual blast furnace were used. The distribution of gas velocity was calculated by the Ergun's equation using the thickness of coke and ore layers and the distribution of particle size measured. The estimation of collapse of coke layer was carried out using the variation of surface profile by laser profile meter and thickness of each layers. The three charging methods are selected for the estimations in two operations of (1) a increasing O/C with constant coke charging and (2) a decreasing coke with constant O/C, in which the O/C is equal to 4.5 as a standard, that is called, I : 0 mode($C_0C_0O_0O_0$), II : A mode ($C_\alpha C_\beta O_0O_0$) and III: C mode ($C_\alpha C_\beta O_0O_\gamma$).

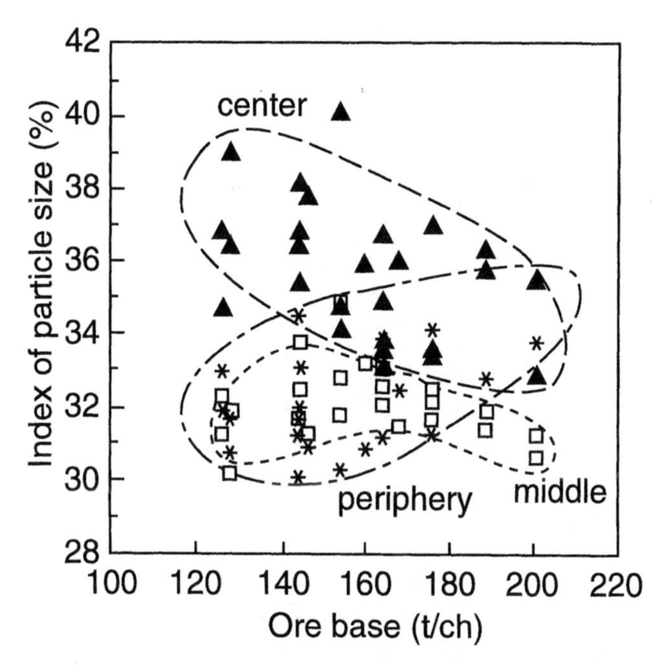

Fig.7-4 Relationship between ore base and particle size index

(1) Distribution of gas flow

Figure 7-3 shows the variation of the distribution of gas flow.[4] In the case of 0 mode, as the lo/lc in the central region decreased over the degree of the collapse of ore layer during charging, the gas flow in the central region increased, when the O/C increased in coke base operation. On the other hand, when the amount of coke decreased with a constant O/C, gas flow in the central region decreased inversely (Fig. 7-3(a)). In this case, the distribution of gas flow in the C mode was the same tendency as that of 0 mode (Fig. 7-3(b)) and the gas distribution of A mode was not changed significantly due to the little increase of collapse of coke layer according to longer terrace of coke layer (Fig. 7-3(b)).

(2) Distribution of particle size

Figure 7-4 shows the index of particle size distribution, which expresses the ratio of average particle size among the peripheral, the central and the middle region in different ore base. [4] When the ore base decreased, the particle size in the central region became relatively coarse and fine in the periphery. This tendency is independent on the charging mode.

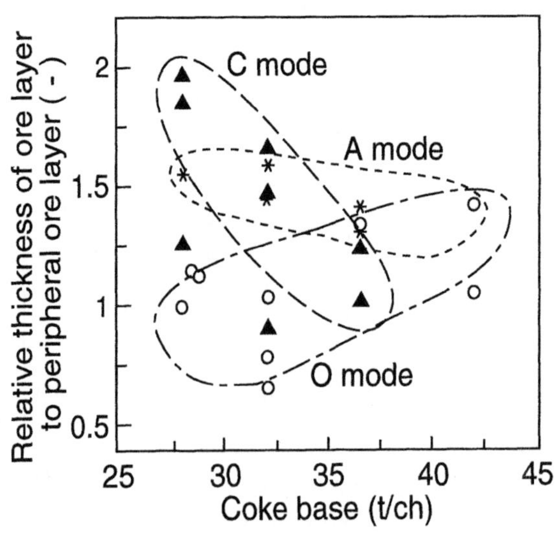

Fig.7-5 Relationship between coke base and relative thickness of ore layer to peripheral ore layer

(3) Influence on the O/C near the periphery

When the thickness of the coke layer decreased, the lo/lc near the periphery decreases relatively in 0 mode, slightly increased in A mode and showed relatively large increase in C mode (Fig. 7-5). From these results, when the O/C increased based on the C mode operation, following phenomena should be noted. [4)]

a) the gas flow in the central region would be inhibited from the reason of the collapse of coke layer and enhanced from the viewpoint of the particle size distribution. Then the gas flow would be control by the coke layer collapse in overall view point.

b) the gas flow near the periphery is easy to be inhibited.

(4) Maximum value of lo/lc [5)]

It is important not to form the high lo/lc region toward the radial direction in the case

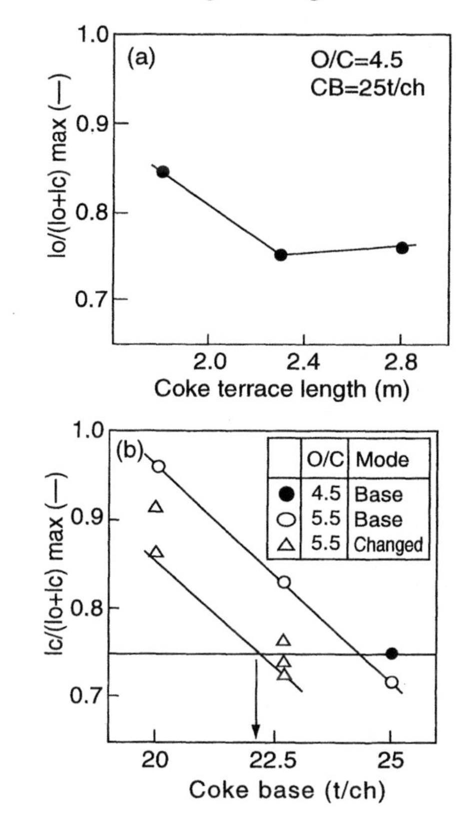

Fig7-6 Influence of coke terrace length and coke base on lo/(lo + lc)

of high O/C operation. Figure 7-6 shows the influence of the length of coke terrace (Fig. 7-6(a)) and the amount of coke base (Fig. 7-6(b)) on the maximum value of lo/(lo+lc). The maximum value decreased, when the length of coke terrace was more than 2m (Fig. 7-6(a)). Although the maximum value increased with decreasing coke base, the influence of charging mode was also significant (Fig. 7.6(b)).

(5) Operation results in actual blast furnace [5)]

Figure 7-7 shows the operation results of the control of the burden distribution in actual blast furnace with the increasing PC rate. The operation was carried out with a constant coke base at 22t/ch((ton per charge), which corresponds to about 260mm of layer thickness in the region of belly). Because of short length of coke terrace at 160kg/thm, a lack of the gas flow in the middle region was resulted in and the fluctuation of gas permeability resistance occurred. And then lower limit of index of gas flow in the middle region was changed from 25.0% to 27.0% and the distribution of gas flow was adjusted by the elongation of the coke terrace length. Afterward, the outside charging of ore and

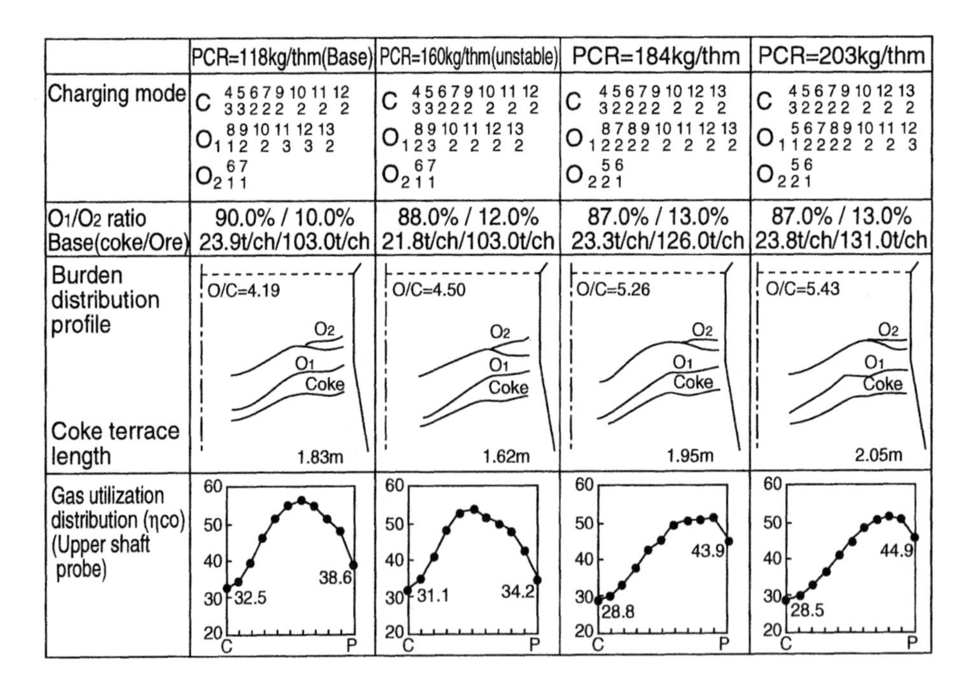

Fig.7-7 Actual results of burben distribution adjustment for high rate PCR

the increase of charge at second batch dumping were carried out as the PC rate increased. Then the adjustment of the burden distribution for the inhibition of gas flow near the periphery has been achieved. Figure 7-8 shows the variation of the index of gas flow distribution measured by gas analysis using upper part of probe. Wide range of gas flow index could be obtained from the periphery region to middle region with the almost same center flow. From this operation, it was possible to maintain a stable gas flow in the upper part of blast furnace.

The results of experiment using hot model of blast furnace were shown in Fig. 7-9 (Cases A to D are illustrated in Fig. 7-10). The lo/lc increased from the center region to periphery and showed large value at the middle region according to the increase of O/C. Figure 7-10 shows the profile of cohesive zone in this operation. It was found that the profile of cohesive zone was changed from the inverse V type to irregular type. The case D shows the profile of cohesive zone having lowest pressure drop in high O/C operation [1], which was obtained in high lo/lc operation in the periphery. Therefore, it can be concluded that the increase of lo/lc in the periphery region is effective for the

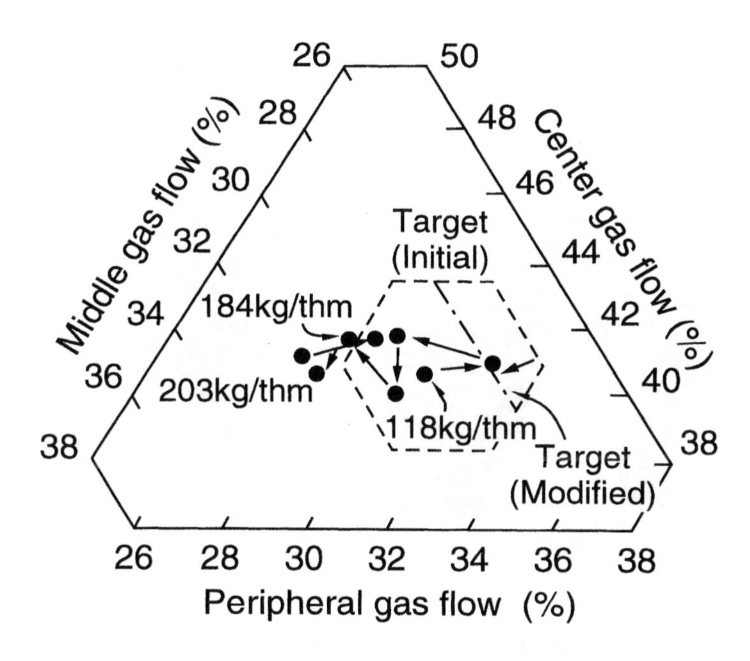

Fig.7-8 Change in gas flow distribution

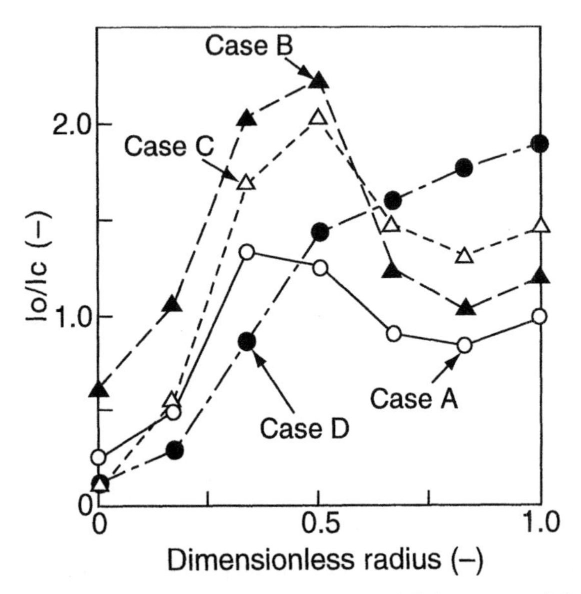

Fig.7-9 Variation of burden distribution with increase of O/C

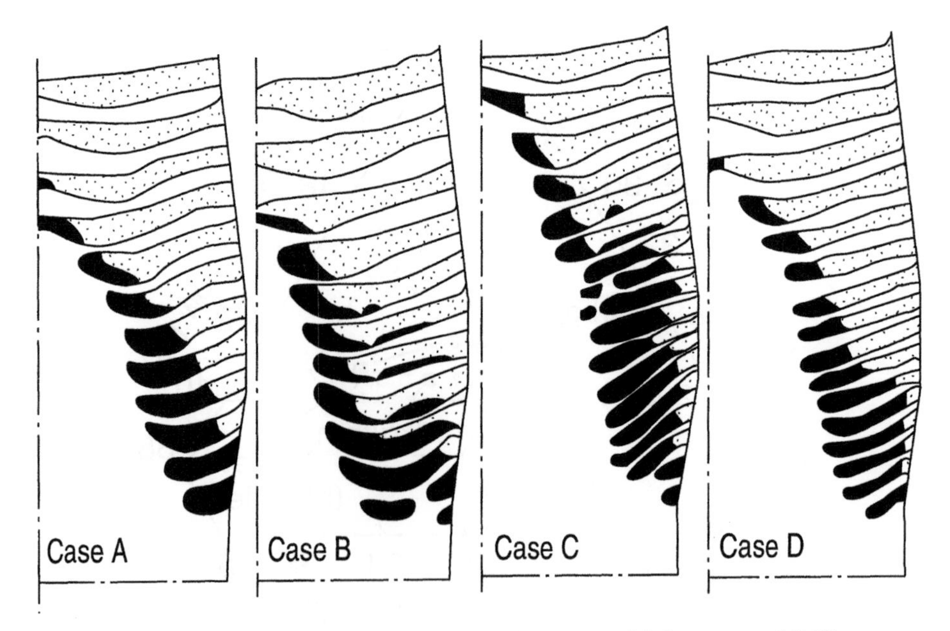

Fig.7-10 Changes in shape of cohesive zone with increase of O/C measured by hot model

stable operation in high rate PCI. Using the No.2 blast furnace in Kakogawa, the trial operations to confirm this result were carried out by (1) the control of the amount of coke in the central region using the center charge of coke, (2) the control of the O/C in the periphery using movable armor and (3) simultaneous injection of PC and heavy oil. The operation result of the average coke rate, 298kg/thm per month (corrected PC rate 220kg/thm) was achieved. [2)]

7.2 Decrease of heat flux ratio

The descending velocity of burden in upper part of blast furnace decreases, as the amount of coke consumption (combustion) at the raceway decreases with increasing pulverized coal rate (PCR). The decrease of descending velocity has a large influence on the temperature distribution in the blast furnace

Generally, heat flux ratio HFR is expressed as following equation .

HFR= (heat capacity flux of solid)/(heat capacity flux of gas)

$= Cp_s \bullet W_s / Cp_g \bullet W_g$

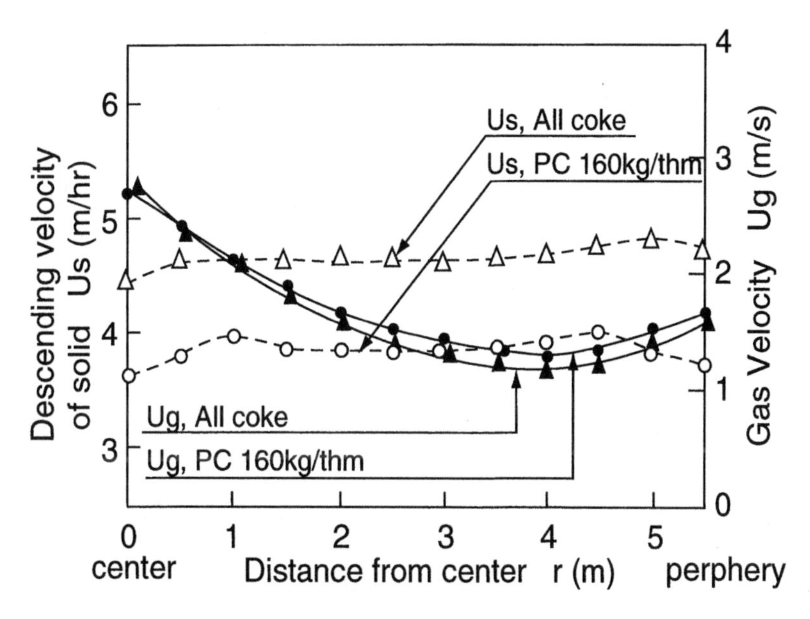

Fig.7-11 Radial distribution of gas and burden descending velocity in middle position of shaft

$$= (Cp_o \bullet W_o + Cp_c \bullet W_c) / Cp_g \bullet W_g \quad (7.1)$$

where Cp is heat capacity, W is mass flux(= V• ω , V: velocity, ω :concentration). Suffixes s, g, o, c express solid, gas, ore and coke, respectively.

Each terms in Equation (7.1) are varied with the temperature and the degree of reaction at a position of blast furnace. Figure 7-11 shows an example of calculation for the influence of PCI on the distribution of gas velocity (Ug) and descending velocity of solid (Us) in the middle position of shaft, which were calculated by Total Model of blast furnace. The distribution of gas velocity has large variation in the radial direction according to the burden distribution, while the distribution of descending velocity of solid does not show big difference. Comparing to the variation in radial direction, the value of gas velocity was almost same between PCI and without PCI. On the other hand, the descending velocity was largely changed with PCI.[4] From these facts, it was found that the descending velocity of solid controls the absolute value of heat flux ratio and the distribution gas velocity controls the heat flux ratio in radial direction. It would be important to control the lo/lc near the periphery not to make a unbalance of heat flux ratio in radial direction, significantly.

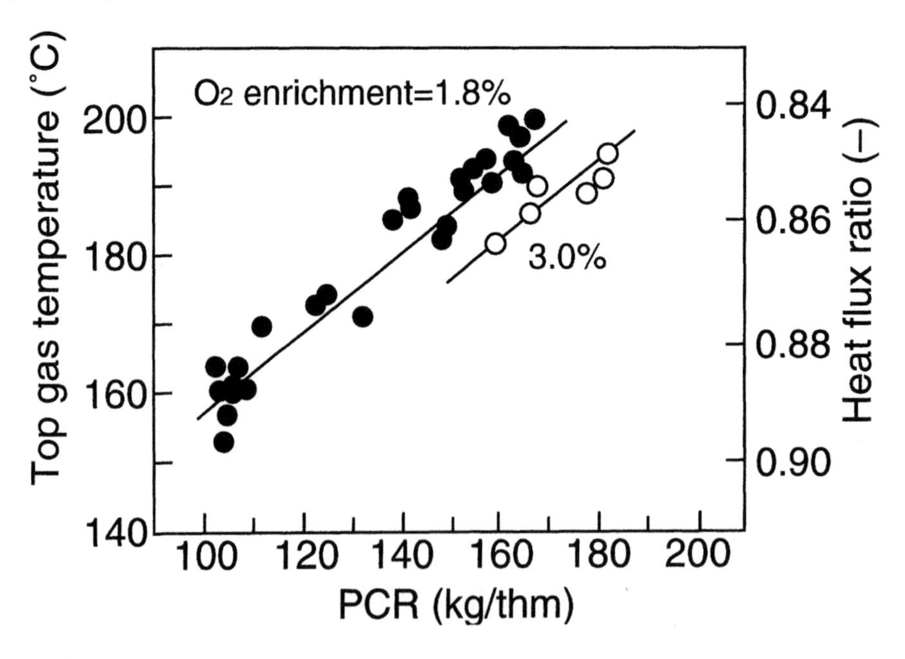

Fig.7-12 Relationship between top gas temperature and PCR

In the region of upper part of blast furnace until the beginning of gasification (≒ thermal reserve zone), the decrease of W_o by the gas-solid reduction controls the HFR mainly as the heat capacity has a low temperature dependence. Therefore, HFR at the top of blast furnace is sometime used as the parameter of the heat balance in the upper part of blast furnace. When HFR decreases, the heating up rate of burden increases and temperature of top gas of furnace increases. In the case of increasing PCR with a constant productivity (a constant W_o charged and replacement ratio is unity), a equivalent coke (ΔW_c) as a increase of PCR would be saved, then the heat flux ratio would decrease due to the change of W_c in Eq.(1) to (W_c-ΔW_c). Figure 7-12 shows the variation of the furnace top gas temperature and heat flux ratio in the actual blast furnace with PC injection. [6) Heat flux ratio decreased linearly with the increase of PCR and the furnace top temperature increased inversely. The result of the 3% oxygen enrichment is also shown in Fig. 7-12. It is found that the increase of 10°C of furnace top gas temperature resulted from the only

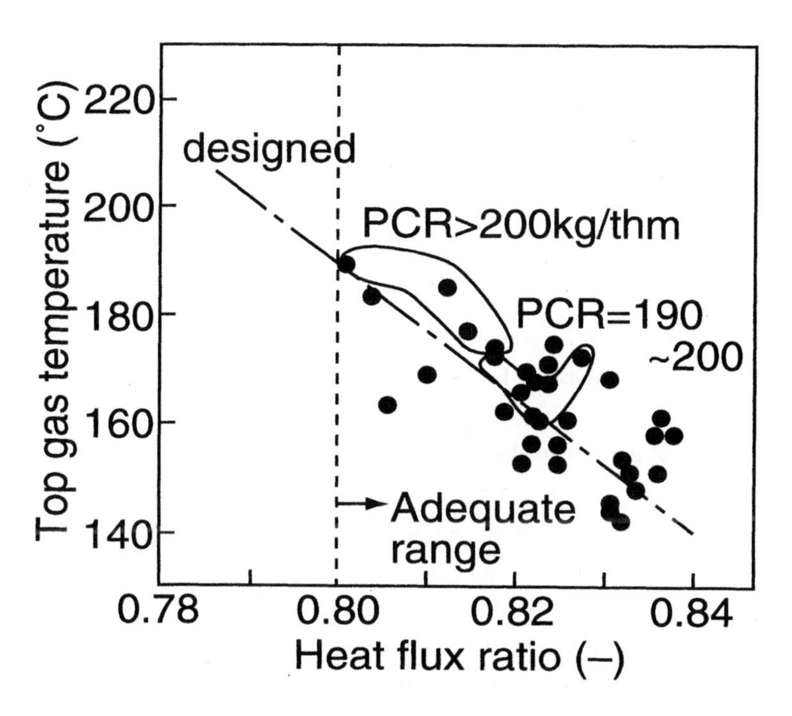

Fig.7-13 Relationship between heat flux ratio and top gas temperature

1.2% oxygen enrichment. It would be explained qualitatively that the decrease of W_g (decrease of Nitrogen) in Eq.(7.1) caused to the increase of heat flux ratio, which shows that the oxgen enrichment is useful for the inhibition of the increase of the top gas temperature.

The heat flux ratio in the actual blast furnace is calculated by Eq.(7.2) more effectively.[5)]

$$HFR = \frac{(0.31 \bullet CR + 0.21 \bullet OR) \times 100}{(0.31 \bullet H_2 + 0.325 \bullet CO + 0.493 \bullet CO_2 + 0.321 \bullet N_2) \bullet V_B} \tag{7.2}$$

V_B in Eq.(7.2) is expressed in Eq.(7.3).

$$V_B = \frac{CR \bullet [C]coke + PCR \bullet [C]pc - [C]pig \bullet 100 - DUST \bullet [C]dust}{(CO + CO_2) \bullet 0.5357} \tag{7.3}$$

Where

CR : coke rate (kg/thm), CO_2 : volume % in top gas

OR : ore rate (kg/thm), N_2 : volume % in top gas

PCR: PC rate (kg-thm) , H_2 : volume % in top gas

DR : dust rate (kg/thm), CO :volume % in top gas

[C] : Carbon concentration in coke (%)

$[C]_{pig}$: Carbon concentration in pig iron (%)

$[C]_{dust}$: Carbon concentration in dust (%)

The heat flux ratio and the gas temperature at furnace top, which calculated by Eq.(7.2) as the PCR was 200kg/thm for the Kimutsu 3 BF, are shown in Fig. 7-13. From the calculated result in Fig. 7-13, it is found that the decrease of heat flux ratio was 0.01, which resulted in the increase of 12.5°C of furnace top gas temperature. [5)]

7.3 Variation of gas permeability resistance

It is considered that the following various factors would widely related to the increase of pressure drop with the increasing PCR.

(1) Decrease of coke layer thickness at high O/C operation.

(2) Increase of ore layer thickness with high gas permeability resistance.

(3) Rising of heat level in the furnace according to the lower heat flux ratio.

(4) Increasing of solution loss reaction around the central region depending on the

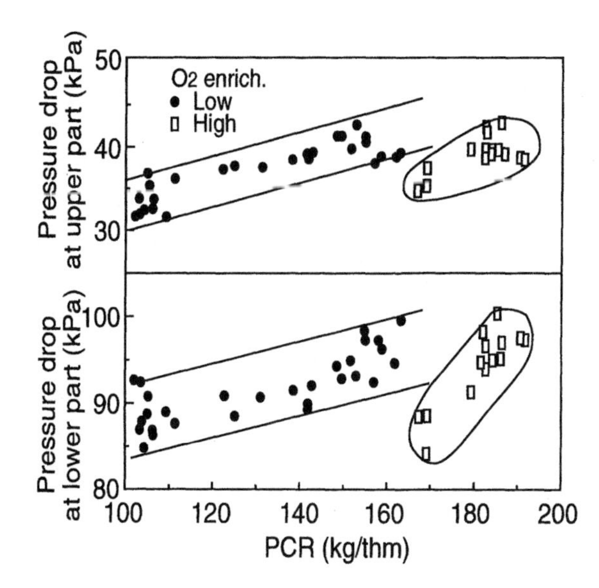

Fig.7-14 Relationship between PCR and pressure drop at lower part of BF

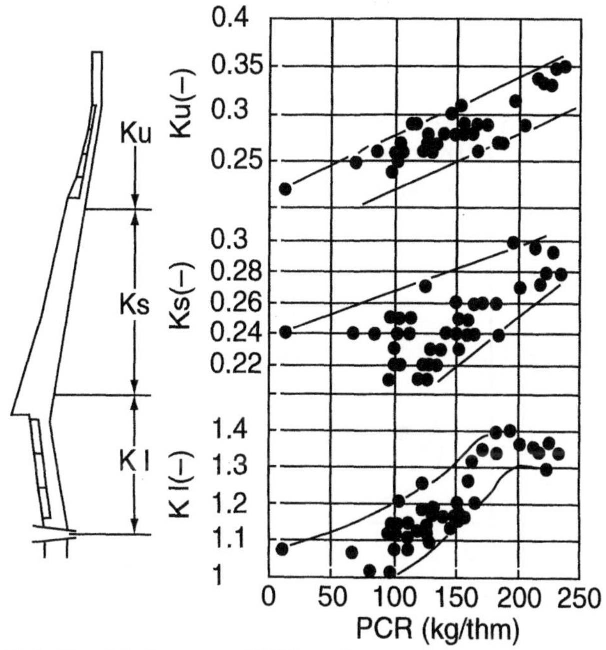

Fig. 7-15 Relationship between PCR and gas permeability resistance

distribution of O/C.

(5) Enhancement of coke degradation with increase of the weight of burden.

Figure 7-14 shows the variation of the measured pressure drop of Kobe 3 BF. The pressure drop both in the upper and lower part increased with the increasing PC.[7] The difference of pressure drop according to the increase of PC rate from 100 to 160kg/thm was 0.15kg/cm^2, which consisted of the increase of gas permeability resistance from the increase of O/C (0.03kg/m^2), the increase of the gas permeability resistance in the tuyere by the combustion of PC (0.05kg/m^2) and the increase of gas volume by the increase of gas temperature (0.07kg/m^2). The increase of gas permeability resistance could be controlled by the inhibition of the increase of the average gas temperature due to the adjustment of the heat flux ratio with a high rate of oxygen enrichment. However, in the case of higher PCI, the effect of these operations disappeared and the pressure drop in the lower part of furnace was back to the original level. This result means that several factors such as the inhibition of gas permeability by the generation and the accumulation of fine coke in the region from the center to the middle due to the change of the packing structure of

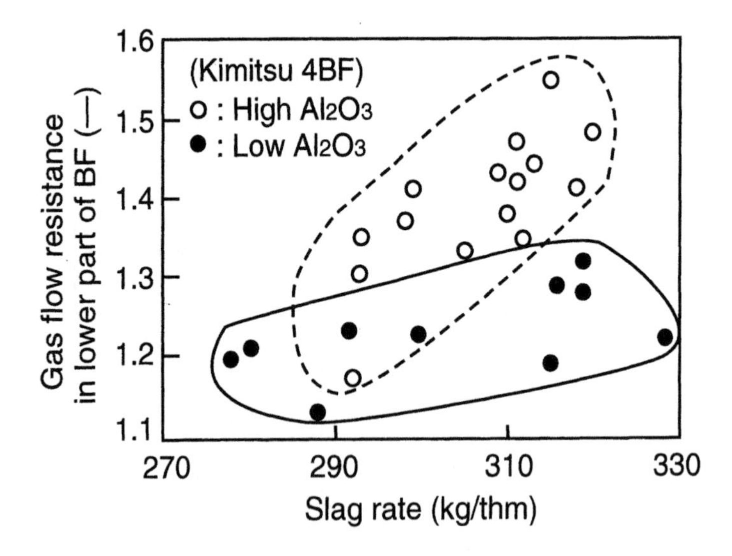

Fig.7-16 Effect of slag rate on gas flow resistance in lower part of BF

deadman, would govern. [7] The details of the generation of fine coke would be mentioned in Chapter 8.

A phenomenon, which the pressure drop in the middle and lower part of shaft varied with some relationship, was often observed in the operation of the PCI.[7] At this moment, the blast pressure increased gradually and decreased rapidly after reaching a given level, when the pressure drop in the upper part varied drastically. It is presumed that a channeling of gas flow caused by the accumulation of fines in the vicinity of the cohesive zone would be occurred. Furthermore, it was recognized that the periodical variation of the blast pressure in unstable condition of furnace roughly corresponds to the variation of a level of slag and metal in the hearth.

Figure 7-15 shows the variation of the index of gas permeability resistance, K value in each part of blast furnace, when the operation condition was moved to 230kg/thm PCI at the Fukuyama 4 BF.[8] K value in the each part of blast furnace was obtained using Eq.(7.4).

$$K=(Pi^2 - Po^2) / U^{1.7} \tag{7.4}$$

Where Pi and Po mean the pressure at the inlet and outlet of the each parts and U means the average velocity of bosh gas.

The gas permeability resistance Ku in the upper part of furnace increased linearly with the increase of the PCI. It is considered that this phenomenon was mainly resulted from the increase of O/C ratio. In order to the larger gas permeability resistance of ore layer, the increase of lo/lc resulted in the high gas permeability resistance. The gas permeability resistance in the lower part of furnace Kl increased monotonously until the 200kg/thm PCI, however, showed almost constant value beyond 200kg/thm. The reason would be explained that the modification of gas permeability in the cohesive zone was designed by the usage of HPS ore having a superior softening and dropping property, and also the gas and liquid permeability was modified by the decrease of the fine ratio due to the high strength coke. These effects would be also obtained by the reduction of slag rate or the change of gangue composition of sinter. Figure 7-16 shows the example of Kimitsu 4 BF, which the index of gas permeability resistance was modified about 20% as the slag rate was changed from 320kg/thm to 290kg/thm. Also the effect of low alumina content in the sinter might be significant. [5]

7.4 Variation of temperature distribution and increase of heat loss

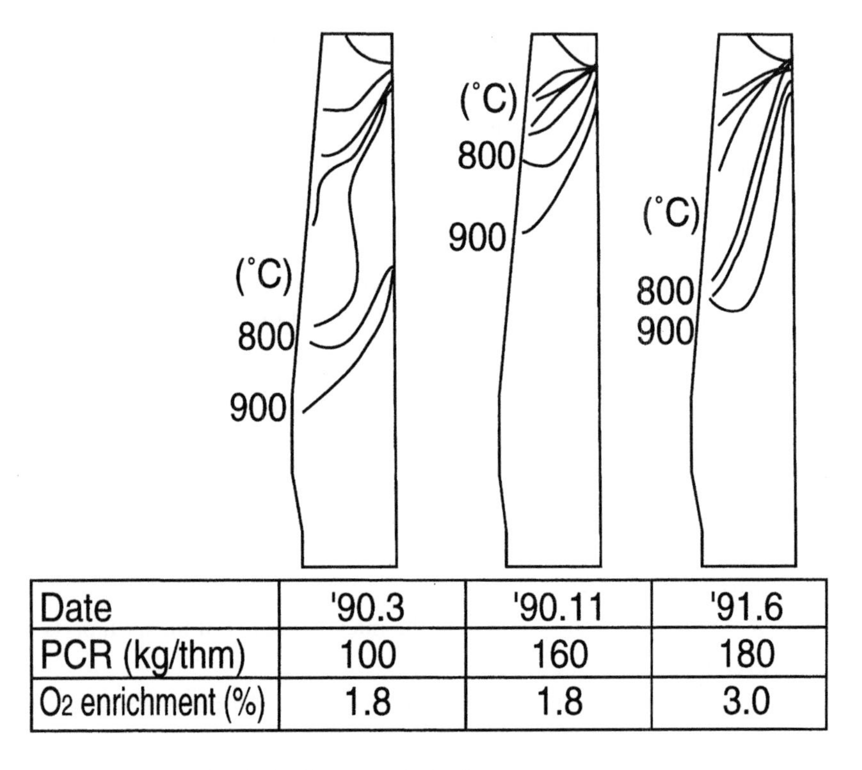

Date	'90.3	'90.11	'91.6
PCR (kg/thm)	100	160	180
O_2 enrichment (%)	1.8	1.8	3.0

Fig.7-17 Changes in in-furnae isotherms

Just as mentioned in Chapter 3 and 4, the combustion focus in a raceway moved toward tuyere side by PC injection. Once the shell was formed around the raceway, which prevented the gas permeation into the deadman, the bosh gas preferred to flow to the peripheral region in BF, and esulted in the increase of the wall temperature of the lower part of BF and the heat accumulation in BF. Since the combustion focus moved toward the tuyere nose in accordance with the PC having high volatile matter, the heat accumulation at the wall in the lower part of BF can be decreased by the adjustment of the position of lance corresponding to the coal quality. The above mentioned decrease of the heat flux ratio introduced the increase of the total heat in BF. The change of the temperature distribution of shaft part with increase of PCR was shown in Fig.7-17. [3] As shown in Fig. 7-17, even at the peripheral region, the sothermal line of 900°C measured by a vertical probe goes up to the upper level of the shaft with the increase of PCR. Due to this change of temperature distribution, the heat loss increased as shown in Fig. 7-18. This tendency was modified to some extent by the increasing heat flux ratio with the

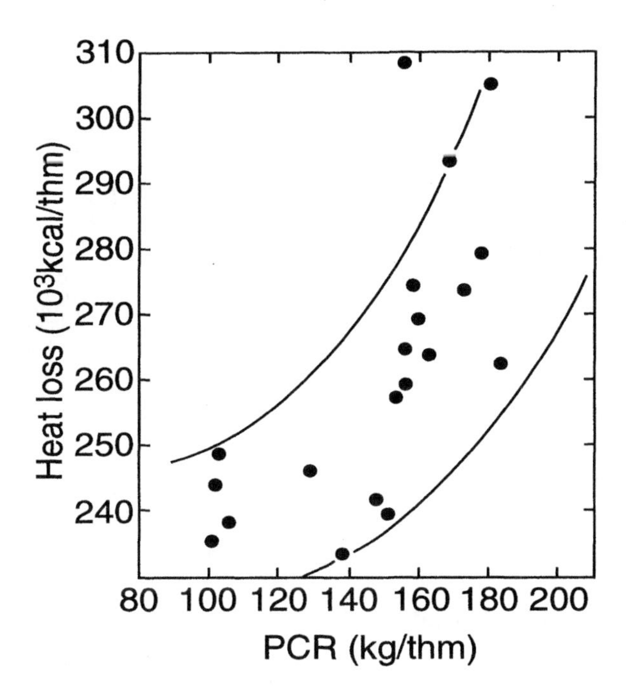

Fig.7-18 Variation of heat loss with PCR

oxygen enrichment and by suppressing the peripheral gas flow with the control of burden distribution. Typical temperature distributions at the center, middle and peripheral region, as a function of PCR (from all coke to 200kg/thm), are shown in Fig. 7-19. The rate of heating up increased with the increase of PCR. The temperature of the thermal reserve zone at the center and the middle region increased from 800°C to about 1000°C with increasing PCR. However, at the peripheral region, the temperature of the thermal reserve zone remained 1000°C, no matter how the temperature goes up early to that of the thermal reserve zone. The gas flow distribution (which was expressed by the gas utilization) measured by horizontal probe and the temperature distribution measured by the movable probe in Fukuyama 4 BF under high PCR were shown in Fig. 7-20. When PCR reached to 230kg/thm, the broadening of the gas flow distribution at the center region was observed, and also the decrease of the isothermal line of 1200°C at the peripheral region and the position of the root of the cohesive zone were significant. According to these phenomena, the pressure fluctuation at the upper part of shaft began to occur. To avoid these situations,

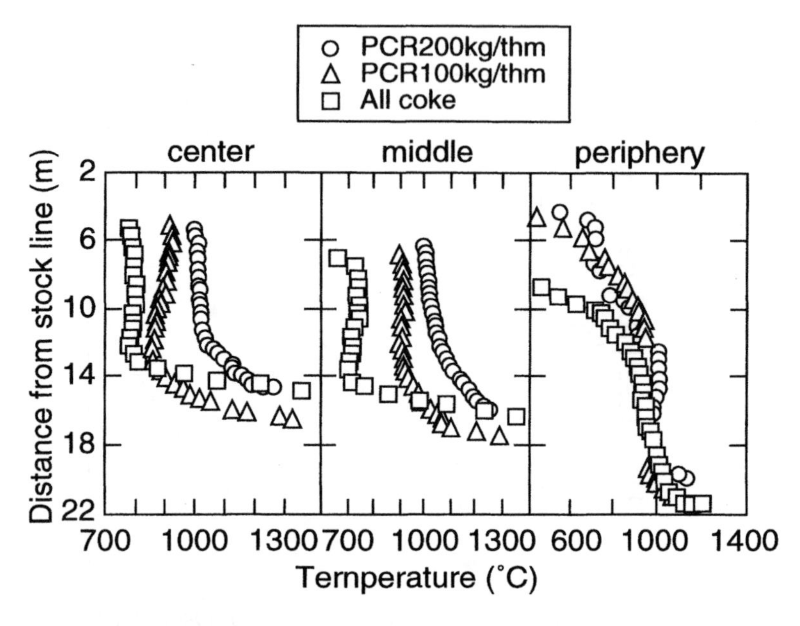

Fig.7-19 Variations of longitudinal distribution of temperature with PCR

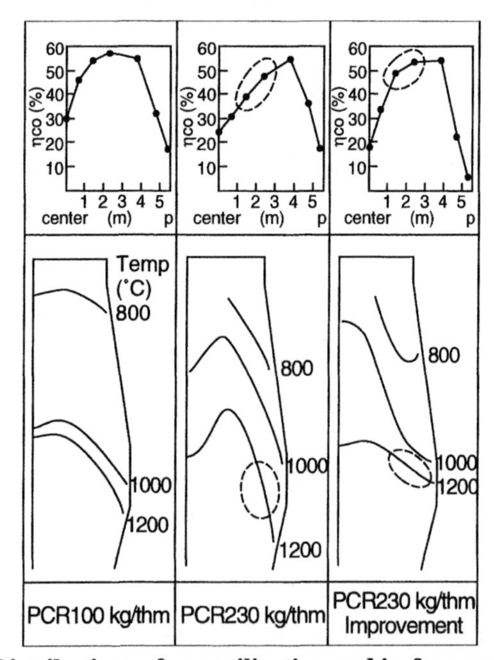

Fig.7-20 Distributions of gas utilization and in-furnace isotherms

the burden distribution control to increase the peripheral gas flow rate was carried out. From this operation, the root of cohesive zone was moved up and the gas flow pattern went back to the original and the inactive situation at the peripheral region was also recovered. The temperature of the upper part of the shaft and that of the bottom hearth as a function of PC rate was shown in Fig. 7-21. [10] Up to 150kg/thm PCR, the temperature of the upper part of the shaft decreased due to the effect of the suppression of the gas flow at the peripheral region, however, once it was beyond 150kg/thm, the temperature increased significantly due to the effect of the decrease of the heat flux ratio. The temperatures at the hearth and the upper part of shaft increased with PCR. These increase of temperatures seemed to be the main reason of the increase of total heat loss in the high PCR operation. It was considered that the increase of the hearth temperature was related

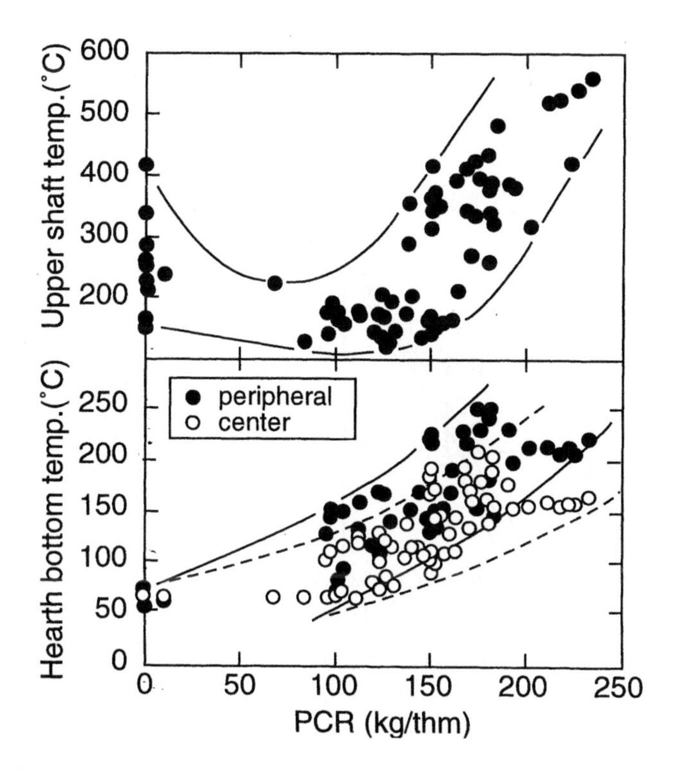

Fig.7-21 Relationship between PCR and peripheral temperature

to the unsound slag and metal tapping, the development of the circulating slag and metal flow due to the accumulation of fine from the middle to the center region of deadman. Depending on a circumstance, the floating-up of deadman by the decrease of void fraction and the formation of the free space in the lower part of furnace would be also related to the increase of the hearth temperature.

7.5 Change of reactions in BF

7.5.1 Rist diagram for PCI

In this section, the effects of the PCR on the process parameter in BF will be discussed based on the Rist diagram. The operation lines of Rist diagram with varying the PCR(H_2

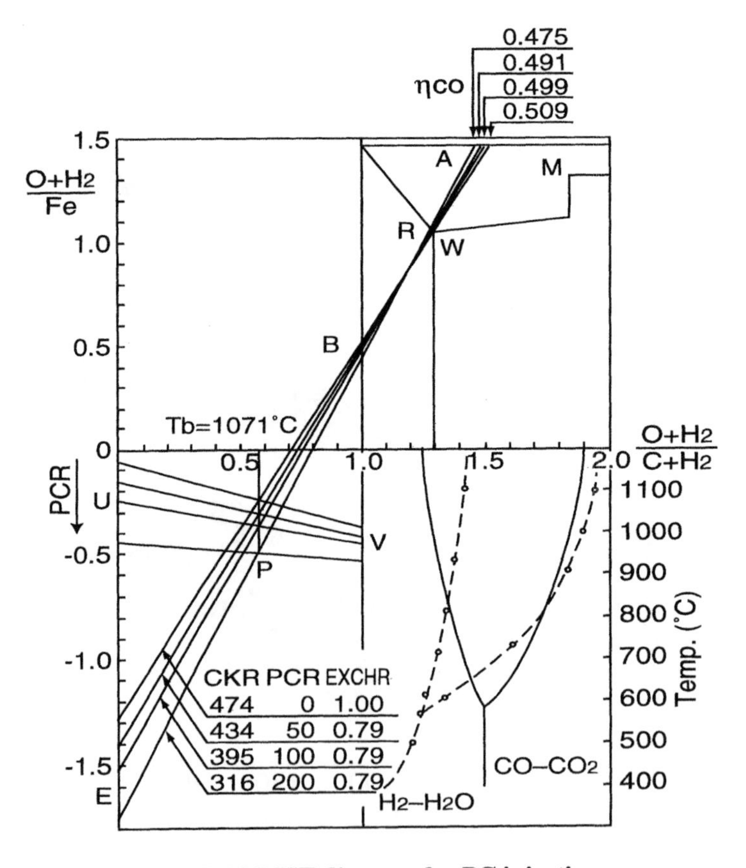

Fig.7-22 RIST diagram for PC injection

≒5%) from 0 to 200kg/thm was presented in Fig. 7-22. [15] The line of heat balance, UV, was moved to negative side with increasing PCR. This means that the increase of the heat of the decomposition of PC and the H_2 content in PC moved the V and U points toward the negative side, respectively. Since the H_2 content in coals was proportional to the heat of decomposition (if the same coals were used), all lines of UV corresponding to

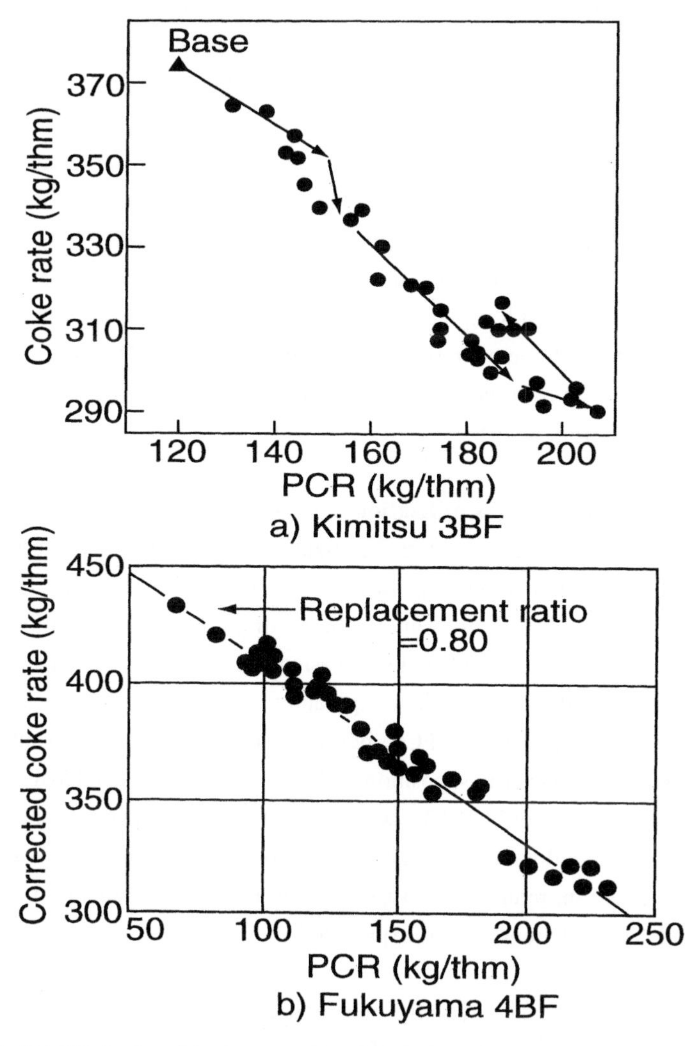

Fig.7-23 Relationship between PCR and corrected coke rate

the various PCR should be passing through the fixed point in the region (X>1, Y<0) of the diagram. This characteristic point represented the relation (in other words, the corrective replacement ratio by the heat production) between the content of H and O in coals and the heat of decomposition. W (Xw, Yw) point in the first quadrant represented the equilibrium composition of Fe/FeO at the thermal reserve zone (chemical reserve zone). W point moved to the right hand side with the increase of PCR, because the equilibrium relation was changed by the increase of H_2 in the bosh gas with PCR. If the operation line was drew through the W point, all operation lines corresponding to the each PCR operation would cross the certain point on the AW line. Since this point did not change with PCR, it can be regarded as new W point corresponding to the coal used. When this operation line was moved to left side, it is easily understood that the shaft efficiency decreased. In Fig. 7-22, under constant shaft efficiency, the operation line AE rotate anticlockwise around R point (increase the gradient) with the increase of PC rate and become a larger gradient. (Since the W point is expressed as a constant for convenience in Figure 7-22, the point R representing the constant shaft efficiency is the fixed point.) In this figure, the replacement ratio was calculated to be 0.79. This is very close to that of 0.80 which was the average replacement ratio at Kimitsu 3 BF(Fig. 7-23(a) [5]) and Fukuyama 4 BF (Fig. 7-23(b) [9]).

7.5.2 Temperature and gas utilization in peripheral region

The gas temperature and the gas utilization (η_{CO}) in the peripheral region were measured by the vertical probe. The results were shown in Fig. 7-24. [5] They showed the similar tendency as Fig. 7-19. Namely, the heating up rate of temperature at the upper part of the shaft increased quickly with increasing PCR and the gas-solid reduction reaction was promoted simultaneously. The temperature of the thermal reserve zone was about 900°C and did not change with PCR, but the end position of the thermal reserve zone which represents the uppermost position of the cohesive zone moved to the lower position of BF with the increase of PCR. The heating up rate of the temperature and the decreasing rate of (CO increased with PCR. The variation of (CO in the furnace was correlated to the both of reaction, reduction and gasification in the mole reaction rate, Ro and Rc, respectively. The relationship was expressed by

$$Ro/Rc > 1 + \eta_{CO} \quad (7.5)$$

Even though the heat exchange and reduction were carried out effectively with increasing PCR, the lower position of the thermal reserve zone simply meant that the

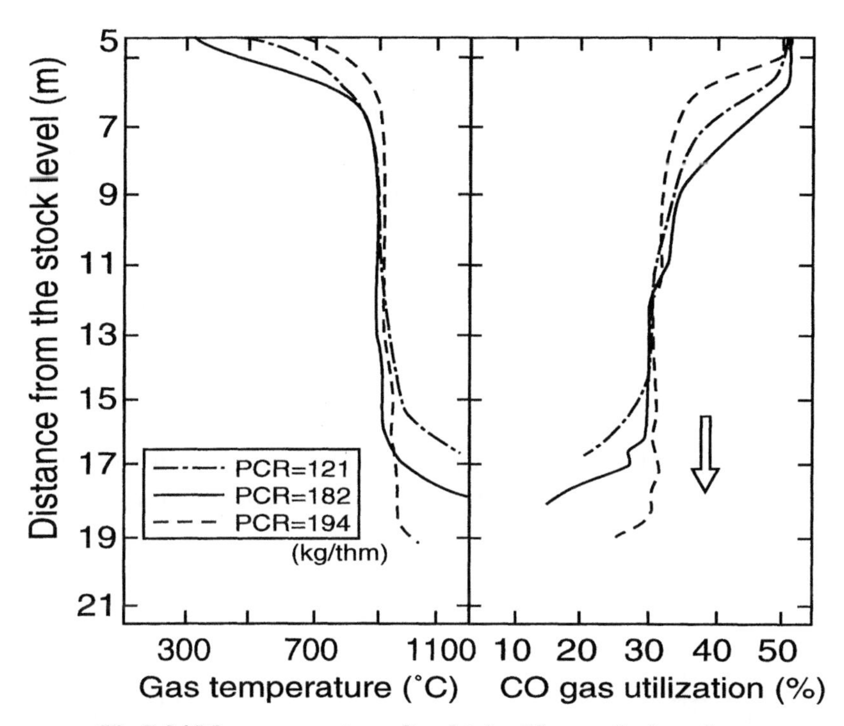

Fig.7-24 Measurement results obtained by vertical probe

distance from the thermal reserve zone to tuyere was short. The abrupt decrease of η_{CO} meant that the quick increase of the rate of solution loss reaction. In other words, it suggested that the increase of PCR introduced the decrease of reduction and melting process of burden significantly in the lower part of furnace and finally caused in unstable condition of BF. These results correspond to the significant increase of the fluctuations of the total time needed for the burden charging (5 times) and the K value at the lower part for every 1 hour, which shown in Fig. 7-25. Especially under the low fuel campaign, the melting ability at the lower part of the cohesive zone could restrict the operation due to following factors;

(1) Decrease in flame temperature

(2) Decrease in bosh gas owing to the oxygen enrichment

(3) Increase in O/C in the peripheral region

7.5.3 Solution loss reaction and PCI

As mentioned in Fig. 7-19, the temperature of the thermal reserve zone at middle and center region increased with PCR. [12] Since it is generally considered that the temperature of the thermal reserve zone was closely related to the reactions (indirect reduction and solution loss reaction) and the amount of the gas flow at that regions, the result shown in Fig. 7-19 suggested that the drastic change of the reduction reaction in the middle and center regions might occur with the increase of PCR. From the temperature and gas composition measured by movable vertical probe, the degree of reduction and oxidation of gas were evaluated. Based on these values the carbon consumed by the solution loss reaction in radial direction was calculated (Fig. 7-26). The amount of carbon by the solution loss reaction was almost constant in the peripheral region, even though the PCR was increased. On the other hand, the amount of carbon at the middle and center region had a decreasing tendency with the PCR. This can be explained by the promotion of H_2 reduction as well as the decrease of the descending speed of burden and the increase of

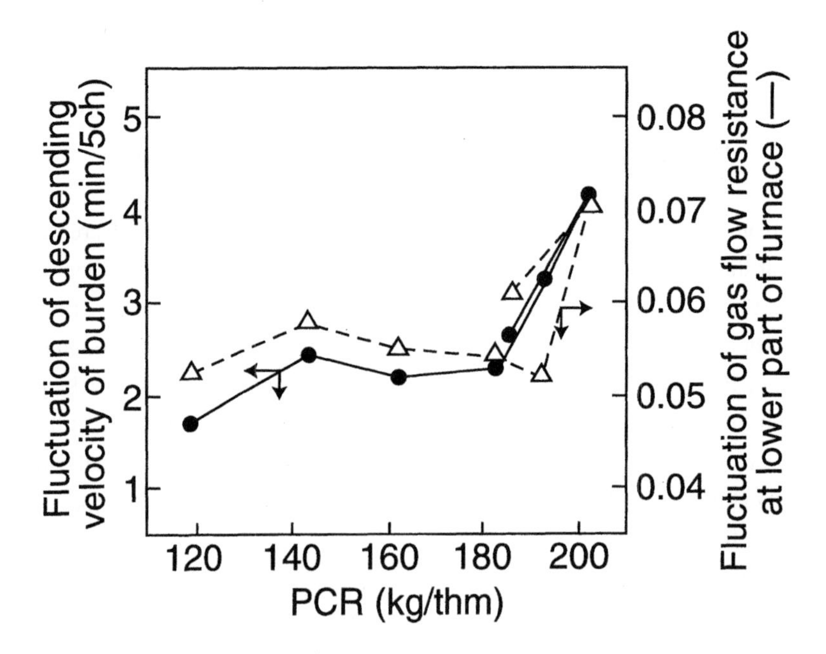

Fig.7-25 Relationship between PCR and K-value at lower part of furnace

the temperature of the thermal reserve zone. The variation of amount of solution loss reaction as a function of PCR is shown in Fig. 7-27, when the PCR was attained to 200kg/thm. The solution loss carbon decreased about 10 kg/thm as PCR increased from 100 to 200kg/thm. The significant decrease of solution loss carbon was found over 200kg/thm PCR, which would be resulted from the usage of HPS ore having high reducibility and small amount slag.

7.6 Phenomena in raceway and lower part of BF

7.6.1 Formation of shell and dripping behavior around raceway

The variation of pressure in blowpipe was measured by a small combustion furnace with PC injection. In the case of all coke operation, the pressure was constant. However, the fluctuation of pressure began to occur just after the starting of PCI and the values of the pressure gradually increased. From the dissection study after experiments, the formation of shell that was strongly bonded by molten ash was found around the raceway. In the actual blast furnace, the thrusting stress was measured by inserting the deadman

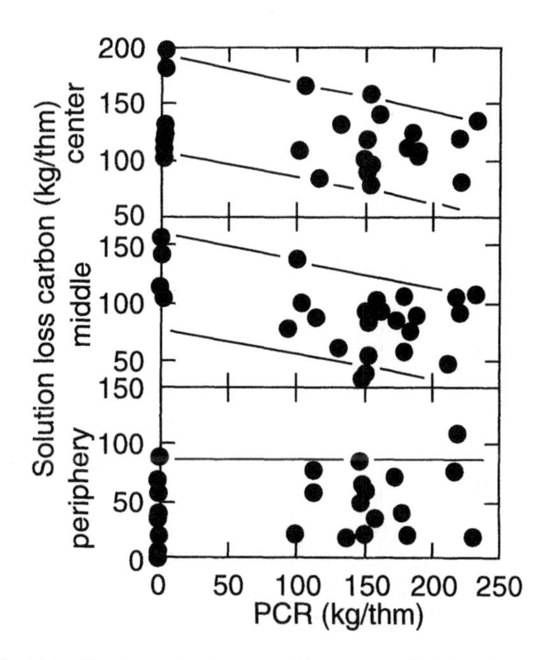

Fig.7-26 Relation between PCR and solution loss carbon

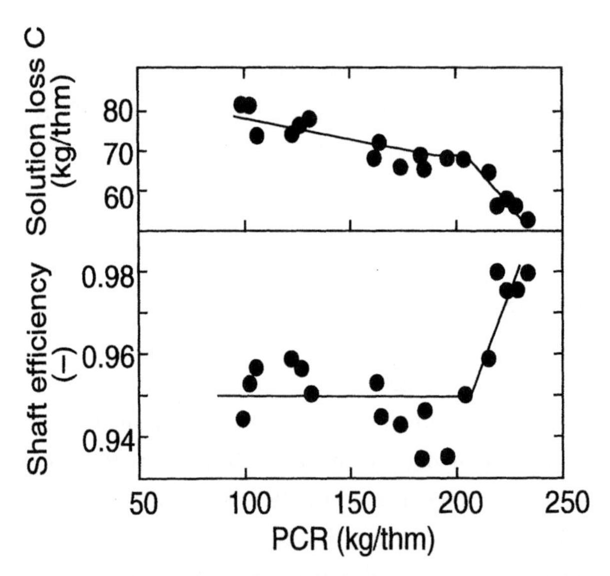

Fig.7-27 Changes in shaft efficiency and solution loss carbon with PCR

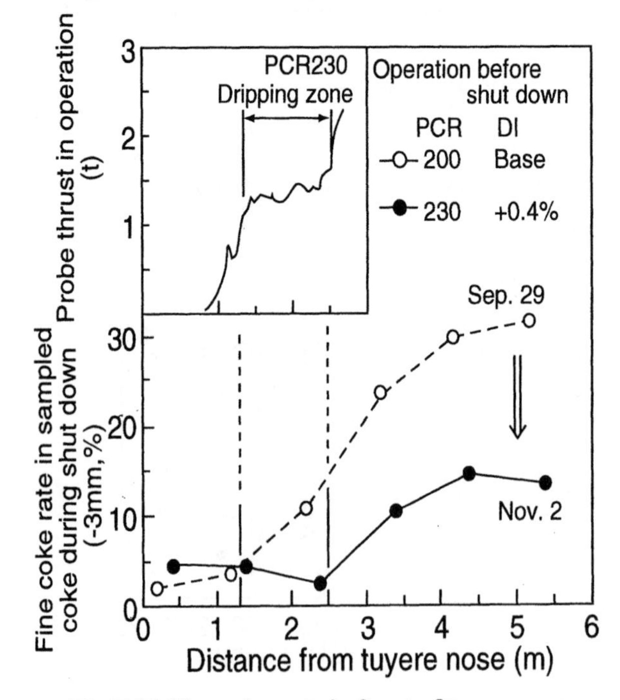

Fig.7-28 Fine coke rate in front of tuyere nose

probe through the tuyere and the result was shown in Fig. 7-28. [17] The thrusting stress began to increase from the position about 1m from the tuyere. Since the temperature was suddenly decreased from the same position, this position could be the border of the raceway. As the coke fine was a little at this region, the increase of the thrusting stress could be resulted by the shell layer. Beyond the shell region, the small thrusting stress region was sometimes observed. The dripping of pig iron and slag were usually observed at this region in the most case. The distribution of the slag and pig iron collected from this region was shown in Fig. 7-29. [17] In the case of 230kg/thm PCR, the position of pig iron and slag was moved about 20cm to deeper position than that of 100kg/thm. It meant that the depth of raceway was increased with PCR. Since the Alumina content in the collected slags was found to be extremely high at the raceway boundary, the ashes from PC generated in the raceway was blown to the boundary region of raceway and combined

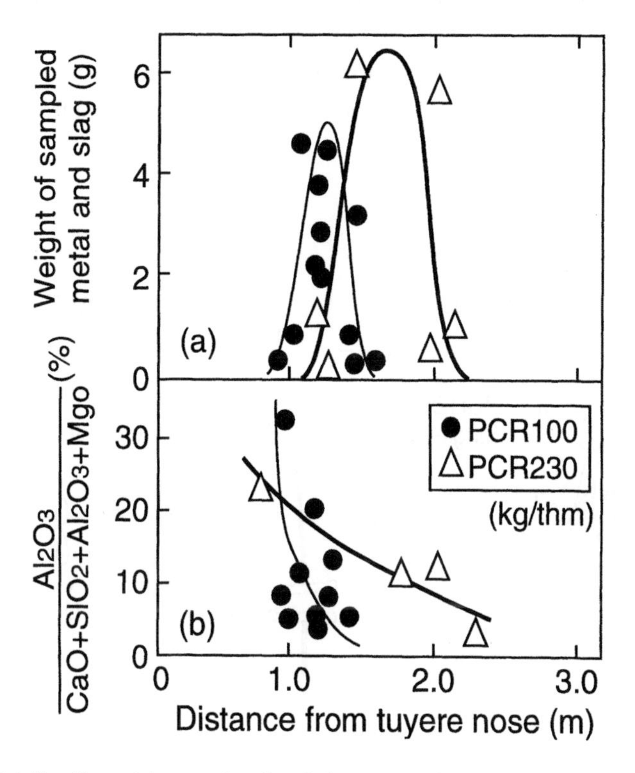

Fig.7-29 Distribution at tuyere level. a) Amount of sampled metal and slag b) Al_2O_3 content of sampled slag

with the high melting point slags and finally formed the shell layer. Therefore, the formation of shell layer could be avoided by applying the burner having a high dispersibility of fine coals. As a fact, the raceway of Fukuyama 4 BF having the eccentric double lance had much deeper raceway than that of 5 BF which was using the single lance. These results were schematically shown in Fig. 7-30.

The formation of shell may be affected by the rate of ash supply and the dispersibility of PC and gas flow pattern in the raceway. The mathematical model of the shell formation is shown in Fig. 7-31. The formation of shell layer was expressed by the balance between the rate of ash supply and its dripping rate. It was considered that the shell layer did not stayed permanently, but it was renewed catastrophically by the coke movement of the surface layer of deadman. However, the exact mechanism was not established and should be clarified in future. Once the shell layer was formed, fine coke was trapped. Then the gas permeability into the deadman was suppressed. As a result of this situation, the depth of the raceway was shortened and elongated vertically, and the peripheral gas flow became dominant. At this case, the liquid phase dripped from the upper part of the raceway has a tendency to flow to the direction of deadman due to the gas pressure from the raceway.

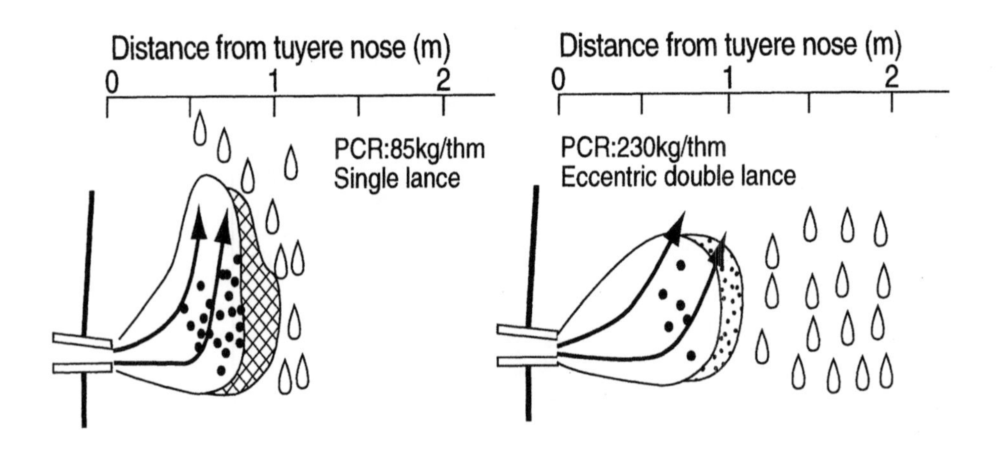

Fig.7-30 Schematic drawing of dripping of pig iron and slag around the raceway

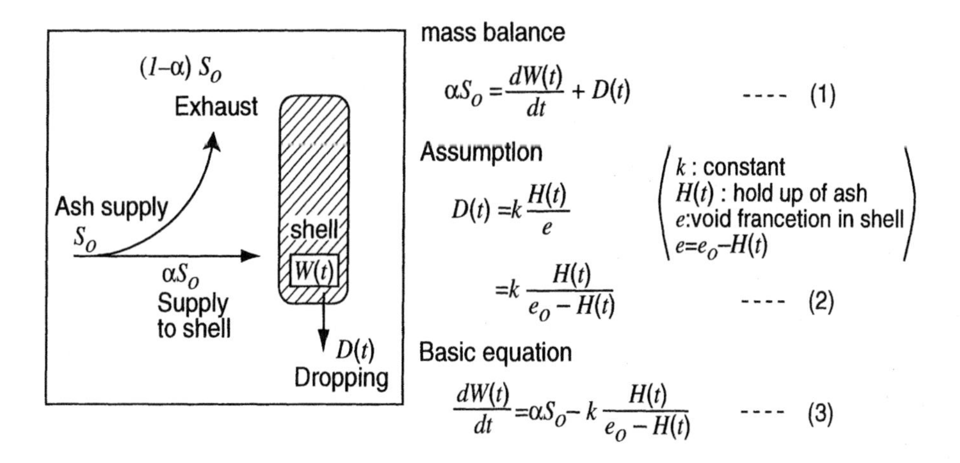

Fig.7-31 Modelling of fromation of shell

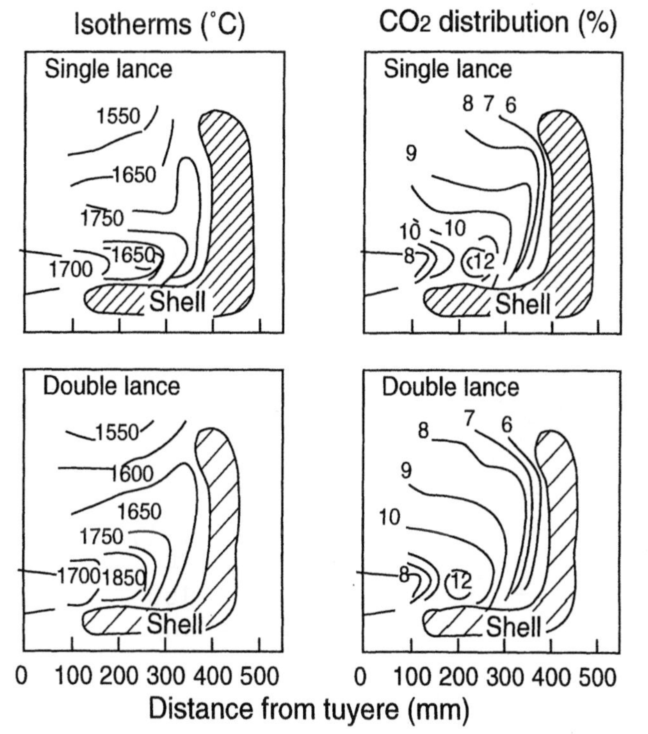

Fig.7-32 Distributions of measured isotherms and CO_2 composition in raceway

This explanation can be supported by the isothermal line and iso-concentration of CO_2 line measured by the small combustion furnace for two kinds of lance (single and double lance) shown in Fig. 7-32. [16]

7.6.2 Fine coke in deadman

The relationship between PCR and the fine coke (-3mm and +5mm) at the middle part of deadman (the average value at the horizontal distance of 4.45m and 3.65m from the tuyere nose) in Kakogawa 1 BF was shown in Fig. 7-33. [11] From these results, the degradation of cokes proceeded with PCR and the coke diameter was decreased, and

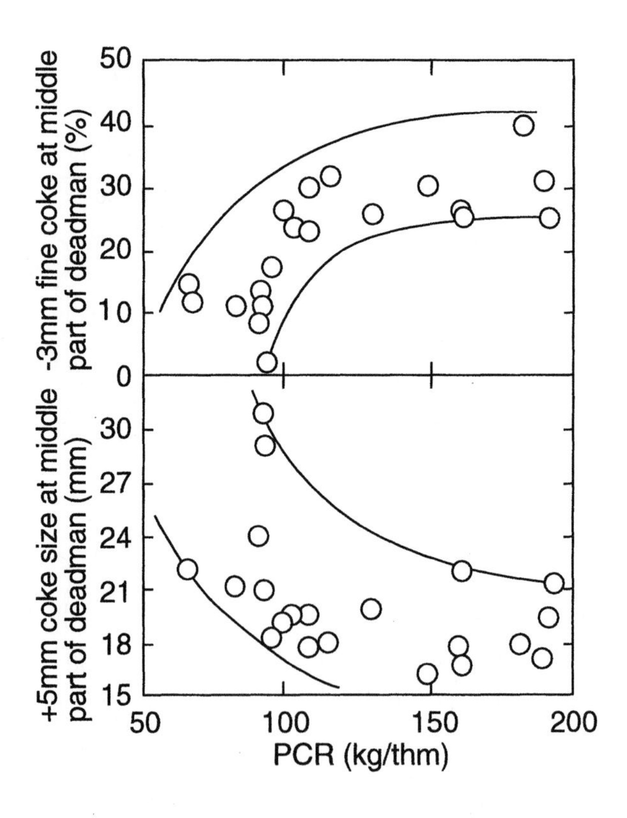

Fig.7-33 Relationship between fine coke in deadman and PCR

then the fine coke rate in the middle region was increased. To estimate the amount of the coke degradation at the high PCR, the carbon consumptions on the solution loss reaction and carburization were calculated and presented in Table 7-1. [11] Since the contribution of carburization was large ((b) in Table 7-1) and combustion of carbon in raceway decreased ((c)), the percentage of carbon solution loss increased (a/(a+b+c)), even though the carbon consumption by the solution loss decreased with PCR ((a)). This situation could be improved by increasing the heat flux ratio with oxygen enrichment and suppressing the peripheral gas flow with controlling the burden distribution. There were several explanations about the generation mechanism of fine coke as follows;

(1) the fine coke generated from degraded coke in the raceway [19]

(2) the fine coke generated in the region from the lower part of lumpy zone to the dripping zone in the center region [18]

(3) the fine coke in the surface part of deadman originated from the race way and the one from the middle to the center region originated from the dripping zone

and so on. However, there was no conclusion at this moment. One of this reason would be a uncertain method to estimate the temperature history of coke, such as X-ray analysis and other method. If the fine coke in the middle and center region were supplied from the upper part of deadman, it is necessary to suppress the coke reaction in the center region by controlling the burden distribution, or improving the coke strength to decrease the fine coke in the deadman. It would be generally considered that the strengthening of coke was effective procedure to suppress the fine generation in the lower part of BF,

Table 7-1 Comparison of coke consumption (kg-coke/thm)

		All coke	PCR=200
Coke rate		515	315
Sol.coke	(a)	110	88
Carburizing	(b)	55	55
Combustion	(c)	340	162
Dust	(d)	10	10
Rate of Sol.C a/(a+b+c)		22%	29%
Rate of carburizing b/(b+c)		14%	25%

regardless the origin of fine.

7.7 Conclusion

With the increasing PCR, many unexpected phenomena have been taken place in BF. Another important subject that was not mentioned in this chapter, such as the relationship of the fine and inactivity of deadman (low permeability of gas and liquid flow and bad tapping ability of pig iron and slag), the heat conduction to the deadman, clean up method of deadman, the quality of pig iron (especially the relationship between the Si content and PCR) were still unclarified. Since these subjects have not been discussed in present research project, it was omitted in this chapter. But these problems could be focused in the future research.

REFERENCES

1) Y. Kimura, M. Shimizu, S.Inaba:CAMP-ISIJ,3(1990),p.80
2) T. Goto: Journees Siderurgiques ATS, Paris (1990) Session 3(7)
3) M. Shimizu :Text of ironmaking, Advanced course of ironmaking and steelmaking seminar-1 , private communication, (1993),Tokyo
4) N. Morishita, M. Higuchi, Y. Inoue, T. Houga, Y. Matsuoka, N. Konno, S. Matsuzaki, S. Naitou, T. Deno: JSPS ironmaking 54th committee, 54-2048, (1995)
5) S. Amano, S. Matsunaga, K. Kakiuchi, H. Ueno, N. Konno, K. Yamaguchi:JSPS ironmaking 54th committee, 54-2025,(1994)
6) T. Kamijo, N. Takahashi, K. Hoshino, Y. Yoshida, R. Itoh, K. Shibata, Y. Miyagawa: CAMP-ISIJ, 6(1993),p.848
7) M. Matsuura, T. Ariyama, T. Satoh, K. Kimura, A. shimomura, K.Mori: CAMP-ISIJ, 9(1996), p.635
8) Y. Kishimoto, H. Kotoh, T. Ariyama, T. Satoh: JSPS ironmaking 54th committee , 54-2038, (1995)
9) K. Mori, K. Tomioka, A. Shimomura, A. Sakai, K. Kimura, A. Maki: CAMP-ISIJ, 8(1995), p.960
10) J. Kiguchi, M. Shimizu, R. Itoh, K. Hoshino: CAMP-ISIJ, 9(1996), p.639
11) K. Mori, N. Takagaki, A. Shimomura, A. Sakai, K. Kimura, A. Maki: CAMP-ISIJ, 9(1996), p.94

12) T. Kamijoh, M. Shimizu, R. Itoh, K. Hoshino: CAMP-ISIJ, 9(1996), p.26
13) T. Sugiyama, S. Naitoh, S. Matsuzaki, T. Kumaoka, T. Nakayama: CAMP-ISIJ, 7(1994), p.46
14) T. Sugiyama: Heat conduction and fluidity of four fluids in blast furnacc. ISIJ, (1996), p.226
15) T. Sugiyama: Flow and heat conduction of four fluids in Blast Furnace, Report of the Research Group for the transport Phenomena of Four Fluid), ISIJ (1996), p.226
16) T. Satoh, R. Murai, T. Ariyama, A. Maki, A. Sakai, K. Mori:Reasearch group of pulverized coal combustion in blast furnace, Rep-36, (1995), JSPS ironmaking 54th committee
17) R. Murai, T. Satoh, T. Ariyama, A. Maki, A. Sakai, K. Mori: CAMP-ISIJ, 8(1995), p.959

CHAPTER 8

Generation of fine in blast furnace at high rate PCI

As mentioned in Chapter 1, with increase of PCI, the significant degradation of coke and the increased unburnt char formation cause the increase of generation of fine in BF and result in the aggravation of BF operation variable such as gas permeability and burden descent. In this Chapter, the origin of generated fine, the effect of PC combustion on fine coke generation, the movement and accumulation of fine and the consumption of fine in BF will be explained.

8.1 Origin of fines

8.1.1 Fine generation phenomena with PCI

Figure 8-1 shows the mechanism of the fine generation in the BF. Generated fines are classified into two types. One is from the degradation of coke. The other is from the injected PC. The location and the causes of the fine generation are thought to be as follows.

(1) Coke degradation

1) In the bosh : solution loss reaction, carburization, reaction with molten FeO and stress acting on coke in the furnace
2) In the raceway: mechanical impact resulting from rotation in the raceway, thermal stress caused by the rapid heating up in the raceway and reactions occurring in the raceway that are the combustion of coke with oxygen and the solution loss reaction of coke with CO_2.

   The degradation in the bosh is mostly the volumetric breakdown, while that in the raceway is mostly surface breakdown.

(2) Fine from PCI

Fine originated from PC is the unburnt char.

8.1.2 Studies of fine generated in blast furnace

(1) Dissection of blast furnace

Several dissection studies of BFs in Japan revealed the deposit distribution of fine and the change of coke mean size in the furnace as a result of coke degradation in the

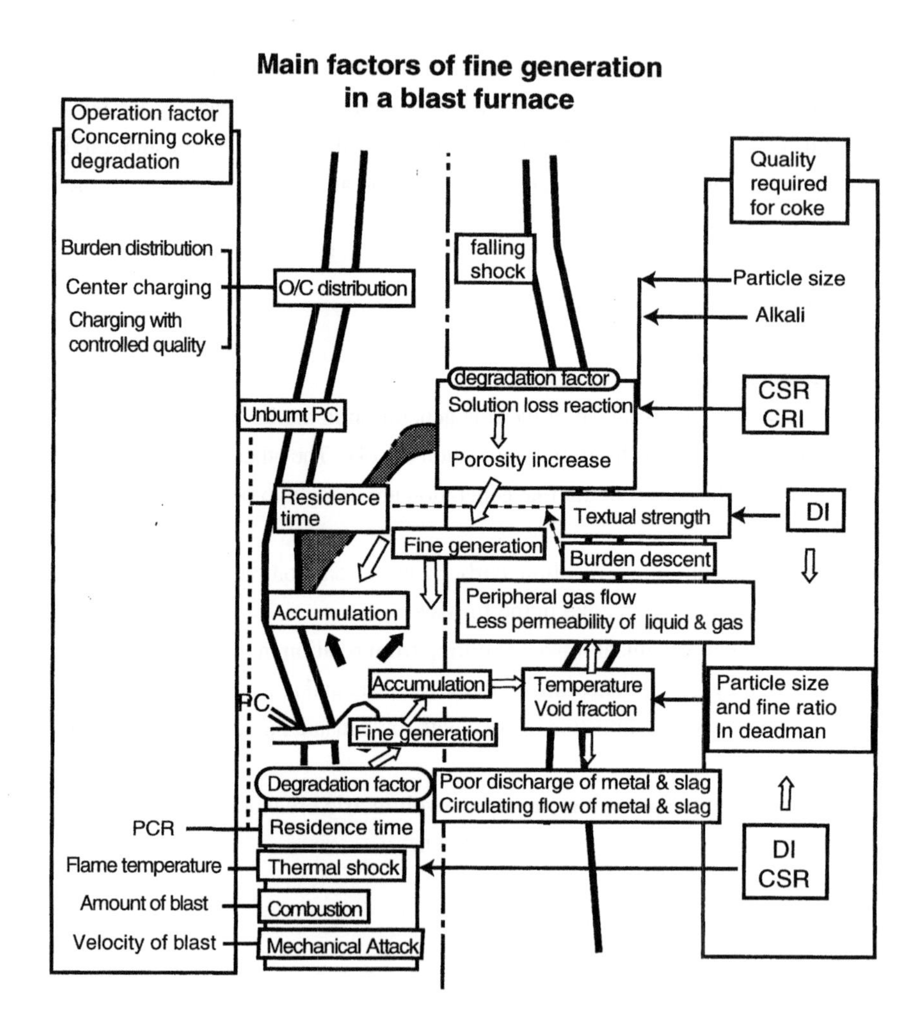

Fig.8-1 Conceptual mechanism of fine generation in BF.

furnace. The main results are summarized as follows.

1) Nagoya 1 BF of Nippon Steel

At Nagoya 1 BF, the dissection study was performed after the BF was quenched by nitrogen. Figure 8-2 shows changes of the coke mean size and the coke strength with change of the temperature estimated through carbon crystallization of coke measured[1].From room temperature to 1000°C, corresponding to upper shaft to middle shaft, no appreciable change of both coke mean size and DI(Drum Index) were observed and the degradation of coke in this region is slight. In the temperature range of 1000-1400°C, corresponding to middle shaft to bottom of cohesive zone, although the coke mean size does not change, DI decreased with increase of temperature. Therefore it is suggested that in this region, though coke surface suffered degradation by the solution loss reaction, fine generation does not occur significantly, because of the small abrasion and impact forces during coke descent in the furnace. In the temperature range of 1400-1600°C, corresponding to the lower shaft to just above tuyere level, coke mean size rapidly decreases with increase of temperature. Therefore, it is suggested that in this region, the degraded coke surface generates coke fine because of the significant abrasion and impact forces in the furnace.

2) Keihin 1 BF of NKK

The proportion of -3mm fine was 2-4% in the shaft part (-10m from the stock line)

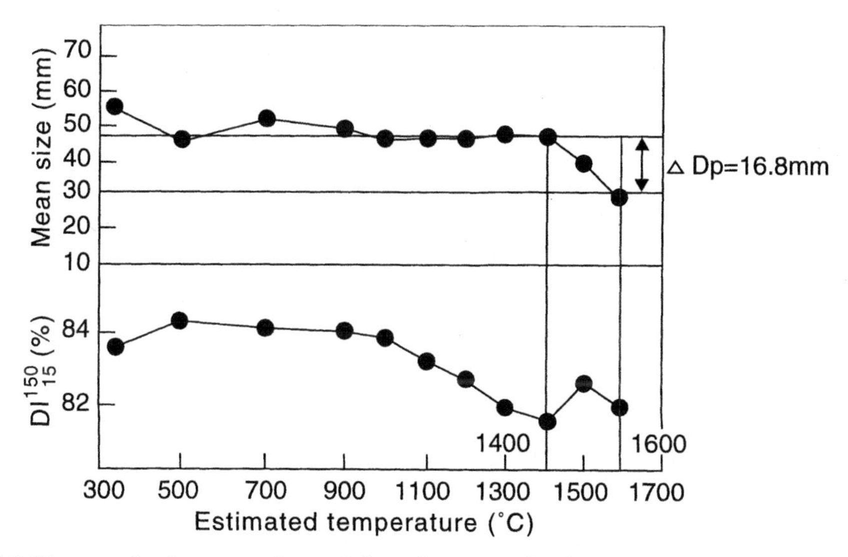

Fig.8-2 Change of coke mean size and the coke strength with temperature estimated[1]

and it significantly increased to 4-14% at the tuyere level.

3) Chiba 1 BF of Kawasaki Steel

Coke mean size did not change from the furnace top down to the middle part of the shaft. Then, it decreased in the bosh part and significantly decreased at 2-3m above tuyere level.

4) Kokura 2 BF of Sumitomo Metals

Figure 8-3 shows the longitudinal distribution of coke mean size, coke drum strength, coke reactivity and content of alkali oxides in coke ash in the dissection study of the BF.[2] The coke mean size was about 40mm constant above 3m of tuyere level. Below that level, it suddenly decreased to about 30mm.

Coke strength, DI_{15}^{30}, gradually decreased from 94 to 86 with its descent in the furnace, with small peaks in reactivity and alkali oxide composition.

Summarizing these results, it is concluded that fine generation suddenly increases 1-3m above tuyere level, that is in the temperature range of 1400-1600°C. The radial distribution of fines is such that it is mostly high in the peripheral region, while it is low

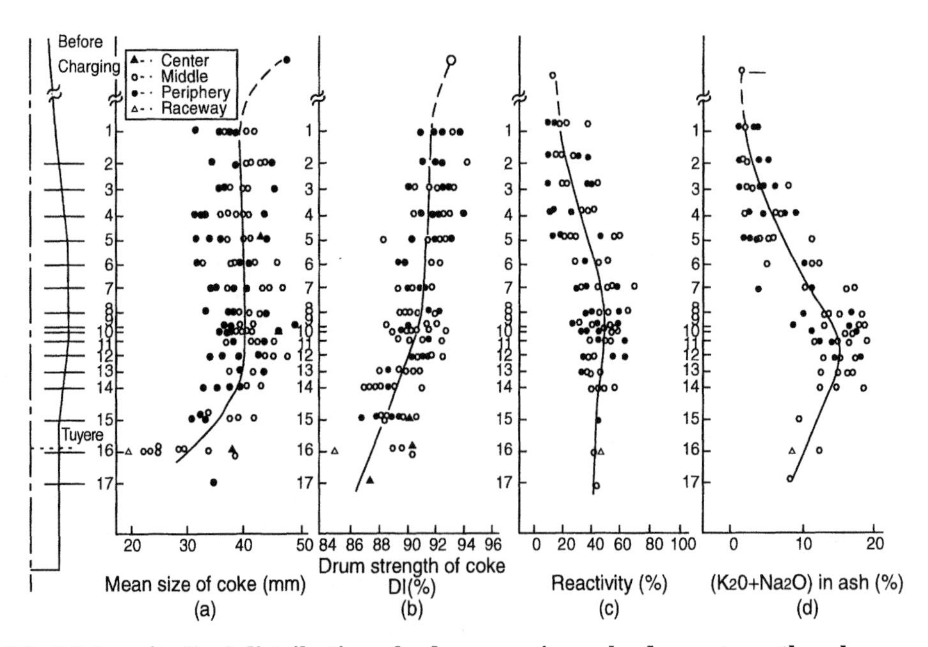

Fig.8-3 Longitudinal distribution of coke mean size, coke drum strength, coke reactivity and the content of alkali oxides in coke ash [2]

in the central region.

(2) Fine sample at shutdown

In the dissection study, it is problem in respect of fine sampling that the generated fine in BF may be washed away and transferred to other places by the quenching process and the reliability of the obtained data may be decreased. Therefore, the fine sampling during furnace shutdowns at the shaft part and the bosh part is also being performed as well as at the tuyere level.

The data of the former were limited because of the cost and the limitation of the facility installation. Figure 8-4 shows the results obtained at Kimitsu 3 BF of Nippon Steel in the test operation with change of CSR(Coke Strength after Reaction). [3] The fine coke content shows its maximum of about 25% in the peripheral to the intermediate region at the belly part. Figure 8-5 shows the data obtained at Kashima 1 BF of Sumitomo Metals in the test operation with change of CSR.[4] The coke fine content shows its maximum of about 20% in the peripheral region at lower shaft and belly parts. Both Figs 8-4 and 8-5 revealed that the "aggravation" of CSR causes the significant increase of coke fine content.

Figure 8-6 shows the change of coke strength and coke mean size in the furnace of Kashima 1 BF.[5] Although coke mean size significantly decreases from belly part (B2) to

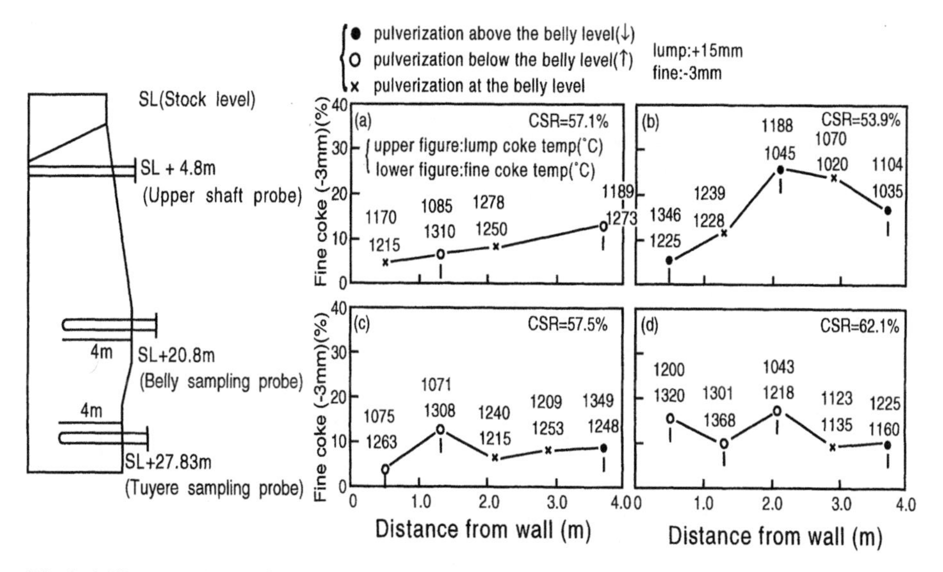

Fig.8-4 Change of coke fine content at belly level with change of CSR at Kimitsu 3 BF [3]

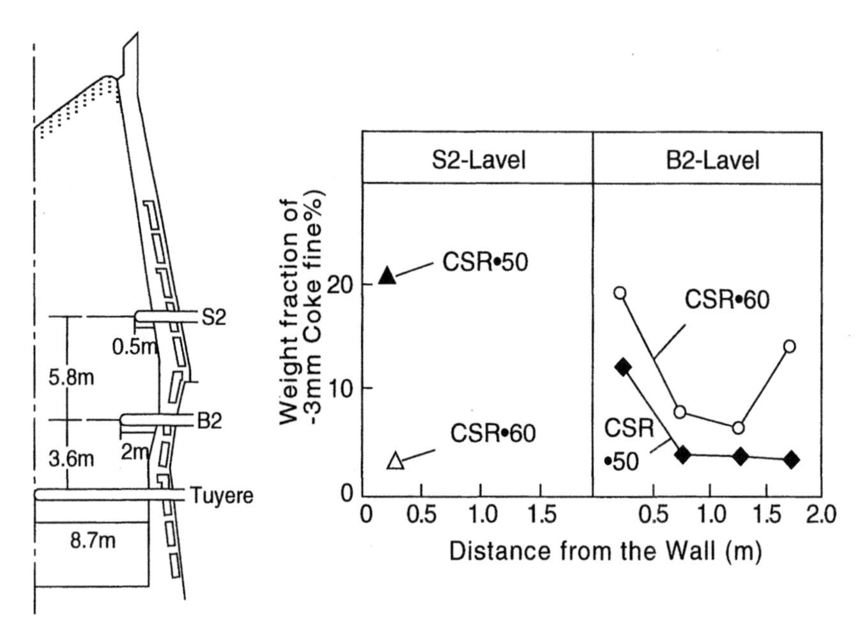

Fig.8-5 Change of coke fine content at middle shaft and belly level with change of CSR at Kashima 1 BF [4)]

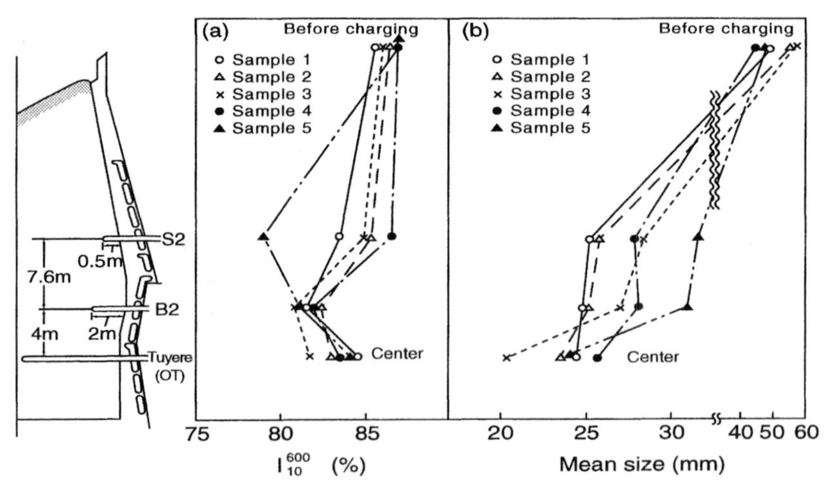

Fig.8-6 Longitudinal change of coke strength and coke mean size at Kashima 1 BF [5)]

tuyere level (OT), coke strength recovers its level as was shown in Fig.8-2. This fact is explained as follows. The peeling off of the degraded part of coke results in the decrease of coke size. On ther other hand, the rest of coke which is still sound shows normal strength.

The sampling of fine at the tuyere level by a deadman sampling probe and a tuyere one has been intensively performed under various operation conditions such as PCI level as well as coke qualities.[6] However, the comparison among different BFs can only be made quantitatively.

Figure 8-7 shows the effect of coke strength on coke fine content at Kobe 3 BF of Kobe Steel.[7] The coke fine content in low drum index operation shows its maximum of about 25% at the intermediate region, while that in normal drum index operation shows a lower value of maximum fine coke content, and the maximum point is shifted to the

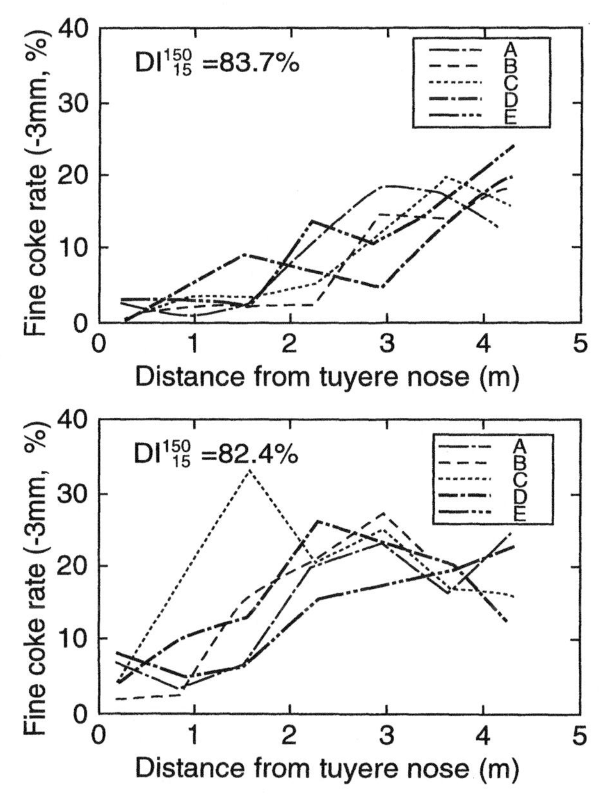

Fig.8-7 Effect of coke strength on fine coke content at the tuyere level in Kobe 3 BF [7]

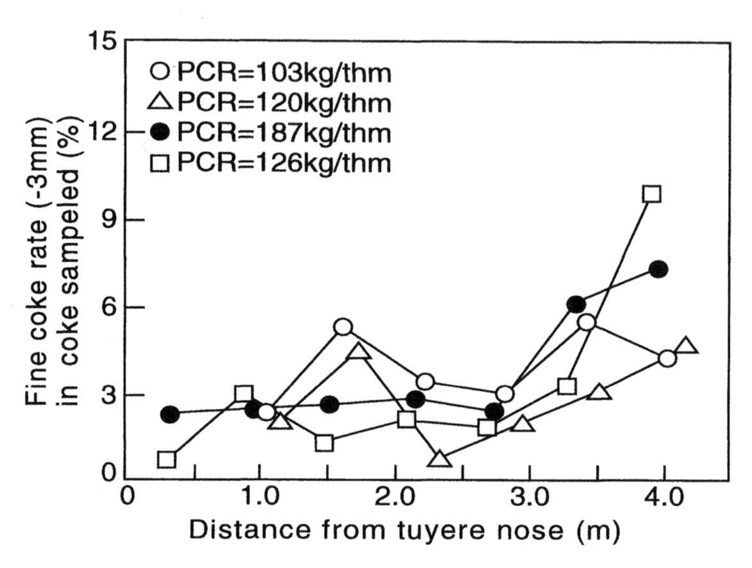

Fig.8-8 Effect of PCR on fine coke content at the tuyere level in Kimitsu 3 BF [9]

central region.

It is also reported that the change of CSR causes the increase of coke fine in the intermediate region and in the central region of the furnace.[8]

Figures 8-8 and 8-9 show the effect of PCI level on the coke fine content at Kimitsu 3 BF of Nippon Steel[9] and Fukuyama 4 BF of NKK[10], respectively. In both cases, the increase of PCR resulted in the increase of the fine coke content. However, at Fukuyama 4 BF at 230 kg/thm PCI, the coke fine content decreased by as a result of countermeasures for the improvement of coke strength.

Although the location of maximum fine coke content is usually observed in the central region, the BF condition is more closely related to the fine coke content in the intermediate region, that is, the surface of the deadman.

The accumulation of coke fine reduces the permeability of gas and liquid in the lower part of the BF and the deadman temperature decreases and the pressure drop in the BF increases.

The measurement of fine portion and the lump coke diameter at the tuyere level qualitatively showed that the fines in the raceway originates from injected PC, while fines outside the raceway originates from coke.[11]

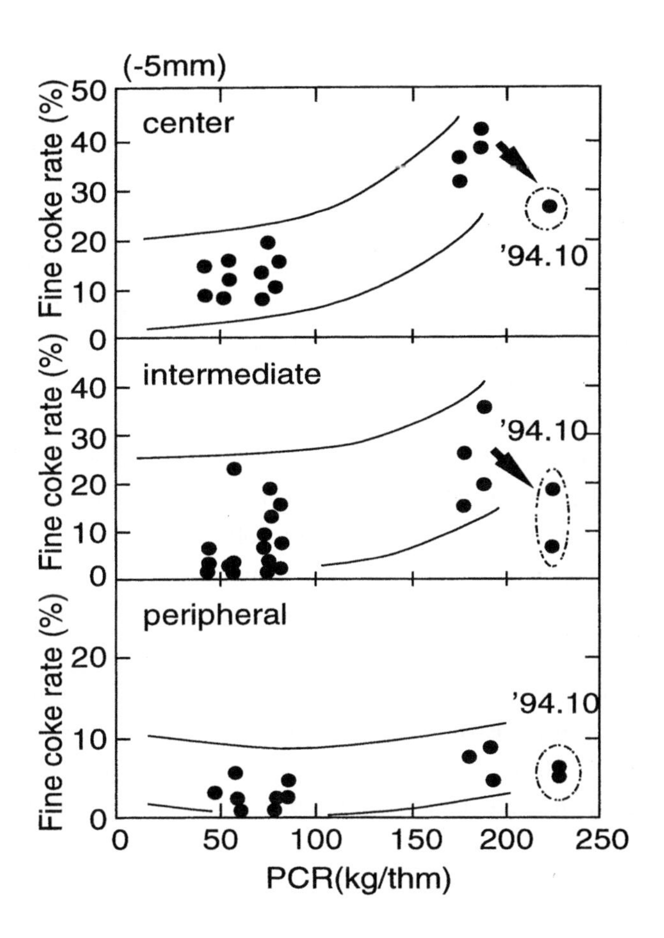

Fig.8-9 Effect of PCI rate on fine coke content at tuyere level at Fukuyama 4 BF [10]

(3) Fine sampled during operation

Fines sampled during operation are unburnt char collected by a raceway probe. The unburnt char is of ballon type shape and has large specific surface area (10^4-$10^6 m^2/kg$). An example of chemical analysis of unburnt char was 92.3% fixed carbon,0.2% total Fe,1.1% CaO,2.8% SiO_2,3.0% Al_2O_3, 0.1% MgO and 0.02% alkali, though the data were obtained in an experimental furnace with a coke packed bed.[12]

8.1.3 Mechanism of fine generation

(1) Coke

1) From furnace top to the bosh part

Based on dissection studies, the fine generation mechanism is thought to be as follows.

From room temperature to 1000°C, corresponding to the upper shaft to the middle shaft, neither the coke diameter nor coke strength change and the degradation of coke is small. In 1000°C-1400°C, corresponding to the middle shaft to the top of the cohesive zone, the coke size does not change, while the coke strength showed a decrease. Although the surface part of coke is degradated by the solution loss reaction, small abrasive forces as well as small impact forces cause no change in the coke size. As a result, coke fine generation does not occur down to this part.

In 1400-1600°C , which corresponds to the lower shaft to just above the tuyere level, coke diameter rapidly decreases and degradation occurs. On the contrary, the coke strength slightly increases. The reason is considered to be as follows.

In this part, although the coke surface suffers solution loss reactions and the strength of surface is decreased, the increase of the abrasive forces and the impact forces causes the separation of the degradated portion of the coke surface and the coke fine is generated. Accordingly, the remaining unreacted coke shows high strength with small diameter.

The breakdown state of coke in BF was investigated in a dissection study of Chiba 1 BF of Kawasaki Steel.[13] Coke sampled from the furnace was provided for both the drum tester and the tumbler tester and the relationships between the coke diameter and the number of revolutions for both test were obtained. Then, the distribution of diameter of sample were compared with those at the same average diameter obtained from both the drum and tumbler tester. As shown in Fig.8-10, no significant change in the average diameter is observed at the upper and middle shaft. The average diameter starts decreasing at the lower shaft by the solution loss reaction. Changes in the coke size distribution in this region agree with the drum results, the dashed line shown in Fig. 8-10. It implies that the coke breakdown occurs in the form of volumetric breakage at the lower shaft. On the other hand, in the lower dripping zone and the raceway, the diameter distribution measured shifts from the dashed line to the solid line, the tumbler test results. This indicates that the coke breakdown proceeds in the form of surface abrasion. The locations of two type of coke breakdown in Chiba 1 BF are illustrated in Fig.8-11.[14]

2) In the raceway

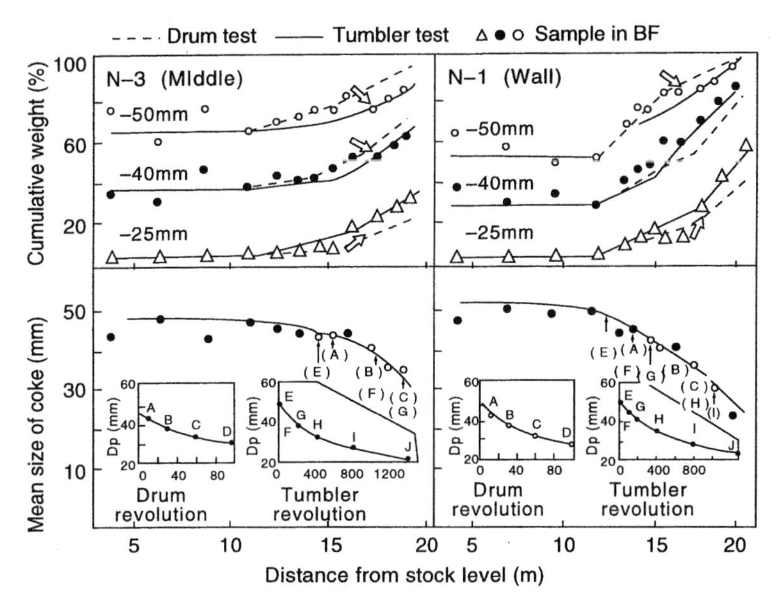

Fig.8-10 Comparison of coke size distributions of sampled coke with those of after the drum and the tumbler tester [13]

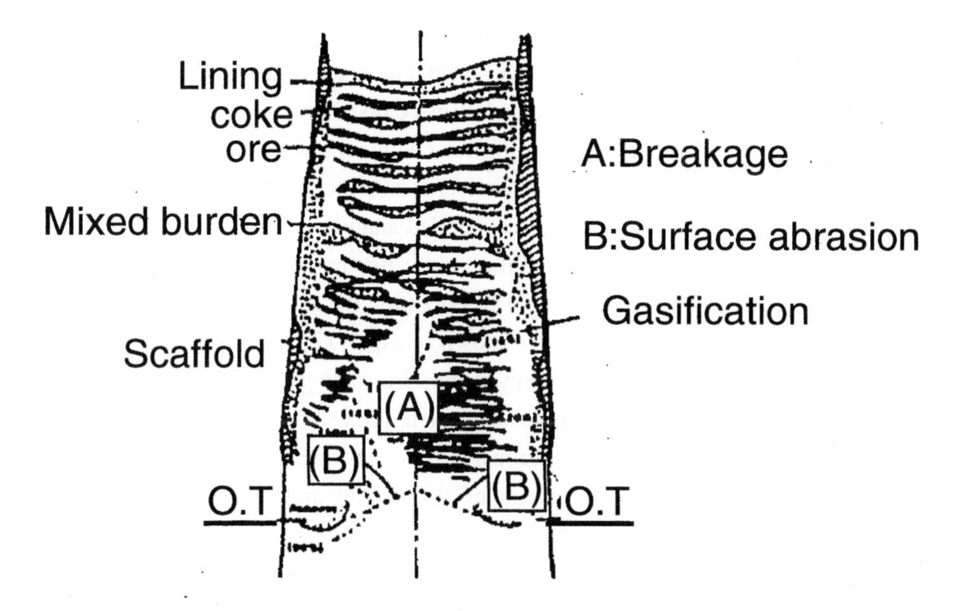

Fig.8-11 Degradation mechanism of coke in BF analyzed for Chiba 1 BF [14]

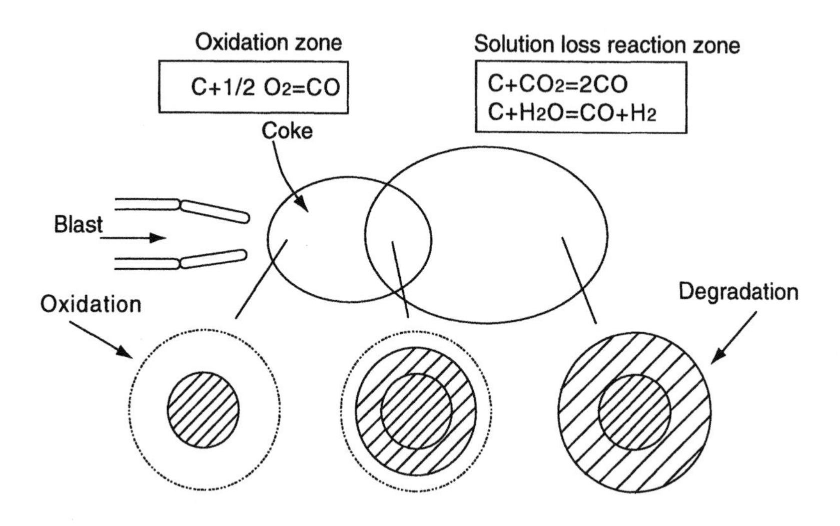

Fig.8-12 Illustration of coke degradation and oxidation [12)]

Probably, the coke degradation occurs due to the mechanical abrasion during rotation in the raceway, thermal stress induced by the rapid heating up and chemical reactions in the raceway. The PC combustion test[15)] revealed that the main degradation of coke occurs by the chemical reactions and the thermal stress causes only slight fine generation. Although from the macroscopic viewpoint the chemical reactions in the raceway are completely expressed by $2C+O_2=2CO$, it consists of both the combustion reaction ($C+O_2$) and the gasification reaction ($C+CO_2$). The detailed analysis of the contribution of each reaction revealed that the contribution of the combustion reaction cannnot be ignored in the all-coke operation, while in high rate PCI operation the gasification reaction mainly contributes to the coke degradation. As shown in Fig.8-12, coke in the solution loss reaction zone in the outer side of raceway suffers from the gasification reactions with CO_2 and H_2O gas, forming porous fragile layers on the coke surface. Then, the part of the fragile layer suffers degradation by abrasion and generates fine coke outside the raceway, while most parts of the remaining fragile layer successively undergoes rapid combustion in the oxidation zone in the internal side of the raceway without causing degradation. Therefore, the degradation of coke in the raceway is presumed to occur outside of the raceway.

For the prevention of the coke degradation, the usage of high reactivity coke has been proposed. In this case, since the gasification reaction is so rapid that the reaction is restricted to occur only within the surface layer and the degradation does not go further into the coke particle and the soundness of the coke particle is maintained. However, there seems to be a certain upper limit in the usage of high reactivity coke, since it has been reported that with overrated progress of the gasification reaction in the use of high reactivity coke, the coke porosity of over 52% may be reached, then the significant degradation of coke would occur.

Since the use of high strength coke requires the production cost increase, the decrease of necessary coke strength in the actual operation is a future task. As a preliminary measure for this, the one method to decrease of load on coke in the BF is by the use of high porosity coke and maintaining sufficient layer thickness of coke in the peripheral region.

3) In the deadman

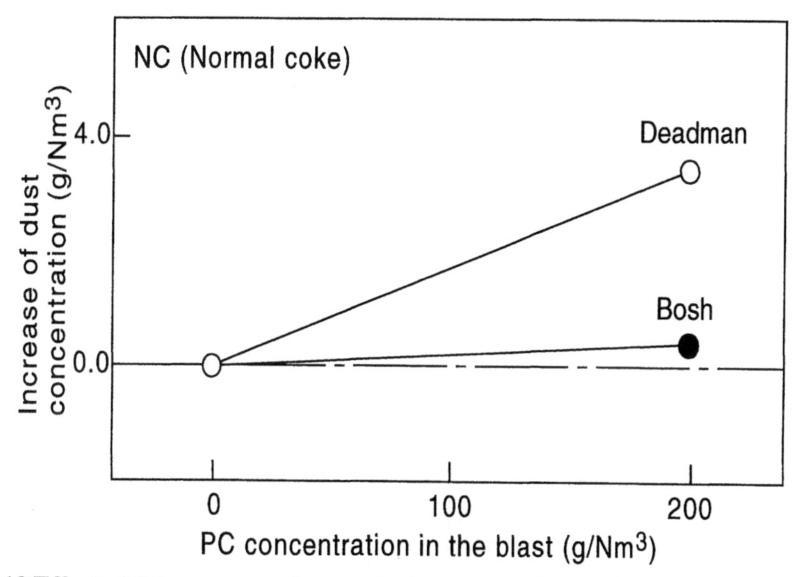

Fig.8-13 Effect of PC concentration on dust concentration in deadman and in bosh part [15]

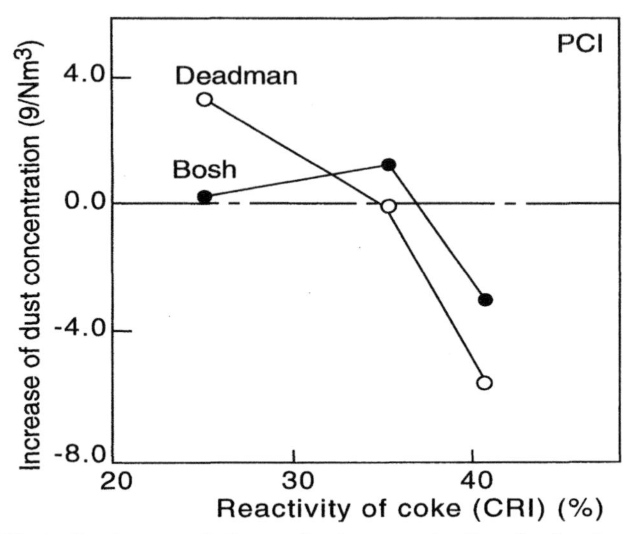

Fig.8-14 Effect of coke reactivity on dust concentration in deadman and in bosh part [15]

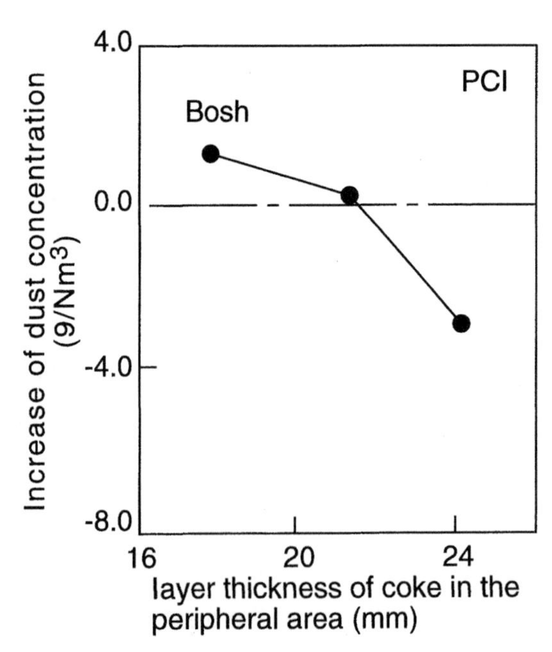

Fig.8-15 Effect of coke layer thickness in the peripheral region on dust concentration in deadman and in bosh part [15]

The effects of PC concentration, coke reactivity (CRI) and the coke layer thickness in the peripheral region on dust concentration, which has a relation to fine coke, in the deadman and in the bosh part were investigated with use of a hot model as shown in Figs.8-13, 8-14 and 8-15, respectively.[15] PC concentration has only slight effect on the fine ratio contrary to previous expectations . On the other hand, the increases of the coke reactivity and the coke layer thickness in the peripheral region decreased the fine ratio significantly and were found to be effective countermeasures against the coke degradation.

The effects of coke drum index (DI) and the coke reactivity (CRI) on both the fine ratio and lump coke size were investigated by the use of a packed bed type combuster as shown in Figs.8-16 and 8-17, respectivey.[11] With use of coke B which has high DI and high CRI, the fine generation is the smallest and the lump coke size is the largest. With use of coke A which has high DI and low CRI, the coke degradation is the most significant. With use of coke C which has low DI but the highest CRI, the degradation was less than is the case with coke A.

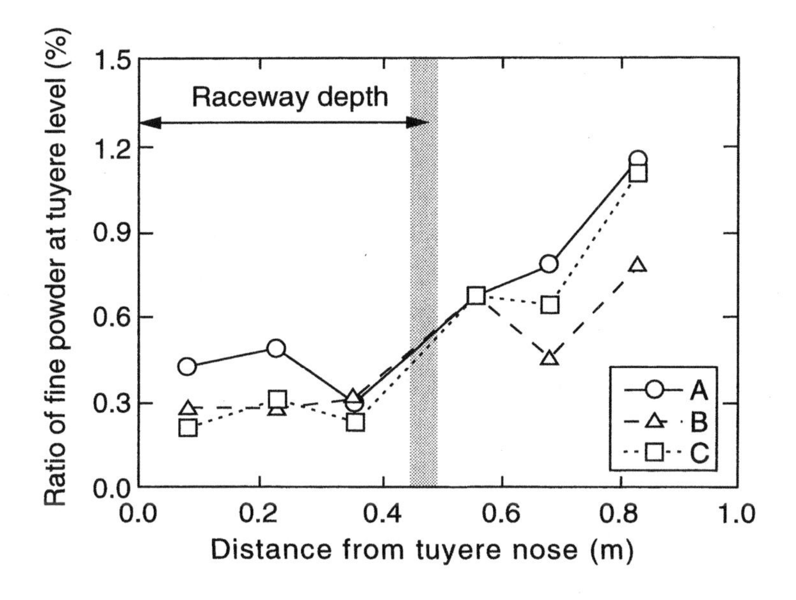

Fig.8-16 Change of coke fine distribution at the tuyere level with change of coke qualities [11].
A: high DI, low CRI, B:high DI high CRI, C:low DI, highest CRI

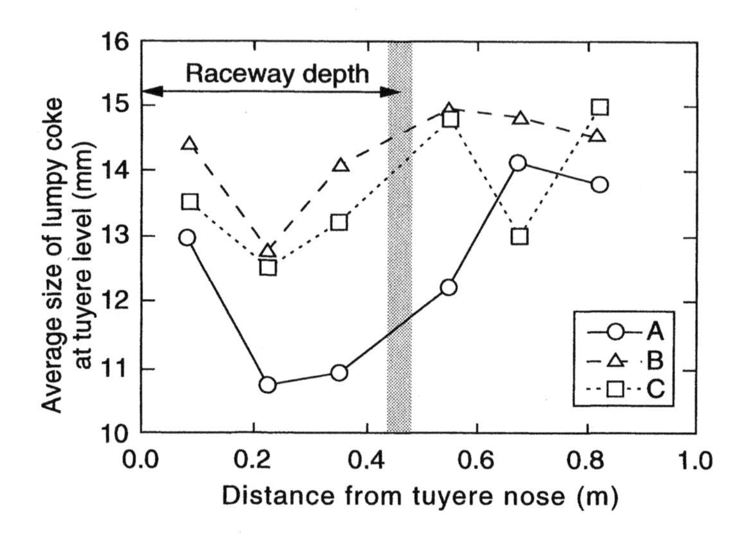

Fig.8-17 Change of coke mean size at the tuyere level with change of coke qualities [11]
A: high DI, low CRI, B:high DI high CRI, C:low DI, highest CRI

(2) PCR

Although in the case of low PCR, all of injected PC is consumed in the raceway without outflow from the raceway, the increase of PCR causes the decrease in the combustibility within the raceway resulting the outflow of unburnt char and fly ash from the raceway as was described in Chapter 3.

8.1.4 Identification of origin of generated fine

(1) Estimation of temperature history of coke

The temperature history of coke at the tuyere level and in the bosh part has been investigated from the carbon crystallization (Lc) and by the laser ramman spectroscopy.

As shown in Figs.8-18 and 8-19, the difference of the estimated temperature from Lc is not observed between fine and lump coke sampled at the tuyere level. On the contrary, a difference of the estimated temperature is observed between the fine and the lump coke sampled in the bosh part. The estimated temperature of fine in the center of deadman is lower than that in the raceway. From these facts, it is estimated that the fine generated at

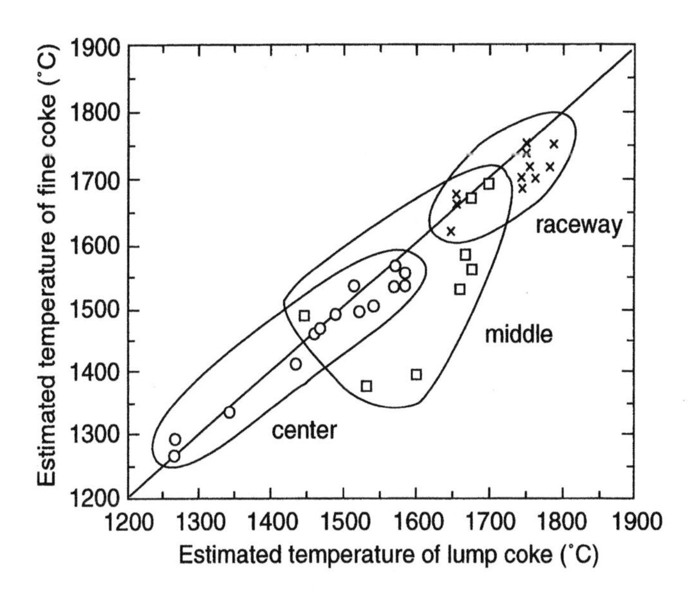

Fig.8-18 Difference between estimated temperature of fine coke and that of lump coke at tuyere level [16]

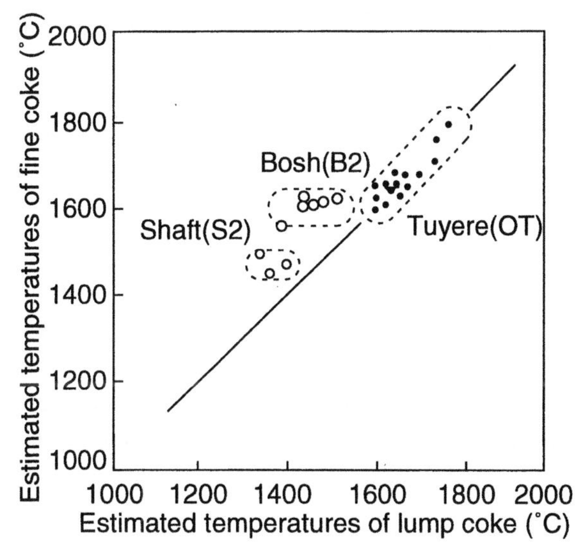

Fig.8-19 Difference between estimated temperature of fine coke and that of lump coke at bosh level [5]

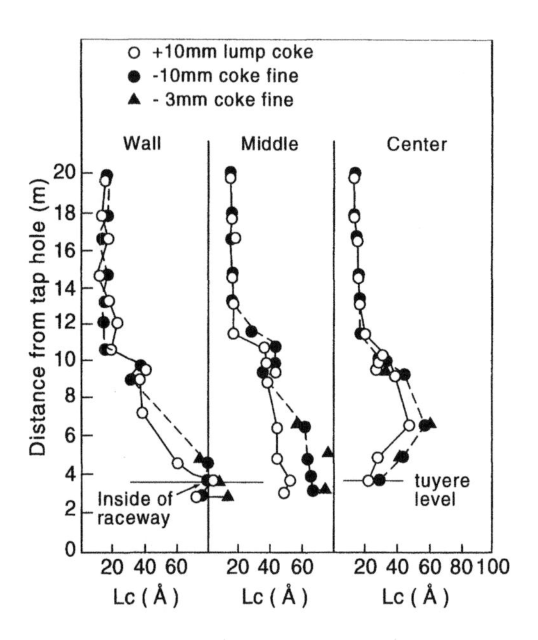

Fig.8-20 Change of Lc of lump coke and fine coke in the blast furnace at Nagoya 1 BF [1]

the tuyere level is transfered upward to the bosh part by the internal gas and the the fine found in the deadman is originally generated in the dripping zone and moved downward into the deadman.[5,16]

The detailed change of Lc in the furnace is shown for Nagoya 1 BF in Fig.8-20.[1] Lc of lump coke and that of fine coke in the periphery of the furnace at tuyere level are almost the same. Lc of lump coke is about 80A and that of fine coke is 80-100A. The estimated temperature of lump coke is 1700°C, which is about 0.75 times of the flame temperature, 2250°C. Therefore, the coke is thought to exist in the raceway. On the other hand, the estimated temperature of fine coke at the same location is 1700-1800°C, which is about the temperature in the raceway. Therefore, it is estimated that this fine coke was generated in the raceway.

The coke temperature history investigated at Keihin 2 BF with use of laser ramman spectroscopy is shown in Fig.8-21.[17] Fine coke in the raceway from 1m ahead of tuyere tip has temperature history of 1800-2000°C which is about the same as that of lump coke. Therefore, the investigation with use of laser ramman spectroscopy has also revealed that this fine coke was generated in the raceway.

(2) Estimation by slag analysis

The unburnt material was sampled during shutdown in front of the tuyere under PCI rate of 29kg/thm and was investigated by the simultaneous analysis of slag and the carbon crystallization and the following results were obtained.[18]

1) Carbon in the unburnt material is found to be fine coke, not unburnt char.
2) Sampled fine carbon contains a lot of slag content. The slag content reached as high as 70% at the boundary of raceway and the deadman. Slag contains high CaO and the slag composition is closer to the final blast furnace slag composition, in which CaO content is higher than that of SiO_2, in a closer position to the furnace center.(Fig.8-22)
3) Coke located closer in a position to the furnace center suffered high ambient

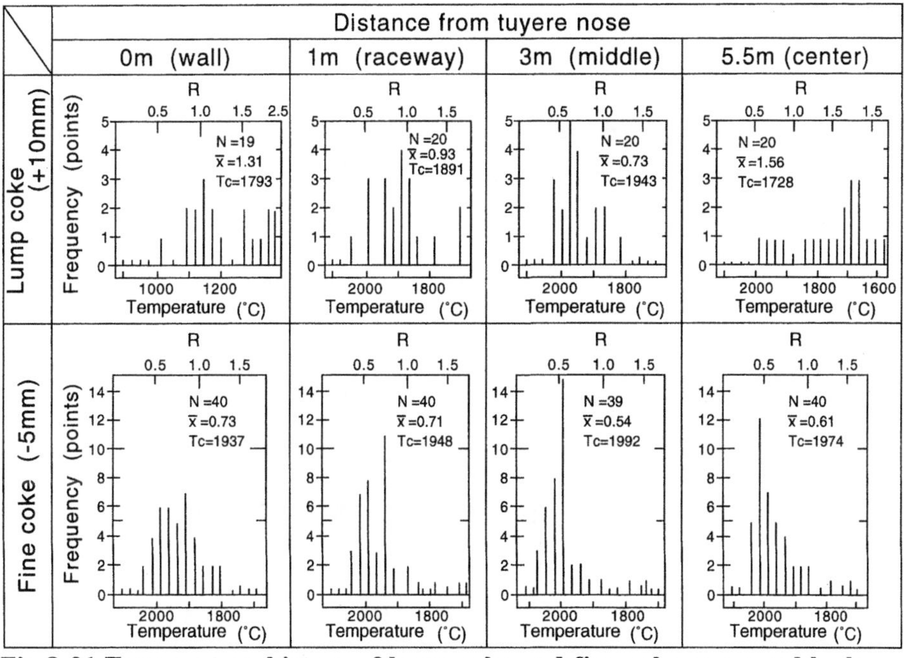

Fig.8-21 Temperature history of lump coke and fine coke measured by laser ramman spectroscopy at Keihin 2 BF [17]

temperature, as is revealed from the carbon crystallization analysis.

These results suggest the possibility that the fine generated at the lower part of cohesive zone flows down and penetrates into the deadman.

Similar investigation for high rate PCI operation will be expected in the near future.

8.1.5 Summary of fine generation

At present, the quantitative contribution ratio of each factor on the total fine generation in the BF has not yet been fully investigated. The factors concerned are fine coke generated in the bosh part, fine coke generated in the raceway and unburnt char of injected PC. However, it is presumed that the main contribution of the fine in the deadman is made by fine coke generated in the bosh part, that in the raceway is made by fine coke generated in the raceway and that in the bosh part is made by both fine flowing upward from the lower part of BF and generated fine in the bosh part.

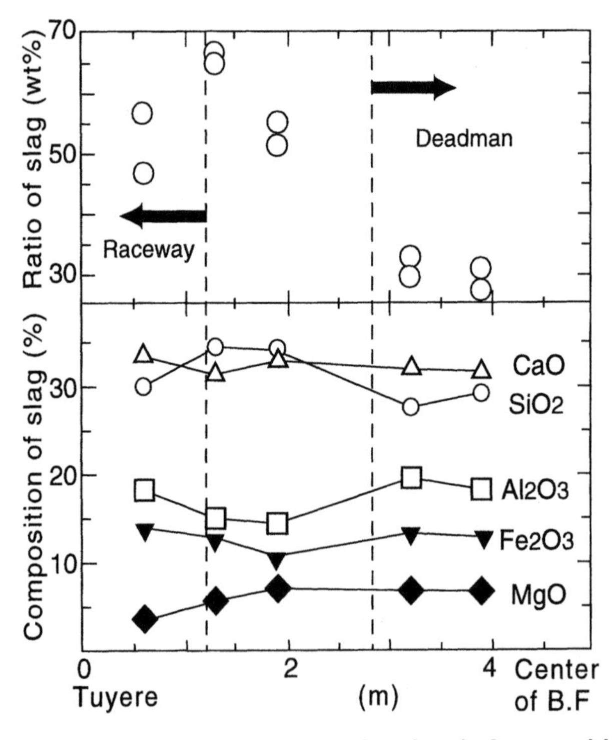

Fig.8-22 Radial distribution of slag amount and slag chemical composition at tuyere level [18]

8.2 Effect of PC combustion on fine coke generation

8.2.1 Effect of PCI on fine coke generation

The effect of the PCI on fine coke generation in the BF is presumed to be as shown in Fig.8-23.

The intensification of PCI decreases the coke consumption rate in the BF, resulting in the increase of the residence time of coke in the bosh part, which means the increase of the solution loss reaction time, and the increase of the residence time of coke in the raceway which means the increase of the mechanical impact time. Accordingly, coke degradates more in high rate PCI operation than all-coke operation and fine generation increases.

The increased hydrogen concentration or decreased CO concentration in high rate PCI operation decreases the total contribution of solution loss reaction and may result in the decrease of fine coke generation. However, the amount of solution loss reaction of each coke particle increases and results in the increase of fine coke generation.

Further, in the high rate PCI, the increase of the generation of unburnt char due to the decrease of the combustibility of PC in the raceway decreases the fine coke consumption.[19] This is because the preferential solution loss rection of unburnt char to fine coke as

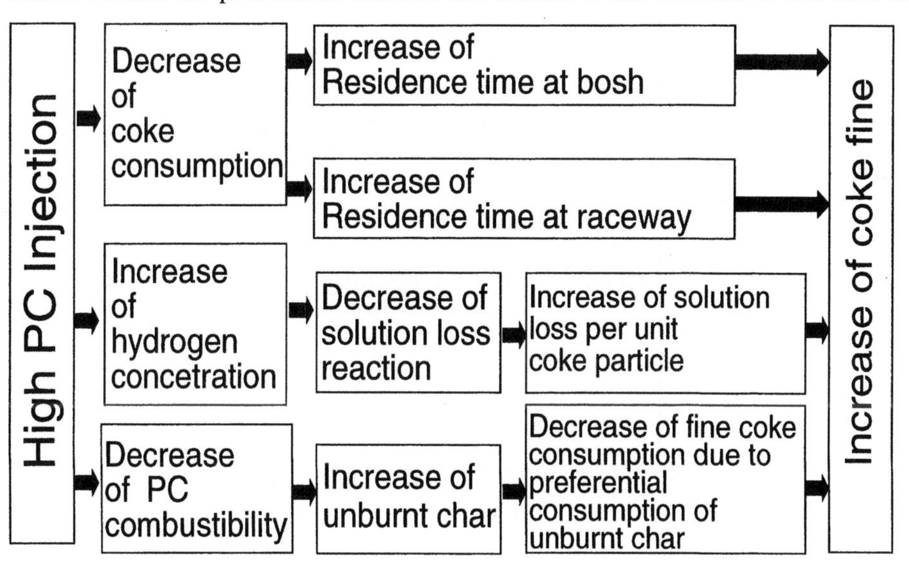

Fig.8-23 Flow diagram of the effect of high rate PCI on fine coke generation in BF

shown in Fig.8-24. As a result, fine coke increases.

The fine coke generation phenomena have been investigated by fine sampling in the actual BF, hot model experiments and the mathematical simulation model analyses as will be explained next.

8.2.2 Fine generation in high rate PCI operation in actual BF

The fine generation behavior in actual BF has been investigated through fine sampling by various companies. Generated fine coke and unburnt char are sampled by a coke sampling probe at the tuyere level during shutdowns. The fine consumption in BF is estimated by sampling of exhausted dust from BF top.

At Kimitsu 3 BF at 187kg/thm PCR, a significant increase of fine at the tuyere level was not observed because of countermeasures taken which both maintain coke strength and the central gas flow through burden distribution control as shown in Fig.8-25.[20]

However, the fraction of unburnt char in fine is as small as 3wt%. On the other hand,

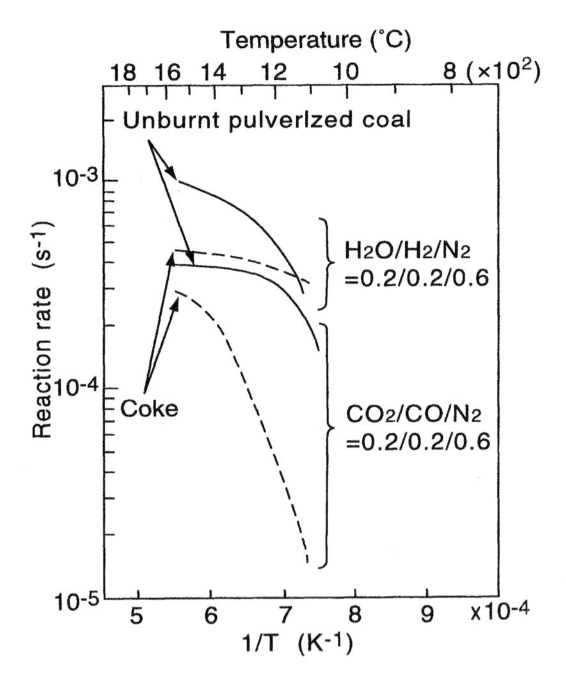

Fig.8-24 Comparison of consumption rate of unburnt char and fine coke by solution loss reaction [19]

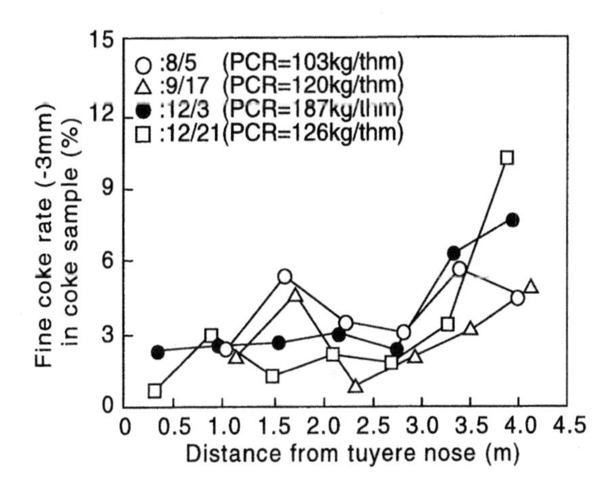

Fig.8-25 Fine coke distribution at tuyere level through tuyere coke sampling at Kimitsu 3 BF [20]

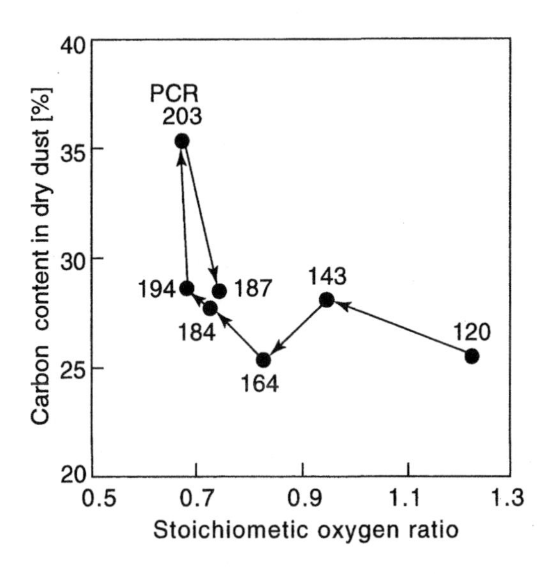

Fig.8-26 Discharged carbon percent in dust from BF top at Kimitsu 3 BF [20]

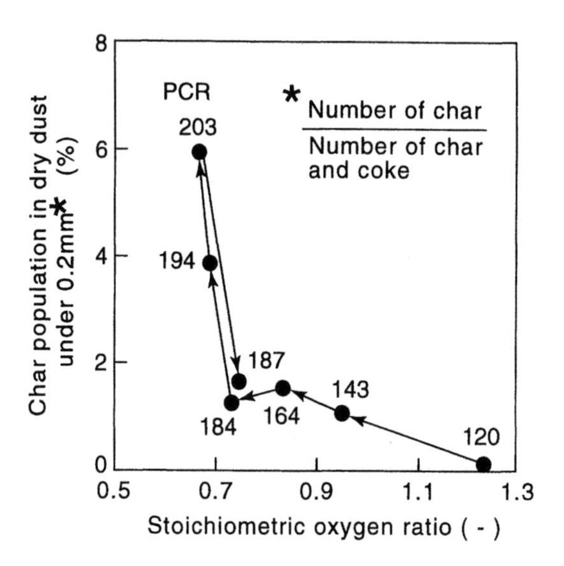

Fig.8-27 Discharged unburnt char percent in dust from blast furnace top at Kimitsu 3 BF [20]

the exhausted dust from BF top significantly increased at about 0.7 in stoichiometric oxygen ratio, which corresponds to 203kg/thm PCI rate, as shown in Fig.8-26. At 203kg/thm PCR the existence ratio of unburnt char in dust increased significantly to 6 % as shown in Fig.8-27, though the level itself is still low. This fact implies that the consumption of unburnt char in the BF may have reached its limit.

At Fukuyama 4 BF, the coke sampled during shutdowns revealed that the fine coke fraction at the tuyere level at 230kg/thm PCR was less than that at 200 kg/thm as shown in Fig.8-28.[21] The reason for this is attributed to the improvements of coke strength by 0.4 % in DI and the reduction of amount of solution loss reaction. Although the existence of unburnt char in fine samples was observed, its content was as small as less than 0.3 % as shown in Fig.8-29. Therefore, it is concluded that in high rate PCI the accumulation of unburnt char in the deadman does not become a serious problem. Further as shown in Fig.8-30, the amount of the exhaust dust from BF top is decreased with increase of PCR and the amount of unburnt char in dust is found to be almost constant as observed by microscope. These results suggest that most of the unburnt char generated is consumed

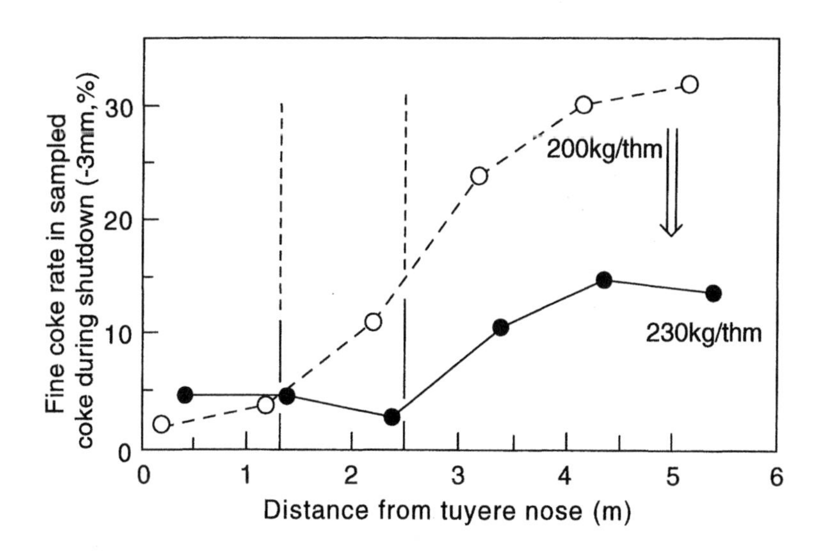

Fig.8-28 Fine coke distribution at the tuyere level through tuyere coke sampling at Fukuyama 4 BF [21]

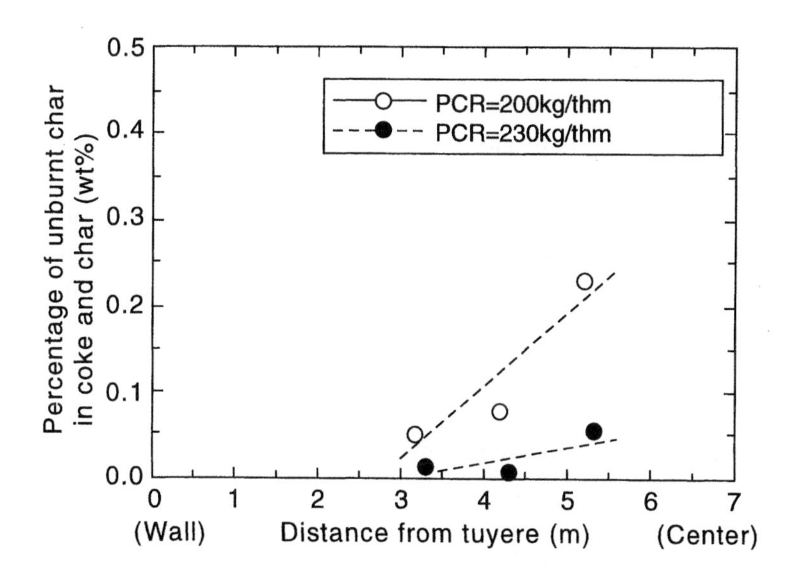

Fig.8-29 Unburnt char percent in fine at tuyere level of Fukuyama 4 BF [21]

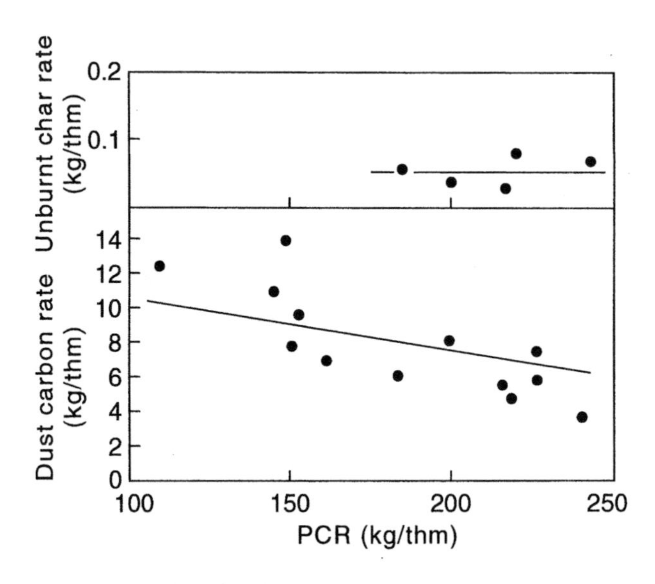

Fig.8-30 Discharged fine from blast furnace top at Fukuyama 4 BF [21]

within the BF.

At Kakogawa 1 BF of Kobe Steel, the 'increase' of fine ratio at tuyere level under high rate PCI rate of 201kg/thm was restrained by the improvement of coke cold strength and the increase of coke size for charging as shown in Fig.8-31.[22] At the BF with use of the center coke charging equipment, coke size for charging into the central region was increased from 45.9mm to 49.4 mm to restrain the aggravation of fine ratio in the deadman, resulting coke for charging into the peripheral region is weakened while improving the coke strength in the central region, no appreciable increase of fine ratio in the deadman was observed. In 201kg/thm PCR, the amount of the exhaust dust from BF top did not increase.

8.2.3 Fine generation by hot model experiments

Since the detailed sampling of fine in the actual BFs is difficult, hot model experiments have been performed .[20]

Based on the coke and ore packed hot model using plasma as blast heating, it is confirmed that at over 245g/Nm^3 PC concentration in blast the rate of the generation of

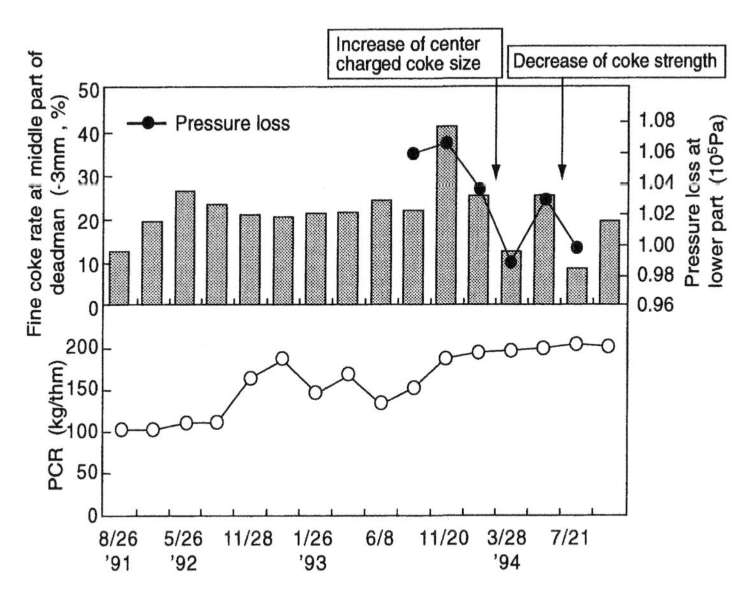

Fig.8-31 Transition of fine coke percent in the middle part of deadman at high rate PCI campaign of Kakogawa 1 BF [22]

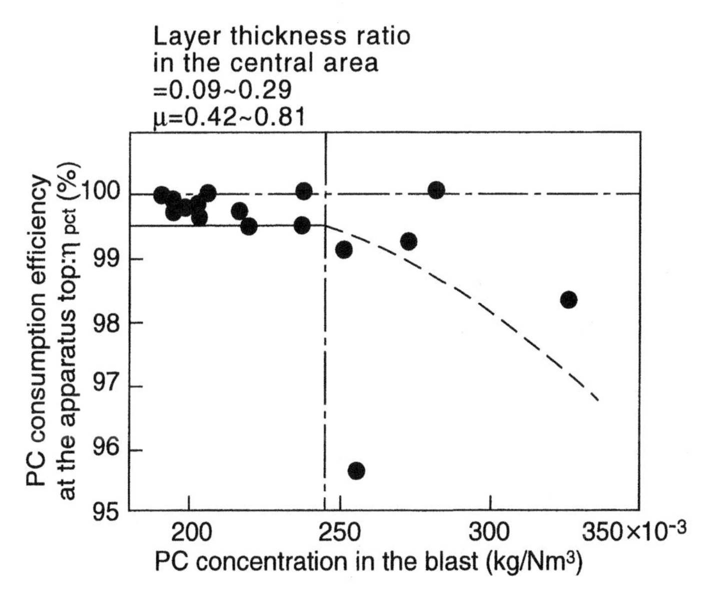

Fig.8-32 Effect of PC concentration in blast on PC consumption efficiency at the apparatus top [20]

fine exceeds the rate of fine consumption in BF and the unburnt char was exhausted from BF top as shown in Fig.8-32.[20]

The generation of unburnt char and coke fine in high rate PCI has been completely investigated with use of a coke packed hot model simulating the lower part of BF.[23] Figures 8-33 and 8-34 show the distribution of unburnt char and fine coke in the model at 200kg/thm PCR, respectively. The unburnt char generated in the raceway moved upward along the gas stream. Just above the raceway, unburnt char accumulates little, while around the raceway there are high unburnt char accumulation regions. The amount of unburnt char from the top of model furnace, which corresponds to the unburnt char carried from the lower part of the BF to the upper part of the BF, was as large as 44%.

The fine coke generated in the raceway accumulates in front of and at the bottom of the raceway and is moved upward by the gas along the periphery of raceway. The high holdup regions of fine coke located near the raceway and the lower part of the model furnace. About 30 % of generated coke fine was exhausted from top of the model furnace.

The fine coke generation and the combustion degree estimated from the material balance of samples are shown in Fig.8-35. With increase of PCI rate, the combustion

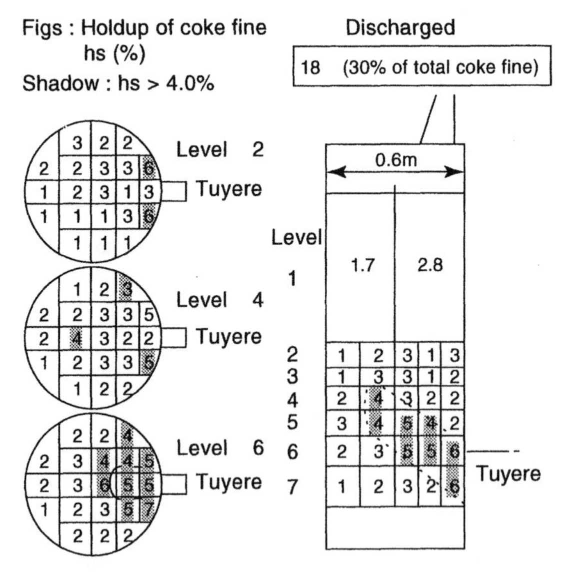

Fig.8-33 Distribution of coke fine holdup at 200 kg/thm PCR in hot model [23]

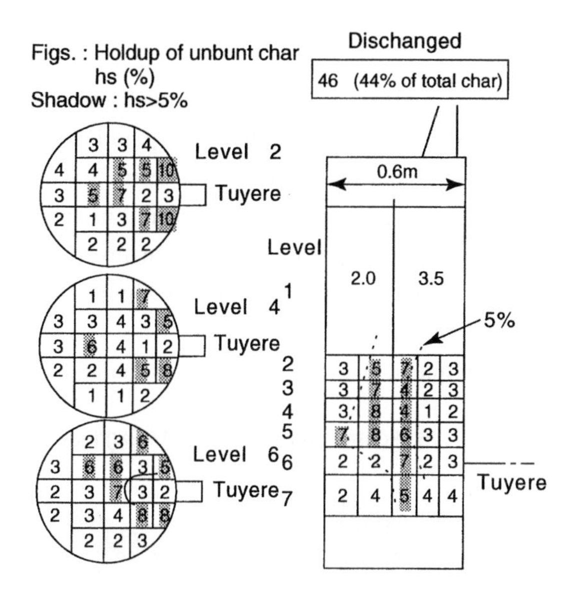

Fig.8-34 Distribution of unburnt char at 200 kg/thm PCR in hot model [23]

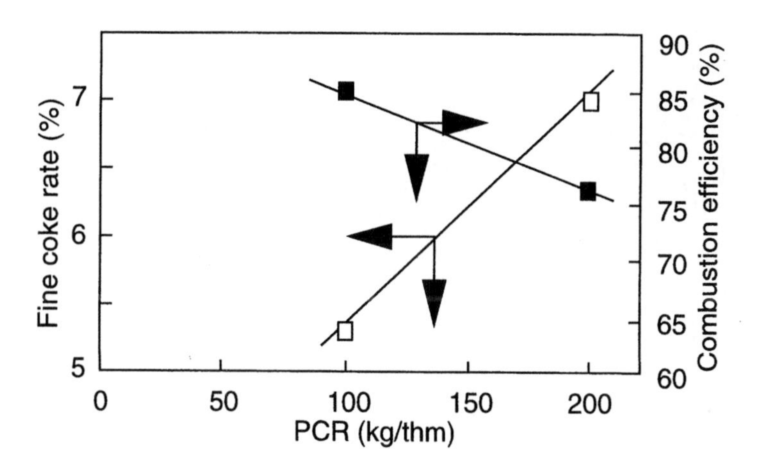

Fig.8-35 Effect of PCR on fine coke generation and combustion efficiency [23]

degree of PC decreases and the fine coke generation increases. It is understood that the high rate PCI is severe for coke degradation.

8.2.4 Fine generation in high rate PCI by mathematical simulation

A total three dimensional BF simulation model incorporating fine coke generation and unburnt char phenomena has not yet developed. Concerning with coke degradation phenomena, only hot model experiments and monitoring in actual BFs have been performed for the understanding of the phenomenon and in this field the research is still at the fundamental stage. Therefore, a simulation model which can totally evaluate the coke degradation phenomenon has not yet developed.

As current research example, a two dimensional BF simulation model considering

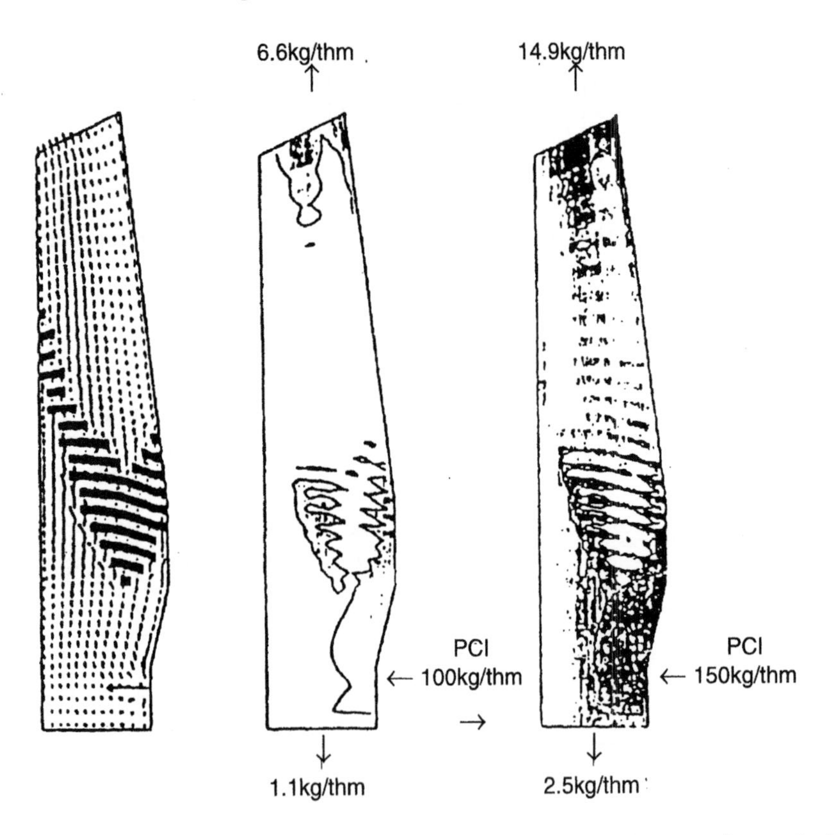

Fig.8-36 Change of behavior of unburnt char in BF calculated by a mathematical simulation model [25]

only the solution loss reaction as consumption of unburnt char has been developed. In the simulation model, the amount of unburnt char in the raceway was evaluated based on the experimental results and the mathematical simulation model on PC combustion and the equation of motion of fine was developed by a gas-solid two phase flow model.[24]

The calculated results by the simulation model is shown in Fig.8-36.[25] The unburnt char is mostly entrained near the cohesive zone and BF top. At 150kg/thm PCR, 66% of unburnt char is exhausted from BF top and 23% of unburnt char is consumed by the solution loss reaction and the remaining 11% of unburnt char moves into the deadman.

However, at this level of PCI rate in the actual BF operation, an increase of dust exhausted from BF top is not observed and the unburnt char is presumed to be entrained in BF and be consumed in BF. Further progress of research in this field is expected.

The analysis of fine movement is performed by a four-fluid simulation model considering solid, liquid, gas and fine, which will be mentioned in detail in the next section.

With simultaneous use of BF simulation model and the consideration of the mechanism of coke degradation and fine consumption, it is expected that internal BF phenomena will be fully analyzed in the near future.

8.3 Transportation and accumulation of fine

8.3.1 Introduction

Since the difficulty of sampling of fine in the actual BF during operation has resulted in a lack of sampling examples except for sampling of unburnt char by raceway probes, deadman probes, a slantingly inserted probe at the Chiba works of Kawasaki Steel and belly probe at the Oita works of Nippon Steel. Although the samplings at the tuyere level during furnace shutdown have been frequently performed, the prediction of the state of fine in operation was difficult. Therefore, cold model experiments as well as mathematical model simulations have been performed to clarify the behavior of fine, that is the movement and accumulation.

8.3.2 Cold model experiment[26]

(1) Fine coke accumulation in deadman

It is presumed that the increase of fine ratio in the deadman inactivates the deadman and aggravates gas and liquid permeability in the lower partof BF. Therefore, the effect of coke particle size on the movement and accumulation of fine in the deadman was investigated by means of a cold model.

The obtained results are as follows:

1) In the case of low gas superficial velocity in the deadman, fine coke generated in the central region not only flow into the raceway but also penetrate into the intermediate region of the deadman by the sieving effect.
2) As shown in Fig.8-37, coke of large size has more significant sieving effect than that of smaller size because of the larger available voidage size.
3) In addition, the sieving effect increases with coke consumption rate in the raceway.

(2) Penetration of fine coke generated from raceway[27]

The effects of packed particle size and fine particle size were experimentally investigated under constant fine injection condition by photography and local pressure measurements.

The obtained results are as follows:

1) The injection of coarse fine causes the accumulation of fine on the surface of the deadman and significantly increases the pressure in front of the tuyere.

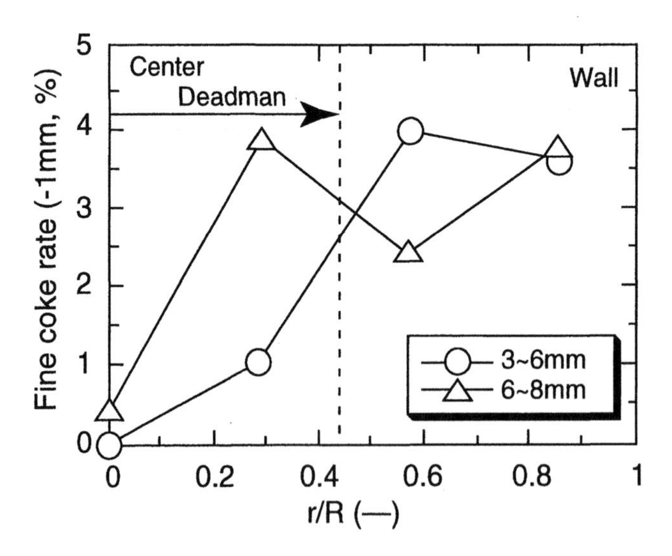

Fig.8-37 Effect of coke size on distribution of fine coke in deadman [26]

2) The passing condition of fine through a packed bed is found to be 1/4-1/5 of fine size to hydraulic diameter of packed particle as shown in Fig.8-38. When the results are applied to the actual BF, the upper limit fine size passing through the coke packed bed is estimated to be 3-5 mm. This value corresponds well to the actual observation that the contribution of fine originated in front of the tuyere significantly decrease of over 4-5 mm in fine coke in deadman.

(3) Formulation of fine coke velocity in packed bed[28)]

The unburnt char generated in the raceway is carried upward by the gas containing CO_2 , with which unburnt char undergoes solution loss reaction in competition with packed coke and fine coke. For the analysis of fine behavior in the BF, the accurate evaluation of the reaction rate and the residence time, that is reaction time, in the BF is indispensable. A formulation to evaluate average velocity (Us) of fine movement in the

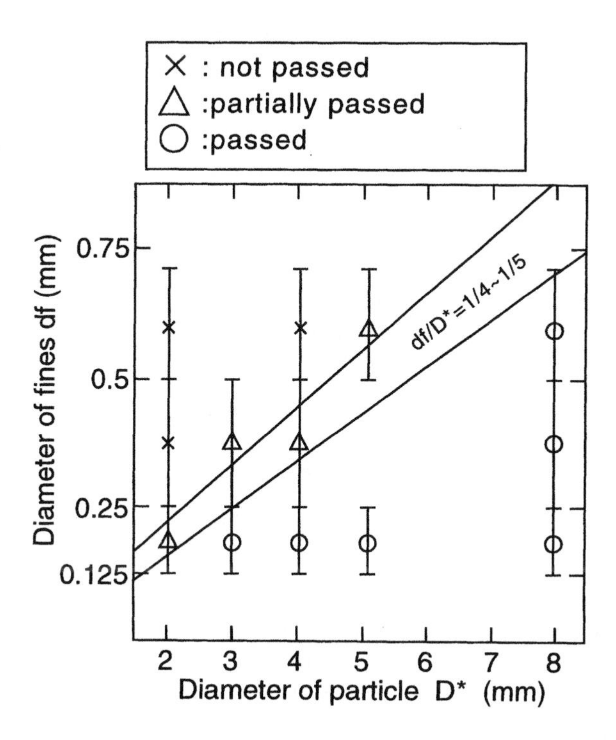

Fig.8-38 Passing condition of fine through packed bed [27)]

packed bed was tried.

Based on the experiment, the accumulation of fine was found to occur in the upper part of packed bed. The entrained part consists of the stationary part of static holdup and the dynamic part which is replaced by newly arriving fine.

The average velocity (Us) of fine movement is related to the residence time in entrained part(td/L) and gas-solid ratio (Gs/Fg) as shown in Fig.8-39, which is expressed by the following empirical relation.

$$Us \cdot (td/L) = -4.39 - 8.641 \cdot \log (Gs/Fg) \qquad (8.1)$$

(4) Fine movement in trickle bed[29]

Since there exist gas, liquid, and three solid phases in the deadman and the dripping zone in the BF, the effect of liquid droplet on the movement and accumulation behavior of fine was investigated by a two dimensional cold model experiment. In comparison with solid-gas two phase case, fine in three phase case accumulates more near raceway and the upper part, while it accumulates less in the exhaust from apparatus top and in the central region of bottom as shown in Fig.8-40. Further, in the lower part of raceway fine

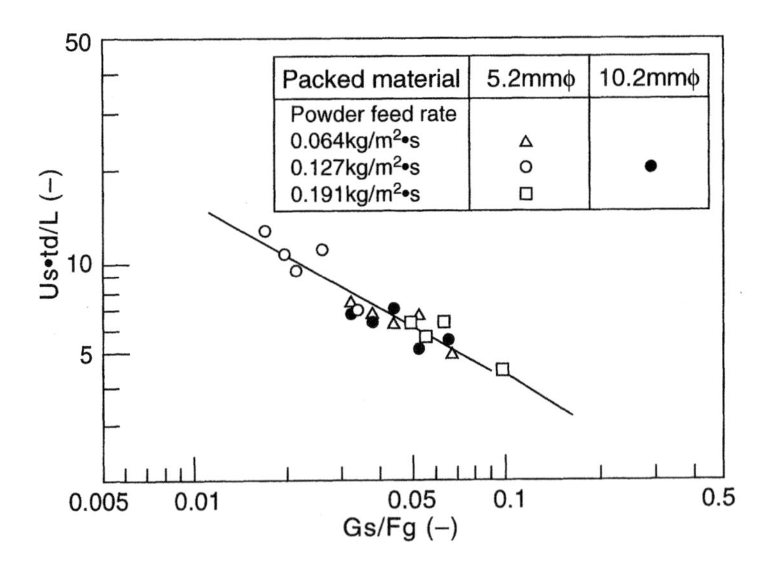

Fig.8-39 Relationship between fine movement velocity and gas-solid ratio[28]

is found to be significant flow down to the bottom part.

(5) Formulation of fine accumulation rate in packed bed[30)]

With an assumption of fine and gas as continuous fluids, the basic transport equations on two phase flow in a packed bed were described as described in APPENDIX. Based on the cold model observation that the accumulated fine in the packed -bed is replaced by a certain rate, the velocity R_f is expressed by Eq.(8.12) as the difference of adhesion velocity, r_s and leaving velocity, r_d and respective rate constants, k_s and k_d were determined based on the experimental results.

(6) Penetration of fine into parallel packed bed [31)]

The packing structure near raceway consists of small particles of low gas permeability in the deadman and sound large particles in the dripping zone. To clarify the effect of the packing structure on the penetration and the accumulation behavior of fine for the cold model experiment , a parallel gas-solid two phase flow was performed with use of an

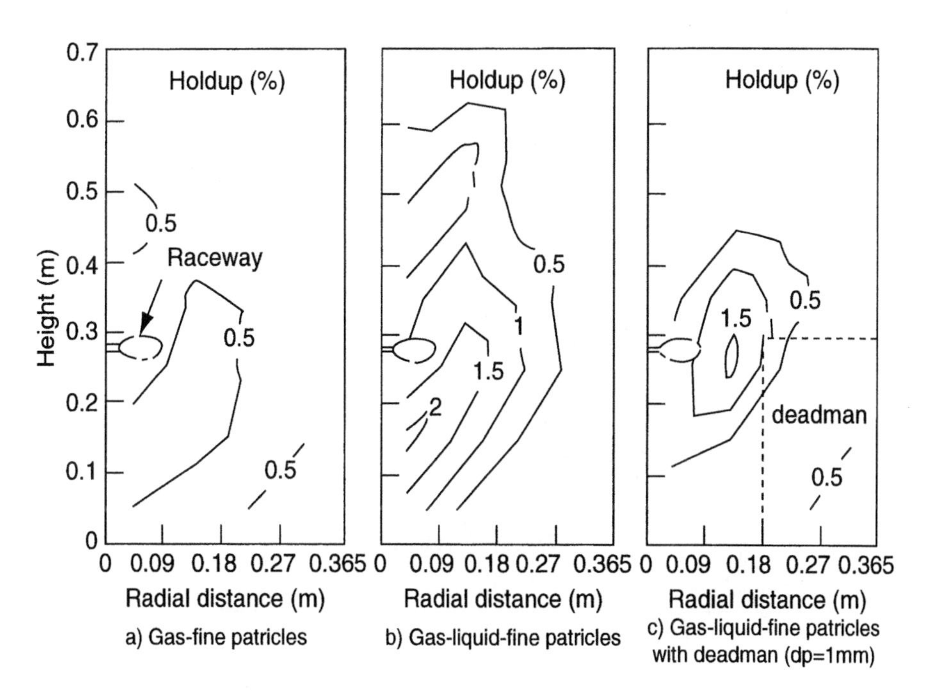

Fig.8-40 Holdup distribution of fine in packed bed[29)]

apparatus shown in Fig.8-41.

The following results were obtained:

1) With increase of particle size, the amount of penetrated fine decreases in a packed bed of small size particles, while it increases in that of large particles.
2) The increase in gas flow rate significantly increases the amount of penetrated fine in a packed bed of small size particles .
3) With enlargement of packed particle size ratio, the distribution ratio of fine to each packed bed also increases.
4) With increase of the amount of fine injection, the deposited amount of fine in small size particle bed significantly increases, which implies the increase of the entrainment of fine in the deadman.

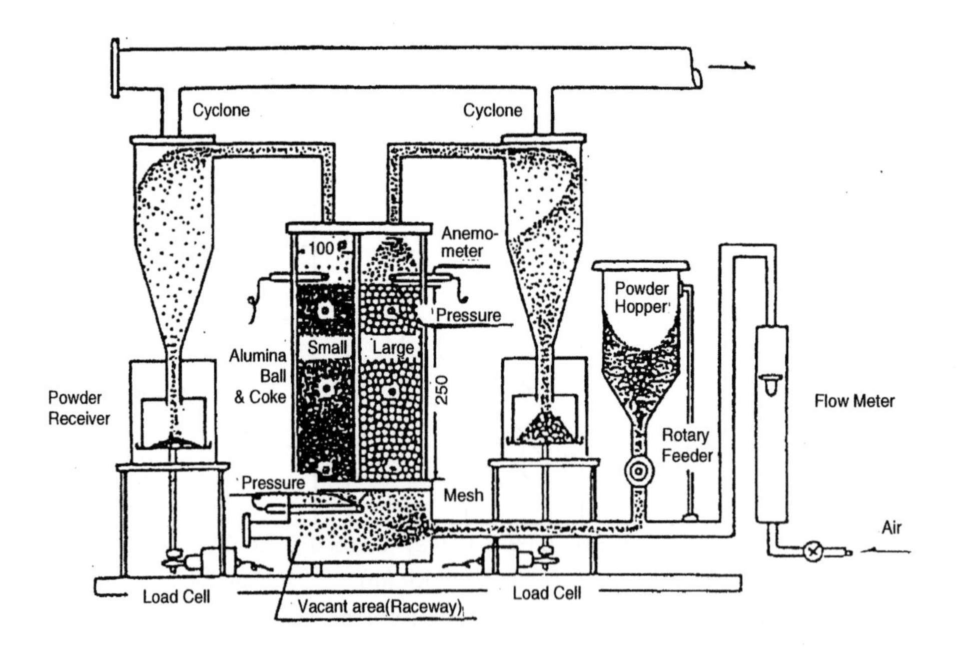

Fig.8-41 Experimental apparatus with parallel beds of different particle sizes [31)]

8.3.3 Mathematical simulation

Several researches using mathematical simulation analysis have been performed and the results were obtained as will be described.

(1) Solution loss reaction of unburnt char

With consideration of unburnt char residence time in a coke packed bed and the solution loss reaction rate of unburnt char, a mathematical simulation was performed to quantitatively evaluate the effect of the bed coke size and bed voidage on the char reaction ratio in a packed bed as shown in Fig.8-42. It has become clear that in the lower part of BF most of solution loss reaction is attributed to unburnt char. For example, in the case of 35mm bed coke size in the lower part of BF, 95 % of solution loss reaction is attributed to unburnt char. Further the simulation was applied to the whole BF and it is concluded that about 80 % combustion amount in the raceway by an improved PC combustion burner is indispensable for the achievement of 250 kg/thm PCRoperation.

(2) Model of movement and dispersion of fine[32)]

A mathematical simulation model considering the movement and the dispersion of

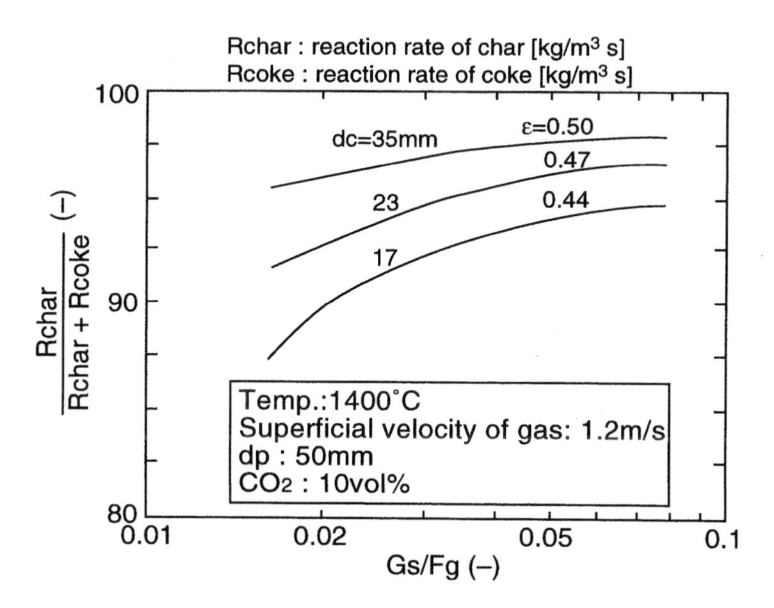

Fig.8-42 Effect of bed coke size and voidage on char reaction ratio in a packed bed.

fine in a packed bed was developed. With use of the particle trajectory method based on the probability process, the average interaction coefficient F_k, which expresses the interaction between the packed bed particle and fine particle, was introduced. The average movement velocity of particle in the packed bed was predicted various packing particle sizes and fine particle sizes and the agreement with the experimental values verified the validity of the model. The average value of the normalized variation of axial particle velocity in free space is about 0.2, while that in a packed bed is about 1.3 as shown in Fig.8-43. The remarkable result that the value of axial velocity variation in a packed bed is larger than that of the average axial velocity shows the significance of fine dispersion in a packed bed.

(3) Simulation of fine movement and accumulation in BF[31)]

Experimental fundamental results on the fine movement and the accumulation in a packed bed were input into a total mathematical simulation model of BF and the fine

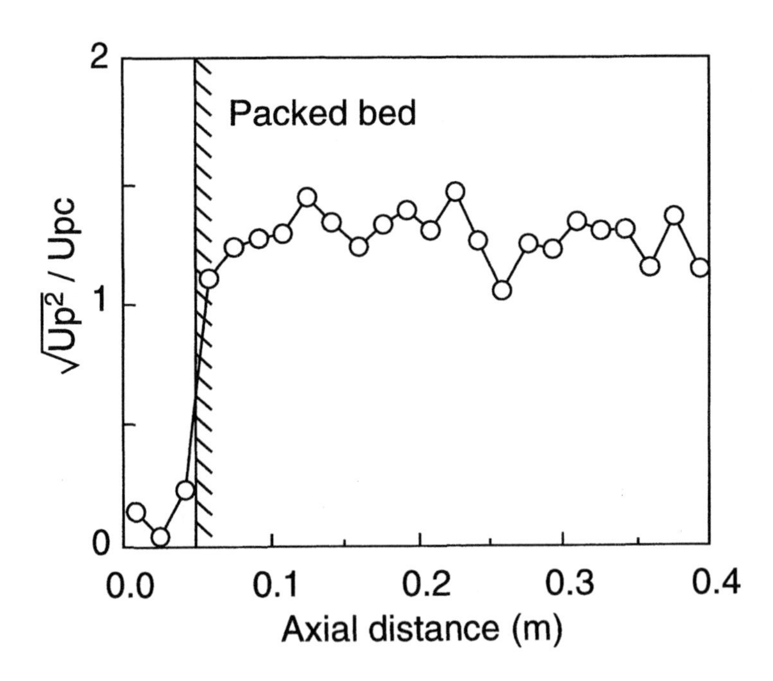

Fig.8-43 Axial variation of normalized standard deviation of axial particle velocity [32)]

movement and the accumulation in BF under different fine sizes were investigated as shown in Fig.8-44. In the case of over 500 μ m, the entrainment of fine initiates at locations not only in the central region but also in the lower part of the BF. In the case of 100 μ m fine size, no entrainment of fine occurs. The fine coke generated in the raceway having 500 μ m to 3mm size tend to be entrained not only in the deadman but also in the dripping zone.

(4) Effect of cohesive zone shape on fine accumulation[33]

The blockage of fine initiates when fine diameter exceeds 1/6 of the hydraulic diameter of packed particle. Based on the experimental results, a two dimensional mathematical simulation model for two phase gas-solid flow in packed bed was developed and was applied to the analysis of the effect of the cohesive zone shape on fine accumulation as shown in Fig.8-45 and 8-46.

In the case of W shaped cohesive zone, fine accumulates significantly at the lowest part of the cohesive zone, while in the case of inverse V shaped cohesive zone it accumulates less than that in W shaped cohesive zone and it accumulates at the root of the cohesive zone. Therefore, for high rate PCI operation, inverse V shaped cohesive zone is preferable, which may be achieved by the burden distribution control. The

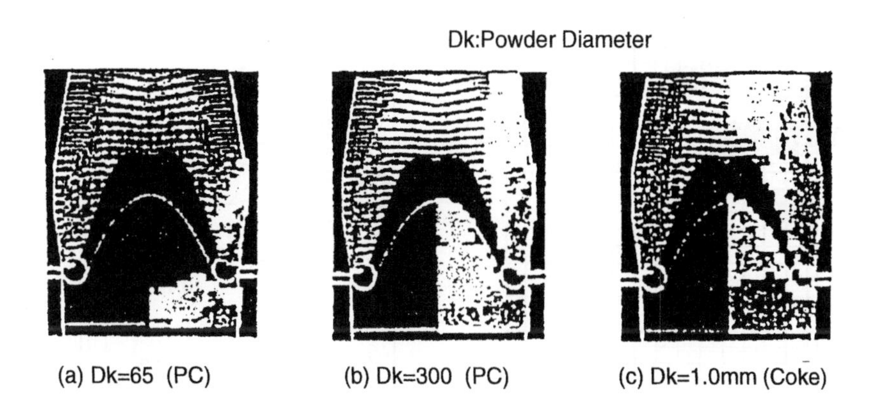

Fig.8-44 Mathematical simulation of fine accumulation in the lower part of BF [31]

maximum PCI rate of 210-240 kg/thm was predicted by the model in respect of fine accumulation.

(5) Prediction of fine accumulation by four-fluid model[34)]

The gas-solid two-fluid model has been developed into a four-phase model with addition of liquid and solid flow. An example of calculated results of streamline flow of gas, solid, liquid and fine is shown in Fig.8-47. Concerning the liquid flow in the lower part of BF, the streamline flow is distorted to a peripheral direction by the effect of the solid flow.

8.3.4 Summary

Summarizing the before mentioned research results both by experiments and the mathematical simulation analysis, it is concluded that fine accumulates at locations where

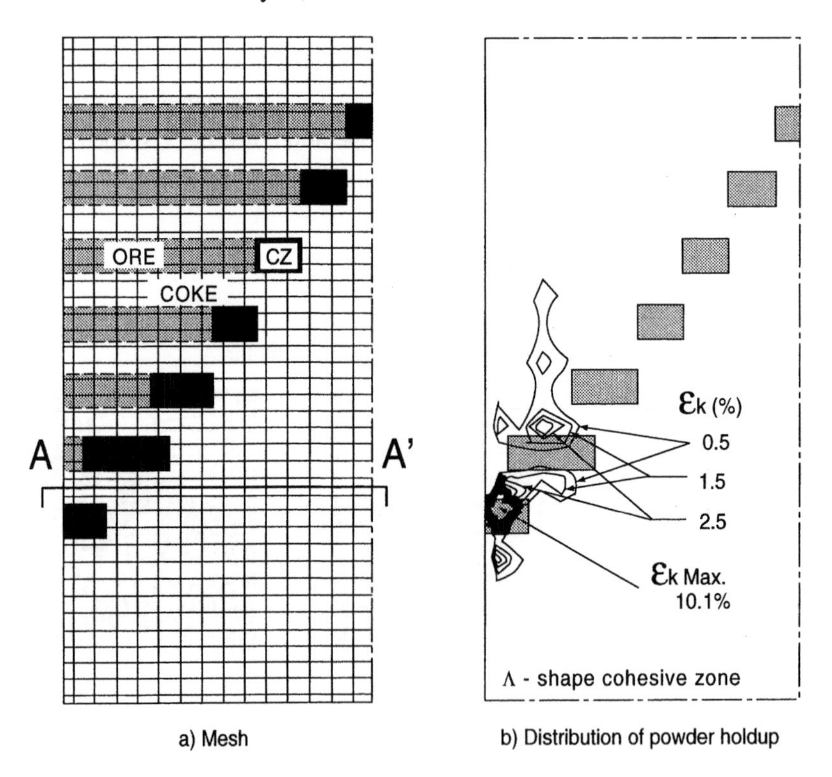

Fig.8-45 Calculated results of fine accumulation for inverse V shaped cohesive zone [33)]
ε_k: Holduped fine

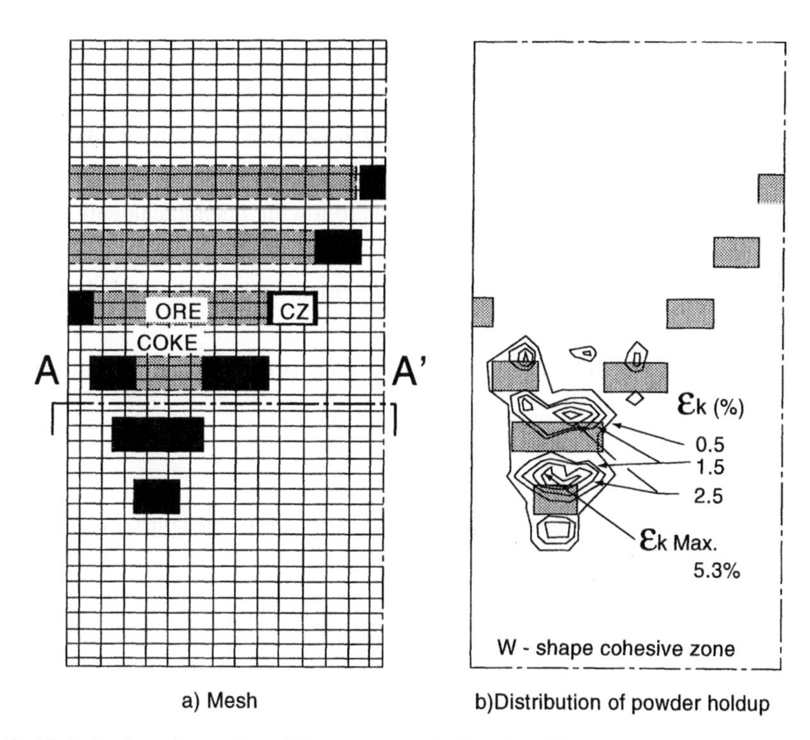

Fig.8-46 Calculated results of fine accumulation for W shaped cohesive zone [33]

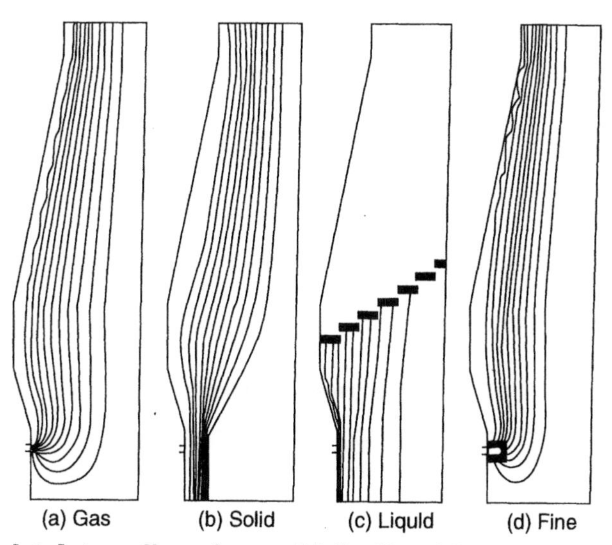

Fig.8-47 Calculated streamlines of gas, solid, liquid and fine by the four-fluid model [34]

gas flow changes its direction and the fine accumulation significantly depends on the cohesive zone shape.

It can be said that the developed mathematical models can precisely simulate relatively simple phenomena of fine movement and the accumulation. In the future, for further precise simulation by mathematical models, the following phenomena will be clarified both by experimentally and theoretically.

1) Not only upward accumulation but also downward accumulation behavior of the generated fine
2) Fine movement and the accumulation in a packed bed with existence of liquid

8.4 Consumption of fine in blast furnace

The following two mechanisms have been mentioned for the consumption of fine.

As described in section 8.3.3, the fundamental research shows that unburnt char reacts more with CO_2 than fine coke does. This fact implies that the unburnt char is responsible for most of the solution loss reaction and the degradation of lump coke by the solution loss reaction is decreased.

It is presumed that unburnt char is partly consumed by the contact with molten iron,

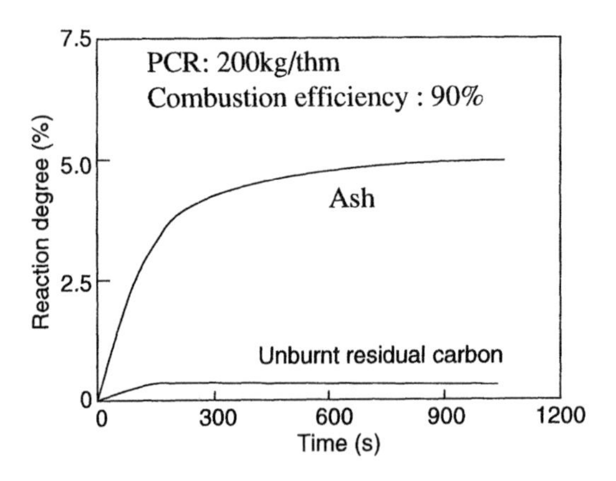

Fig.8-48 Reaction rate of unburnt char with molten slag [35)]

that is the carburization reaction, or slag, that is the assimilation reaction. An experiment into the unburnt char reaction with molten iron and slag was performed with use of a rotating crucible. As shown in Fig.8-48 [35)], the contribution of unburnt char carburization is less than 1.5% and that of unburnt char assimilation to slag is about 4%. From these results, it is presumed that the amount of unburnt char consumption in the dripping zone in BF is about 5%.

These facts show the possibility that PC is not completely combusted or consumed within the raceway and it can be consumed in the BF.

However, the maximum possible amount of the consumption of unburnt char remains as future task to be quantitatively investigated.

APPENDIX

FORMULATION OF MATHEMATICAL MODEL

It was observed that as fine particles ascend in the packed bed, some of them deposit on the surface of the packed particle or in the dead space to form static powder, which was called the "static" holdup of powder. They were apparently "retained" there while some of them adhered and others departed at the same time. For the two-phase unsteady flow of gas and powder in the packed bed, the accumulation of powder will reduce the voidage of the packed bed and result in a higher pressure drop. Powder in the gas stream, dynamic holdup, will decrease because of the deposition of powder in the packed bed. Therefore, the influence of the powder accumulation on the voidage of the packed bed and the reduction of powder in the gas stream must be taken into account in computations. Table 8-A1 lists the related conservation of variables in fundamental equations considered here. If "i" refers to gas or powder phase, the two-phase flow of gas and powder can be described via the continuity and momentum equations in cylindrical coordinates.

The variables Φ_i, Γ_ϕ and $S_{\phi i}$ represent the dependent variable, and diffusion coefficient and source terms for Φ_i. The interaction force F^i_j will briefly be discussed below. The interaction force between gas and packed particles was assumed to be evaluable by Ergun's equation considering the effect of the static holdup of powder on the voidage of the packed bed (Eq.(8.5)). The pressure drop of the two-phase flow of gas and powder

Table 8-A1 Transport equations of two phase flow in a packed bed

Conservation of	ϕ_i	$\Gamma_{\phi i}$	$\varepsilon_i S_{\phi i}$
Gas phase			
Continuity	1	0	0
Momentum	V_g	0	$-F_g^{\,p} - F_g^{\,f} - \varepsilon_g \,\mathrm{grad}\, P$
Powder phase			
Continuity	1	0	$-R_{pw}$
Momentum	V_{pw}	0	$-F_g^{\,f} - F_f^{\,p}$

$$\frac{\partial(\varepsilon_i \rho_i \phi_i)}{\partial t} + \mathrm{div}\,(\varepsilon_i \rho_i V_i \phi_i - \varepsilon_i \Gamma_{\phi i}\, \mathrm{grad}\, \phi_i) = \varepsilon_i S_{\phi i} \tag{8.2}$$

$$\varepsilon_g + \varepsilon_{pw} + \varepsilon_{pws} + \varepsilon_p = 1 \tag{8.3}$$

$$\frac{\partial(\varepsilon_{pws} \rho_{pw})}{\partial t} = R_{pw} \tag{8.4}$$

$$F_g^p = 150.0\, \mu_g \left[\frac{\varepsilon_g + \varepsilon_{pws}}{(1 - \varepsilon_p - \varepsilon_{pws})\phi_p d_{pw}}\right]^2 \left(\frac{V_g}{\varepsilon_g}\right) + 1.75\, \rho_g \left[\frac{\varepsilon_g + \varepsilon_{pws}}{(1 - \varepsilon_p - \varepsilon_{pws})\phi_p d_{pw}}\right] \left(\frac{1}{\varepsilon_g}\right)^2 |V_g| V_g \tag{8.5}$$

$$F_g^f = \varepsilon_g^{-4.65} C_d \frac{3\rho_g \varepsilon_f}{4\phi_f d_f} \left|\frac{V_g}{\varepsilon_g} - \frac{V_{pw}}{\varepsilon_{pw}}\right| \left(\frac{V_g}{\varepsilon_g} - \frac{V_{pw}}{\varepsilon_{pw}}\right) \tag{8.6}$$

$$C_d = 24.0 / Re_{pw} \qquad Re_{pw} < 1.0 \tag{8.7a}$$

$$C_d = 24.0\,(1.0 + 0.15\, Re_{pw}^{0.687}) / Re_{pw} \qquad 1.0 < Re_{pw} < 10^3 \tag{8.7b}$$

$$C_d = 0.44 \qquad 10^3 < Re_{pw} \tag{8.7c}$$

$$Re_{pw} = \frac{\phi_{pw} d_{pw} \rho_g \rho_g}{\mu_g} \left(\frac{V_g}{\varepsilon_g} - \frac{V_{pw}}{\varepsilon_{pw}}\right) \tag{8.8}$$

$$F_{pw}^p = \frac{F_k}{2D^\cdot} \varepsilon_{pw} \rho_{pw} \frac{|V_{pw}|}{\varepsilon_{pw}} \frac{V_{pw}}{\varepsilon_{pw}} \tag{8.9}$$

$$F_k = \begin{cases} 10.50 / Fr^{1.33} & 0.003 < Fr < 0.05 \text{ (vertical)} \quad (8.10a) \\ 14.98 / Fr^{1.33} & 0.03 < Fr < 0.5 \text{ (horizontal)} \quad (8.10b) \end{cases}$$

$$D^\cdot = \frac{2\phi_p d_p (1 - \varepsilon_p)}{3\varepsilon_p} \tag{8.11}$$

r_s = *(rate parameter)* × *(inflow rate of powers)*
r_d = *(rate parameter)* × *(deposited amount of powders)*

$$R_{pw} = r_s - r_d = k_s \varepsilon_{pw} \rho_{pw} |V_{pw}| - k_d \rho_{pw} \varepsilon_{pw} \tag{8.12}$$

was studied in the previous work.[33] The pressure decreased with decreasing gas velocity. The interaction force, Eq.(8.6), between gas and powder was evaluated based on the drag force equation for a single particle with Richardson and Zaki's voidage function, Fo = $\varepsilon_g^{-4.65}$, modifying the drag coefficient. The quantity C_d is related to the modified Reynolds number in terms of the particle diameter(Eqs.(8.7) and (8.8)).

The interaction between powder and packed particles, including the gravitational force on the powder and the collision and friction between powder and the packed particles, was determined by a Fanning type equation. According to previous works, the coefficient was correlated by Eqs.(8.9)-(8.11). The relationship in Eq.(8.10a), was obtained from steady state experiments.

The variation of the static holdup of powder with time was assumed to be equal to the deposition rate of powder, and can be written by Eq.(8.4). Formulation of the rate of powder depositions, Rf , is discussed elsewhere.[30]

REFERENCES

1) H. Haraguchi, T. Nishi, Y. Miura, Y. Ushikubo and T. Noda: Tetsu-to-Hagane, 70(1984), p.2216.
2) K. Sasaki, M. Hatano, M. Watanabe, T. Shimoda, K. Yokoya, T. Ito and T. Yokoi: Tetsu-to-Hagane, 62(1976),p.580
3) Y. Ishikawa, M. Kase, Y. Abe, K. Ono, M. Sugata and T. Nishi: 42th Ironmaking Conference Proceedings, AIME, (1983), p.357.
4) I. Kurashige, F. Nakamura, M. Hatano, Y. Iwanaga, H. Nomiyama, M.Kojima and H. Aminaga: JSPS 54th Committee No.54-1689(1984).
5) H. Shimizu, K. Sato, M. Kojima, Y. Aminaga, F. Nakamura and Y. Iwanaga: Tetsu-to-Hagane, 72(1986),p.195.
6) M. Ichida, K. Kunitomo, Y. Fujiwara, H. Kamiyama and Y. Morisane: CAMP-ISIJ, 6(1993),p.860
7) S. Kitayama and T. Kamijyo: private letter.
8) H.Tsukiji, M.Hattori, A.Yamaguchi, A.Shimomura K.Ishii and S. Itagaki: CAMP-ISIJ,6(1993),p.892.
9) K. Yamaguchi, H. Hino, S. Matsunaga, K. Kakiuchi and S. Amano: ISIJ Int., 35(1995), p.148.
10) A. Sakai and M. Sato: private letter.
11) A. Kasai, K. Miyakawa, T. Kamijyo, J. Kiguchi and M. Shimizu: Tetsu-to-Hagane,

83(1997), p.239.
12) Y.Iwanaga: Tetsu-to-Hagane, 79(1993),p.927.
13) H. Itaya and H. Nishimura: private letter.
14) K. Takeda, M. Igawa, S. Taguchi, M. Kiguchi, Y. Imoo and S. Tamura: CAMP-ISIJ, 3(1990), p.10.
15) K. Yamaguchi, T. Uno, T. Yamamoto, H. Ueno, Y. Konno and S. Matsuzaki: Tetsu-to-Hagane, 82(1996),p.641.
16) Y. Yoshida, S. Kitayama, S. Ishiwaki, M. Isobe and Y. Miyagawa: CAMP-ISIJ, 6(1993),p.100.
17) Y. Yamaoka, Y. Suzuki, S. Itagaki, I. Shimoyama, K. Fukada, M. Hattori and H. Tsukiji: JSPS 54th Committee No.54-2006(1994)
18) K. Ishii: Research Group of Pulverized Coal Combustion in Blast Furnace in JSPS 54th Committee Rep-44(1996).
19) Y. Iwanaga: Tetsu-to-Hagane, 71(1991),p.71.
20) K.Yamaguchi, H.Ueno, S. Matsunaga, I. Kakiuchi and S. Amano:ISIJ Int.,35(1995),p.148.
21) S. Kishimoto, H. Mitsufuji, T. Ariyama and M. Sano: private letter.
22) M. Shimizu, A. Kasai, H. Iwakiri and T. Kamijyo: private letter.
23) K. Takeda and N. Ishiwata: Research Group of Coke Fundamentals in JSPS 54th Committee Rep-54(1996).
24) R. Ito, T. Yabata, T. Goto, K. Kadoguchi, T. Kamijyo and K. Shibata: private letter.
25) S. Nagami, T. Murai, Y. Shimoda, S. Komatsu and T. Inada: CAMP-ISIJ,6(1993), p.26
26) A. Kasai, H. Iwakiri and T. Kamijyo: CAMP-ISIJ, 7(1994),p.91.

27) H. Ohkushi, F. Shimoshige, Y. Tomita and K. Tanaka: CAMP-ISIJ,6(1993),p.40
28) T. Ariyama, M. Sato and Y. Asakawa: Chemical Engineering, 22(1996), p.171
29) T. Ewatari, T. Uchiyama, K. Takeda and H. Itaya: CAMP-ISIJ,7(1994), p.92

30) J.Z. Chen, H. Nogami, T. Ariyama, R. Takahashi and J. Yagi:Advances in Mutiphase Flow, Elsevier Science(1995), p.619
31) T. Sugiyama: Tetsu-to-Hagane, 82(1996),p.29.
32) K. Takeda and F.C. Lockwood: Tetsu-to-Hagane, 82(1996),p.492.

33) K. Shibata, R. Takahashi, M. Shimizu and J.Yagi: Tetsu-to-Hagane, 75(1991), p.1267.
34) H. Nogami, P.R. Austin and J.Yagi: CAMP-ISIJ,8(1995),p.30.
35) S.Inaba: Tetsu-to-Hagane, 81(1995), p.271.

NOMENCLATURE

A_v =surface area normal to flowing direction per unit volume (m^2/m^3)

C_d =drag coefficient for a single particle

d_{pw}=diameter of powder (m)

D^* =equivalent diameter of the packed bed for powder-packed particle interaction (m)

F_k =interaction coefficient between powder and the packed particles (-)

Fr =Froude number ($V_{pw}(D\dot{g})^{-0.5}$) (-)

g =gravity (m/s^2)

P =pressure (Pa)

R_{pw}=deposition rate of powder (m^3/s)

Re_{pw}=modified Reynolds number defined by Eq.(8.8)

U_g,U_{pw}=axial superficial velocity of powder (m/s)

Ut=terminal velocity of powder (m/s)

Vg,Vp=radial superficial velocity of gas and powder (m/s)

Greek Letters

ε_i = volume fraction of 'i' phase in the packed bed (i=g,pw,p) (m^3/m^3bed)

ε_{pw}, ε_{pws}= volume fraction of the dynamic and static holdups of powder in the packed bed (m^3/m^3bed)

$\Gamma_{\phi i}$= diffusive transport coefficient of ϕi (i=g,pw) (kg/m^2s)

ϕ_{pw}= shape factor of powder (-)

ϕ_p= shape factor of packed particles (-)

Suffix pw=powder

CHAPTER 9

Burden properties suitable for high rate PCI

In the present research activity in "Research Group of Pulverized Coal Combustion in Blast Furnace", detailed examination has not been directly made as to a change in reduction behavior of agglomerates in a blast furnace relating to high rate pulverized coal injection (PCI), the measures for improving the properties of agglomerates and so on. However, such changes, e.g., an increase in Ore/Coke ratio and a decrease in heat flux ratio with an increase in the PCI rate, have a large effect on permeability, temperature profile of burden, distribution of reduction degree (reduction rate), etc. As described in Chapter 7, these effects are controlled by various actions taken regarding the operation including the burden distribution control. But, at the same time, it is also necessary to take a corresponding step to improve the agglomerate properties so that they fit the high rate PCI. In this chapter, an attempt has been made to clarify the appropriate properties of sinter for high rate PCI operation by comparing the recent information on the practical sinters, referring mainly to the special proceedings of ISIJ in the last five years.

On the other hand, a problem has clearly come up in regard to the coke, that is, coke fines inside the raceway increase with an increment in PCI rate. This phenomenon has been made clear by measuring the amount of fines in the coke sampled from the inside of the tuyere level at the shutdown and a tendency that permeability at the lower part of the blast furnace deteriorates has been accompanied. And, it has been reported that when the cold strength (Drum Index; DI) of charging coke is raised, the amount of coke fines decreases[1)], and then, at almost all blast furnaces having the PCI equipment installed, corresponding measures have been taken which control the generation of coke fines by raising DI.[1-5)] However, the mechanism as to how the coke is degraded and fines are generated is not yet clear. In this chapter, an attempt has been made to clarify the coke properties controlling the coke degradation and generation of fines based on the latest literature information and supplemented by the description in Chapter 8 of the phenomenon of coke degradation and generation of fines during high rate PCI.

However, since the experience in the operation performed with high rate PCI has been acquired merely over a few years, there is a limit to the phenomenon described herein, and thus lots of problems related to this subject which need to be further studied are indicated herein.

9.1 Agglomerate properties for high rate PCI

9.1.1 Sinter properties required for high rate PCI

(1) Reduction degradation at low temperatures (RDI)

It has been generally considered that the control value of RDI (Reduction Degradation Index) can be raised since a high temperature region at the upper part of the blast furnace is enlarged and a low temperature thermal reserve zone of 500 to 700°C is contracted because of a heat flux ratio being lowered in line with the increase in PCI rate.[6-7] This fact leads to a relaxation of the restrictions on properties in the sintering process and makes it possible to increase reducibility (JIS Reducibility Index; JIS-RI) easily. Therefore, it also gives an advantage from the viewpoint of raw material conditions, including a flexibility in the selection of ore brands.

Whilst, since Ore/Coke ratio in the radial direction in the blast furnace becomes less uniform, a burden descending rate is lowered and residence time at a low temperature zone is prolonged as PCI rate increases, a zone with a high heat flux ratio is locally formed at the peripheral part in the blast furnace. Therefore, a raise in the control value of RDI seems to be difficult because reduction degradation will be significant in such a zone.[8]

Also in the operation with high rate PCI, especially under the condition of high productivity, oxygen enrichment in the blast is made aiming at suppression of an increase in pressure drop in the furnace and/or improving the combustibility of PC. However, it also reported that the oxygen enrichment procedure raises the heat flux ratio and makes the low temperature thermal reserve zone larger.[9] The oxygen enrichment, as shown in Chapter 7 (Figs.7-12 and 7-13), is another effective action means for decreasing the top gas temperature and preventing the abnormal burden descending.

As described above, it is quite important to take the change required for RDI into account based on the understanding (prediction) of the shape and size of the low temperature thermal reserve zone in consideration of radial distribution in the blast furnace.

(2) Reducibility (JIS-RI)

As stated before, with increasing PCI rate, the temperature at the lumpy zone rises and the upper part of the lumpy zone tends to become heat excessive. This implies the enlargement of the high temperature thermal reserve zone of about 1000°C, which gives an advantage in general to gaseous reduction of iron ore.

However, to cope with a phenomenon that a radial distribution of Ore/Coke ratio in the blast furnace becomes less uniform with an increase in PCI rate, in other words, a local increase in the thickness of the iron ore layer, it would be indispensable to improve JIS-RI.

It is already known from the reduction test result that the reduction degree of iron burden till reaching the high temperature thermal reserve zone of about 1000°C (cohesion starting position) shows nearly the same value independent of the value of JIS-RI (Fig.9-1).[7] JIS-RI, which is a value under the gaseous reduction condition at 900°C, seems to have no significant influence on the reduction degree at the lumpy zone of the blast furnace. However, since a porosity, pore size distribution and mineral structure, which are determinants of JIS-RI, have a close relation to the reducibility in a cohesive zone at over 1000°C, it becomes necessary to control JIS-RI comprehensively taking the internal conditions of the blast furnace into account. As shown in Fig.9-1, sinter having a high

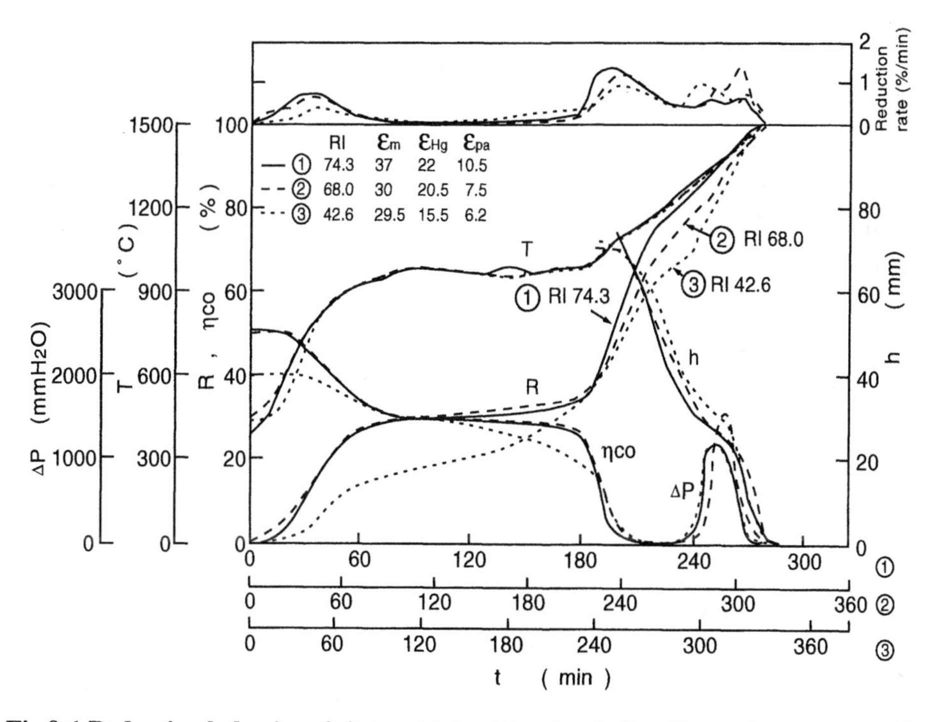

Fig.9-1 Reduction behavior of sinter obtained by simulating thermal reserve zone[7]

value of JIS-RI gives a high reducibility of above 1000°C.

Additionally, the concentration of hydrogen in the bosh gas of the blast furnace increases with increasing PCI rate, but a solution loss reaction is restrained by a relative decrease in the amount of coke. This allows the possibility of a reduction of gas utilization. Further, to perform a low fuel rate operation, it is absolutely necessary to improve JIS-RI as there is a limit to how high a blast temperature can be increased.

(3) Softening and melt-down properties at high temperatures

It is supposed, as mentioned before, that the temperature at the upper part of the furnace goes up with the increase in PCI rate, but the temperature around the raceway goes down due to a pyrolysis heat of PC and the temperature in the deadman is also estimated to decrease.[10] Therefore, the lower part of the furnace is put wholly in a heat short condition. This results in the top face of the cohesive zone rising and the bottom face lowering, which will probably make the width (thickness) of the cohesive zone larger and increase a pressure drop in the lower part of the furnace. Accordingly, for a stable operation under the condition of high rate PCI, how to control high temperature properties of agglomerates may be one of the most important problems to be resolved in the future.

It may be considered that the permeation resistance of the cohesive zone is primarily controlled by the softening and melt-down properties of the iron ore (agglomerates); however in the actual operation, the thickness of the iron ore layer and/or an interaction with the coke layer also becomes a factor exerting such an influence. The former depends on such properties as a reduction rate and permeability being lower at the upper part of the ore layer compared with the lower part contacting a high reducible potential gas at all times, and it has already been verified that the starting temperature of pressure drop increase falls due to the influence of a part of the ore layer which has been thickened by PCI.[7] Fluctuation in burden descending and permeation resistance at the lower part of the furnace which have occurred in the case of a PCI rate being larger than 200kg/thm at Kimitsu 3 BF shown in Chapter 7 (Fig.7-25) can also be ascribed to the above reason.

9.1.2 Change in sinter properties during high rate PCI in BF

A comparison of sinter properties at PCI rates of 100 and 200kg/thm is shown in Table 9-1. These properties are prepared based on the published operational data and figures[11-15] obtained by actual furnace operation with PCI rate of 200kg/thm. Some information listed in Table 9-1 was obtained via a private communication. However for Kokura 2 BF, data acquired when the PCI rate was 170kg/thm have been adopted.

From Table 9-1, the sinter properties in the case of high rate PCI at 200kg/thm have been changed as follows relative to the operation with PCI rate of 100kg/thm.

(1) Improvement in JIS-RI

For achieving higher JIS-RI values, the amount of SiO_2 or Al_2O_3 in sinter has been decreased. As a result, the slag volume of the blast furnace has been lessened.

However, for Kobe 3 BF, since high JIS-RI sinter with a low SiO_2 content has already been used in the case of operation at 100kg/thm, SiO_2, Al_2O_3 and JIS-RI for the operation at 200kg/thm are almost fixed and the slag volume of the blast furnace is at a low level.

Also, as for Kakogawa 1 BF, because of pellet being blended about 35% in burden, the slag volume of the blast furnace is at a low level although the amount of SiO_2 in sinter is relatively high.

As to Kokura 2 BF, since the PCI rate is as low as 170kg/thm, the slag volume of the blast furnace is still at a high level.

Table 9-1 Sinter properties during high rate PC injection

Blast furnace	Month	PCR	SV	RDI	JIS-RI	SiO_2	Al_2O_3	SI	TI	Remarks
		kg/thm	kg/thm	%	%	%	%	%	%	
Kimitsu 3BF	93.9	118	295	32.6	66.2	5.14	1.89	88.6	—	
	93.11	203	278	31.2	69.1	5.17	1.61	90.2	—	
Kobe 3BF	90.1	104	283	43.8	70.4	4.69	1.74	86.0	69.6	
	96.1	204	279	38.3	69.6	4.91	1.71	87.7	70.4	
Kakogawa 1BF	93.6	134	276	22.5	65.9	5.60	1.88	87.0	68.9	PR=36.0%
	94.6	204	283	27.1	67.2	5.60	1.86	89.0	69.8	PR=33.0%
Fukuyama 4BF	94.8	112	322	44.3	69.4	5.10	1.80	—	67.8	
	94.10	218	274	40.0	70.7	4.70	1.71	—	66.5	
Kokura 2BF	95.3	132	312	37.2	67.3	4.77	1.99	—	77.8	
	96.3	171	320	38.1	70.1	4.69	1.94	—	79.1	

PR : Pellet Ratio

(2) Aggravation of RDI

Lowering of RDI was attained by improving the operation condition, including adjustment of SiO_2 and Al_2O_3 in sinter.

However, as for Kakogawa 1 BF, the amount of SiO_2 in sinter is high and therefore RDI is at a low level.

For Fukuyama 4 BF, a high level of RDI is still kept although some decrease thereof is made. This phenomenon will be described later.

As to Kokura 2 BF, the level of RDI is high since PCI rate is as low as 170kg/thm.

(3) Improvement in cold strength (SI and TI)

The cold strength has been raised successfully by improving the operation conditions in addition to the adjustment of SiO_2 and Al_2O_3 in sinter.

However in the case of operation at 200kg/thm for Fukuyama 4 BF, Tumbler Index (TI) is lowered and its level is also low compared to the other case of PCI rate below 200kg/thm. The reason why there is no effect on the operation is unclear; however, there is a possibility that such a low level has been compensated by the decrease in the amount of SiO_2 in sinter and/or the raised cold strength of coke.

It is to be understood from the above actual furnace operation results that for achieving the operation at the PCI rate of 200kg/thm, the following sinter properties seem to be required; JIS-RI ≥ 70%, RDI ≤ 38%, Cold strength ≥ +1.0% and Blast furnace slag volume ≤ 280kg/thm.

9.1.3 Effect of sinter properties on gas permeability

In this section, the effect of the sinter properties on the blast furnace operation will be introduced, along with an analysis of the results of actual furnace operation.

(1) Slag volume (content of SiO_2 in sinter)

It can be assumed that the slag volume of blast furnace depending on the content of SiO_2 in sinter is controlling the permeability at the lower part of the furnace in the case of high rate PCI. When Fukuyama 4 BF in Chapter 7 (Fig.7-15) is taken as an example, it can be seen that the permeation resistance at the lower part of the furnace (K1) which increases with the increment of PCI rate reaches the maximum value at PCI rate ≥ 200kg/thm and this is due to the decrease in the slag volume of the blast furnace by a decrease in SiO_2 content in sinter, as shown in Table 9-1.

A similar phenomenon is observed in Kimitsu 3 BF (Fig.9-2[11]). The actual permeation resistance at the lower part of the furnace shows a lower value than the operation design line. In this case, this permeation resistance is dependent on the decrease in the slag volume of the blast furnace resulting from the decrease in Al_2O_3 content in sinter. Also at Kashima 2 BF whose PCI rate is as low as 100kg/thm, the permeation resistance at the lower part of the furnace has been lowered successfully owing to a decrease in the slag volume as the result of lowering SiO_2 content in sinter (Fig.9-3[16]).

Further, the decrease in the slag volume of the blast furnace can be attained by the decrease of SiO_2 and Al_2O_3 contents in sinter, as seen from the example of Kimitsu 4 BF (PCI rate = 80kg/thm) in Chapter 7 (Fig.7-16). However, even when the slag volume of the blast furnace remains the same, a smaller content of Al_2O_3 provides a greater improving

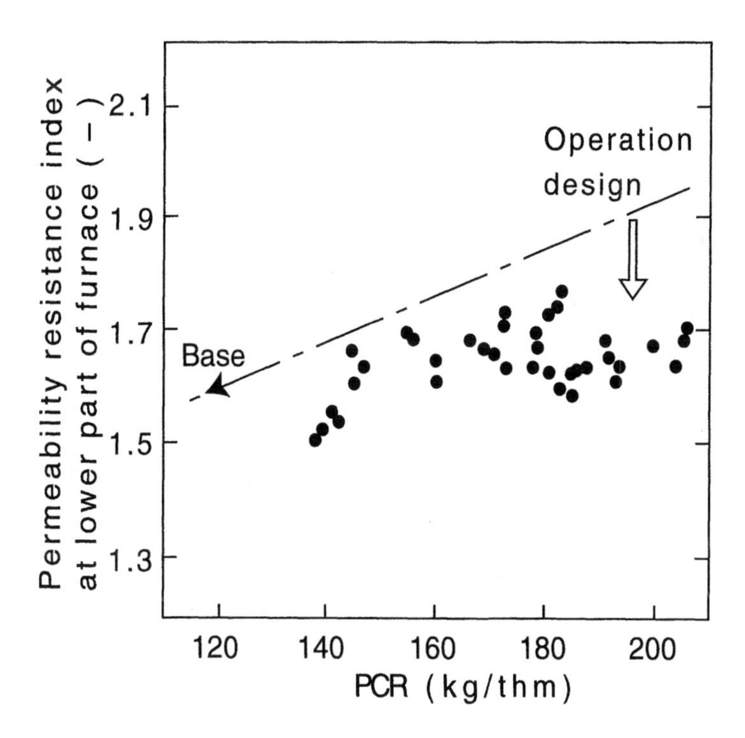

Fig.9-2 Relation between PCR and permeability at lower part of furnace (Kimitsu 3 BF)[11]

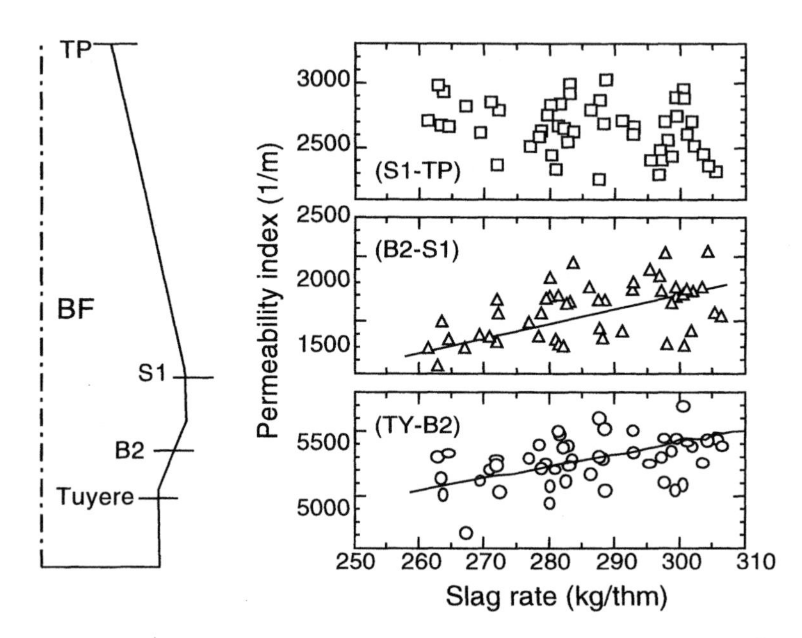

Fig.9-3 Effect of slag rate on permeability of furnace (Kashima 2 BF)[16]

effect on permeation resistance at the lower part of the furnace. This implies that the effect of Al_2O_3 itself also cannot be ignored. However, since the effect of Al_2O_3 lessens when the slag volume is smaller than 290kg/thm, decreasing SiO_2 content in sinter will be a main subject regarding which future research will have to be conducted, considering the future prospects of raw materials.

Improvement in permeability at the lower part of the furnace due to the decrease of the slag volume (decrease in SiO_2 and Al_2O_3 contents in sinter) is the result of a decrease in the permeation resistance of the cohesive zone. As shown in Fig.9-1, since FeO containing melt is generated in the sinter from around 1100°C, and pores are filled up in turn[17], the pressure drop of the sinter layer increases progressively. The extent of this increase in pressure drop controls the permeability of the cohesive zone, that is, the permeability at the lower part of the furnace.[18] Defining a S value[19] obtained by integrating a pressure drop-time curve shown in Fig.9-1, this value has been used as an index available for evaluating the permeability of the sinter layer (permeability at the lower part of the blast furnace). An index (KS) of permeation resistance at high temperatures obtained by integrating a pressure drop-temperature curve is also a similar kind of index.[20]

KS decreases with a decrease in SiO_2 content in sinter (Fig.9-4 [21]). Considering the phenomena of improvement in JIS-RI owing to the decrease in SiO_2 content (Fig.9-5[22]) and lowering of KS due to the rise of JIS-RI (Fig.9-6[21]), it appears that the reduction degree attained at a high value before the formation of FeO containing melt together with low SiO_2 lead to a decrease in the amount of formed melt. Also, according to Fig.9-4, the effect of MgO content is greater than that of Al_2O_3 content for KS, and therefore it is necessary to use the SiO_2-MgO flux like a serpentine and the MgO flux like a dolomite effectively.

(2) RDI

As shown in Fig.9-5, RDI generally deteriorates due to decrease in SiO_2 content in sinter. Taking account of the fact that maintaining or strengthening a central gas flow is essential for stabilizing the blast furnace operation (Fig.7-10 in Chapter 7), a region where a heat flux ratio is relatively high may form because of an increased Ore/Coke

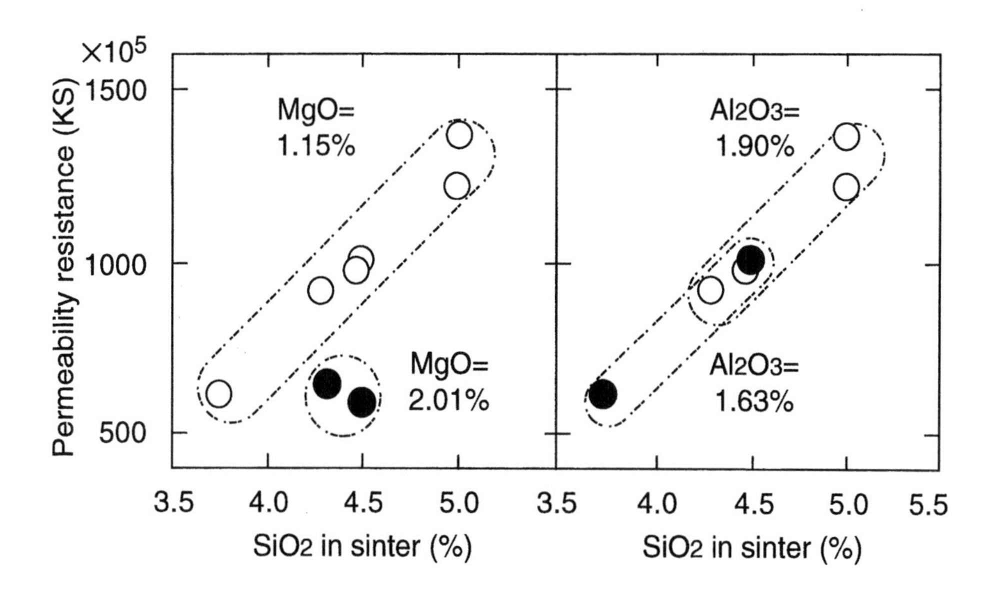

Fig.9-4 Effect of chemical compositions of sinter slag on permeability resistance (CaO/SiO_2=1.8)[21]

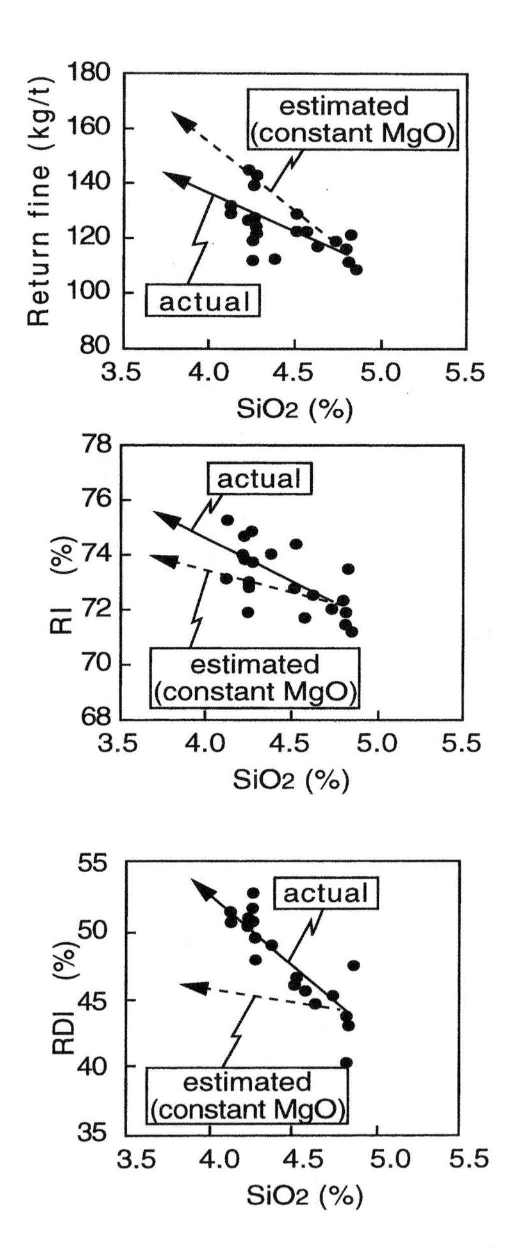

Fig.9-5 Effect of SiO_2 and MgO in sinter on yield, reducibility (JIS-RI) and reduction degradation (RDI) (Fukuyama 5 DL)[22]

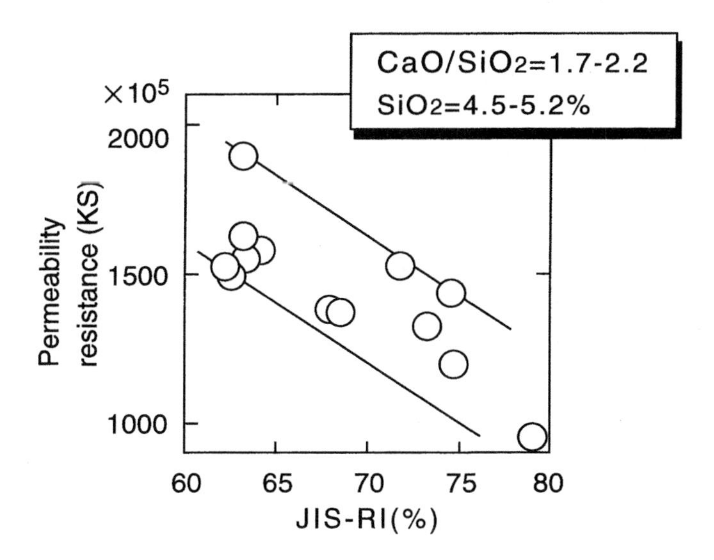

Fig.9-6 Effect of reducibility (JIS-RI) on permeability resistance[21]

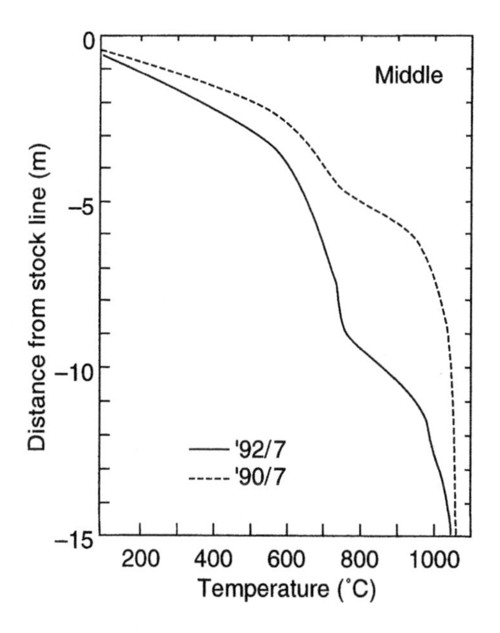

Fig.9-7 Temperature profile measured by vertical probe (Kokura 2 BF; July/ '92: PCR=150kg/thm, July/'90:PCR=75kg/thm)[23]

ratio and ununiformity in radial distribution of Ore/Coke ratio with the increase in PCI rate. This inevitably results in the operation being in trouble along with a deterioration of RDI.

In Kokura 2 BF (RDI = 40%) shown in Fig.9-7[23], a low temperature thermal reserve zone of 500 to 700°C has formed at the middle part of the furnace, which is measured by a vertical probe, and there is an increase in PCI rate from 75kg/thm (July/90) to 150kg/thm (July/92). It is estimated, from the calculation results obtained by a simulation model, that sinter degradation in the furnace has been accelerated by about 10%. This is because a zone where the heat flux ratio is high has been formed in the middle part of the furnace.

Also in Kobe 3 BF (RDI = 40 to 42%) shown in Fig.9-8[6], the low temperature thermal reserve zone is found to have contracted at the peripheral part of the furnace, which is measured by the vertical-horizontal probe, when PCI rate is raised from 100kg/thm to 180kg/thm. However, an increase in RDI value up to 50% results in the low temperature thermal reserve zone being enlarged. The low temperature thermal reserve zone becomes large because of a peripheral gas flow having been suppressed excessively. This cause lies in a retardation of the sequential reduction from magnetite due to a shortage of reducible gas volume. It has thus been stated that RDI must be kept less than 40% at a level where PCI rate is 200kg/thm.[24]

As seen in Table 9-1, in Fukuyama 4 BF, the operation with PCI rate of more than 200kg/thm has been accomplished with the use of sinter with RDI of more than 40%. According to the temperature distribution determined by the horizontal probe in Chapter 7 (Fig.7-20), there is a possibility of a zone where the heat flux ratio is high being formed to range from the middle to the peripheral area of the furnace. But, according to the results obtained by the vertical probe in Chapter 7 (Fig.7-9), it has been verified that the operation was possible because a low temperature thermal reserve zone has not been formed at the middle and peripheral areas of the furnace. The fact that a fuel rate is high (532kg/thm) might be one reason for no low temperature thermal reserve zone; however, this fact suggests that a decrease in SiO_2 content in sinter (decrease of slag volume in the blast furnace) leads to the possibility of RDI being relieved with a high rate PCI.[25]

(3) Unburnt char

In the case of high rate PCI, unburnt char flows out of the raceway and goes up in the blast furnace. When the unburnt char has been captured by the cohesive layer of iron ore, it contributes to reduction of FeO in the iron ore, whereby the high temperature properties are improved (Fig.9-9[26]). According to the results of a reduction test under load performed by mixing char in reduction gas, the pressure drop of the ore layer is sharply lowered at 1250°C.[27] This implies that softening and melt-down properties vary due to the existence of char having higher reactivity than fine coke. Further, at the temperature over 1250°C, the consumption rate of char is greater than the accumulation rate.[27] This shows that the char might be consumed by an active solution loss reaction in the cohesive layer.

There is also a report stating that the reaction of lump coke at the lower part of the furnace is controlled by such preferential reaction of unburnt char and that fewer fines

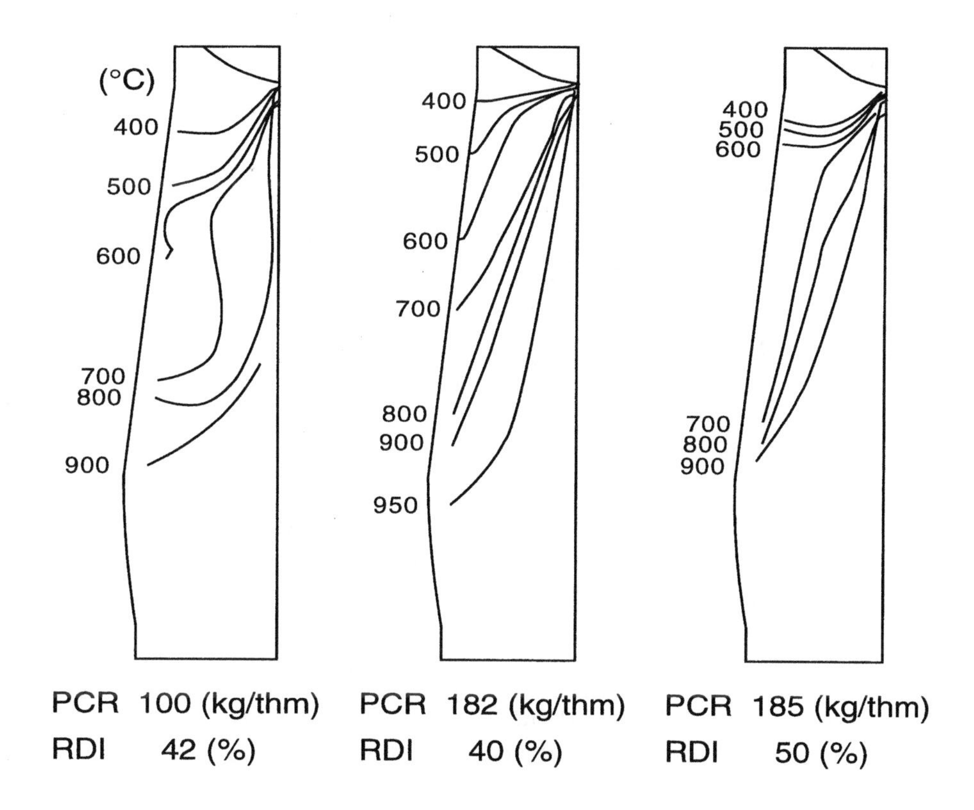

Fig.9-8 Change in temperature distribution with PCR and RDI (Kobe 3 BF)[6]

are generated.[26] However, as stated in Chapter 8, viewed from the necessity for the operation to control the fine generation of coke in the raceway by increasing the combustibility of PC, it would be hardly possible to adopt a means which uses the unburnt char for promoting the high temperature reduction of iron ore.

(4) Mixing nuts coke into iron ore layer

It has been verified through both a model experiment and an actual furnace test operation that the permeability of the blast furnace can be improved and the reduction efficiency can be increased by mixing small size coke of about 20mm in average particle size (5 to 25mm, as an example) into iron ore. This is due to the facts that small size coke acts as a resistance to shrinkage of the iron ore layer and CO_2 generated by the iron ore

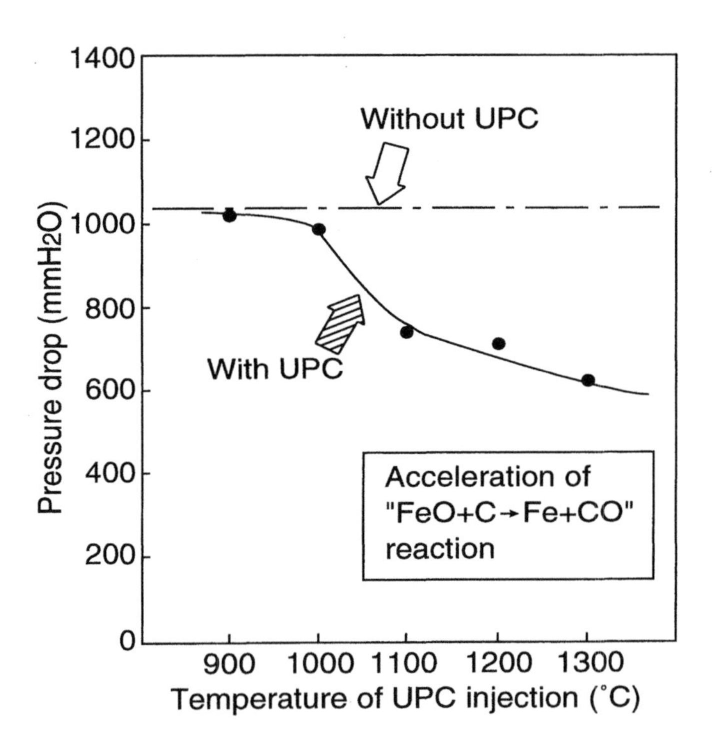

Fig.9-9 Effect of temperature of unburnt PC injection on pressure drop[26]
UPC: unburnt PC

reduction reacts with the mixed coke whereby CO is regenerated. There is also the effect that the large size coke is restrained from deteriorating because the small size coke reacts preferentially.[28)]

Although there are some restrictions, e.g., decrease in dropping temperature due to acceleration of metal carburization[28)] and an upper limit of the amount of use from the aspect of burden distribution control (100kg/thm, 26% of coke rate of 390kg/thm)[29)], it may be possible to adopt high rate PCI as a countermeasure for accelerating the iron ore reduction.

Lately, a means has also been developed aiming at further increasing the reduction efficiency by providing the small size coke with high reactivity (from 22% in usual JIS-RI to 61%) (Fig.9-10[17)]). This means lowers the temperature of the thermal reserve zone where the temperature is usually about 1000°C and at the same time, shifts the starting

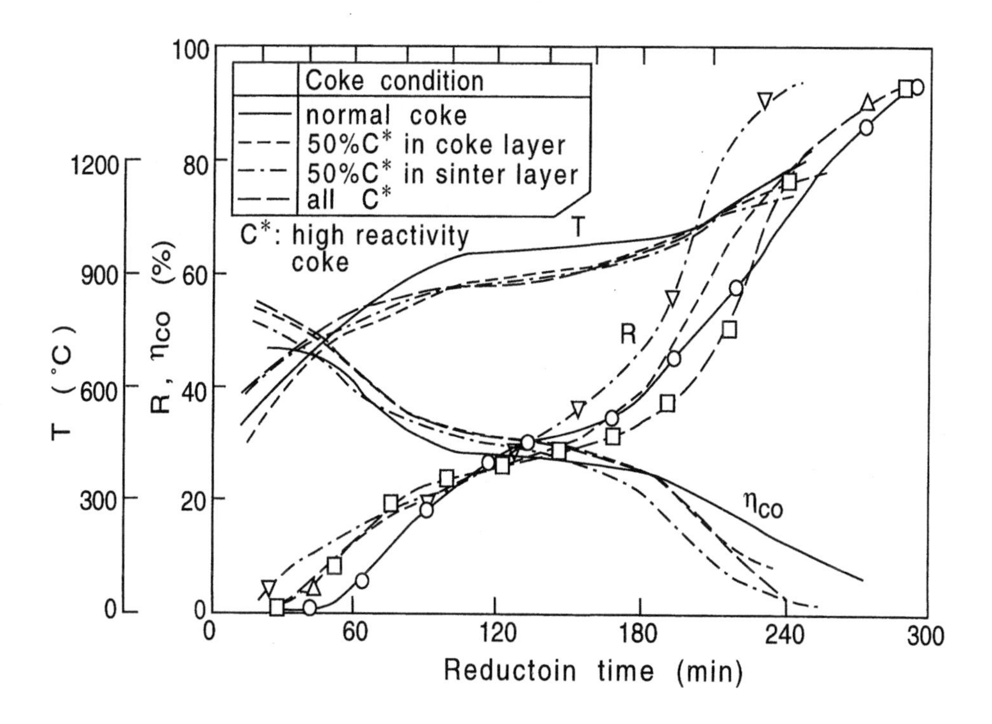

Fig.9-10 Reduction behavior of sinter obtained by using high reactivity coke[17)]

temperature of reduction of wustite and calcium ferrites to iron to a lower temperatures, to thereby cause an increment of the volume of pores in the ore by the reduction of wustite and calcium ferrites, a decrease of FeO amount at a high temperature zone and reduction of formed melt.

However, use of the coke producing technology available currently is accompanied with a reduction in strength (hot and cold) by providing the coke with high reactivity. Thus, the requirements for use of this technology are to react the coke preferentially by mixing it into the iron ore layer, and further, not to allow the fines to remain after consumption of almost all mixed small size coke.

9.1.4 Future work

The internal conditions of the blast furnace significantly change with an increase of the PCI rate and accordingly, the properties required for the sinter also vary. However, since such properties vary even due to the production conditions, including productivity, fuel rate and so on, or an alteration in the operation way accompanying the increase in PCI rate, it is difficult to discuss the properties without setting the operation conditions.

Up to the preceding section, (1) control of reduction degradation at low temperatures (RDI of less than 38% is a numerical value obtained from the actual operation result) and (2) decrease in slag volume of blast furnace due to the decrease in SiO_2 content in sinter (slag volume of less than 280kg/thm is a numerical value obtained from the actual operation result) have been described. In consequence of (1) and (2), (3) improvement in sinter reducibility (JIS-RI of more than 70% is a numerical value obtained from the actual operation result) has been gained and along with the decrease in SiO_2 content, the softening and melt-down properties at a high temperature zone over 1000°C have been improved. Also, these improvements in the sinter producing process lead to, (4) an increase in cold strength (SI or TI of higher than +1.0% is a numerical value obtained from the actual operation result). The effect of cold strength on the actual blast furnace operation seems to be large, although there is no example of an analysis of the operation available so far.

Generally available measures to produce the sinter providing such properties are, e.g., increase in porosity, decrease in FeO content and increase in melting temperature by controlling slag basicity of product sinter. However, attention should be paid to the fact that maintaining RDI value is difficult under the application of such measures. In addition, the restrictions on the raw materials selection are becoming severer, including the increasing amount of use of a high Al_2O_3 and/or a high combined water containing iron ores such as Pisolitic ores. Thus, it is necessary to make precise design on the properties and establish a sophisticated control technology taking these measures into account synthetically.

For realizing these measures, one idea would be to produce a high-reducible sinter retaining the characteristic feature of a pellet having a low reduction degradation property (RDI) at low temperatures, providing a high porosity, and especially containing many micro pores and less slag content.[7)] HPS having a low SiO_2 content and a high reducibility (JIS-RI) is supposed to be one of the processes corresponding to this idea. It has been reported that HPS greatly contributes to high productivity in the actual operation.[30)]

Further, a sinter produced from fine magnetite ore containing many micro pores and having a good reduction property is thought to have properties similar to those mentioned above. Therefore, it is necessary to establish an agglomerating technology applicable to a wide range of raw materials.

On the other hand, there is a report on the operation that by blending a large amount of high Al_2O_3 and combined water containing iron ores, it has been possible to succeed in making sinter having low SiO_2 content to improve the softening and melt-down properties (reduction of permeation resistance of cohesive zone) without a significant rise in cost.[31)] This measure can be considered to be one of the solution measures.

Finally, since the sintering process uses many kinds of raw materials by blending them, a granule design of a raw material mixture is a key technology for enabling the property control of product sinter by enabling the properties of these raw materials to be understood and utilizing them effectively. Together with the granulation process producing granules of the raw material mixture which are hard to collapse during conveying and charging, overall progress in the development of pretreatment technology of raw materials is required.

9.2 Coke properties for high rate PCI

As already described in Chapter 8, the main coke fine generating area in a blast furnace is a dripping zone at the lower part of the furnace and an inside of the raceway. Degradation of coke and generation of fines occurs due to the bulk breakage and surface abrasion at the dripping zone at the lower part of the furnace and the surface abrasion inside the raceway. Also, there are three probable factors controlling the degradation of coke and generation of fine, at the dripping zone at the lower part of the furnace, that is,

(1) gasification reaction by CO_2 (H_2O), (2) contact reaction with molten FeO and (3) contact carburization reaction with molten metal, and two factors inside the raceway, that is, (1) gasification reaction by CO_2 (H_2O) and (3) contact carburization reaction with molten metal.

In this chapter, focusing on the phenomenon of coke fine generation, an attempt has been made to derive the properties necessary for controlling the coke degradation and fine generation.

9.2.1 Effect of fine coke on BF operation

As shown in Chapter 8 (Figs.8-4,8-5,8-7 and 8-8), the amount of coke fines (-3mm and -5mm mass%) was measured by inserting a probe into the furnace and sampling the coke at the shutdown. From the results acquired then, it can be seen that quite a big dispersion is found in the amount of coke fines; that is, about max. 25mass% at the lower

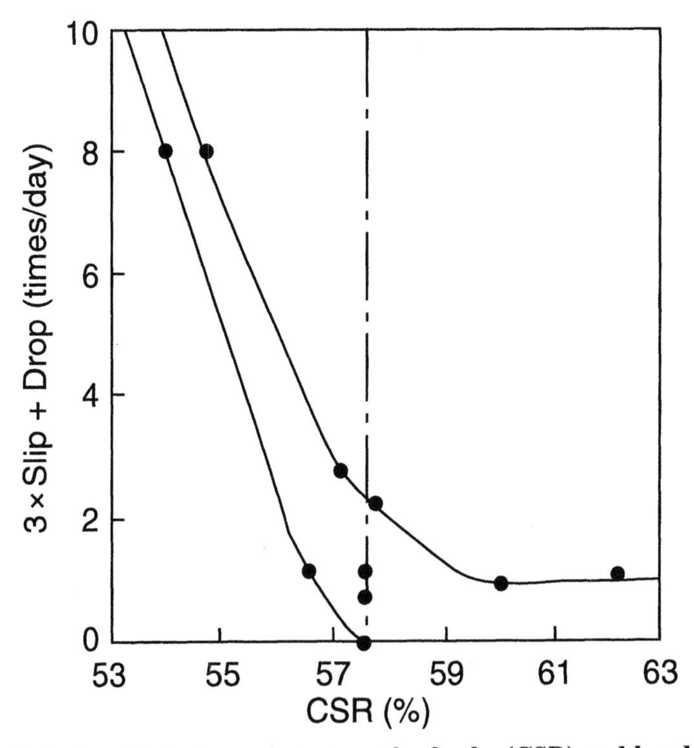

Fig.9-11 Relationship between hot strength of coke (CSR) and burden descent (Kimitsu 3 BF)[32]

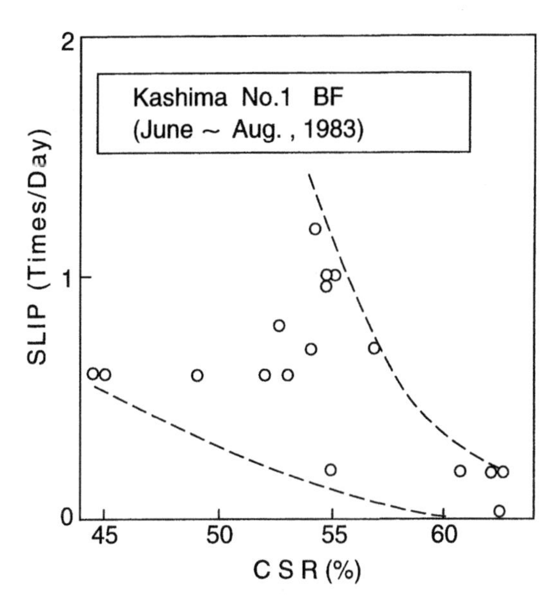

Fig.9-12 Relationship between hot strength of coke (CSR) and burden descent (Kashima 1 BF)[33)]

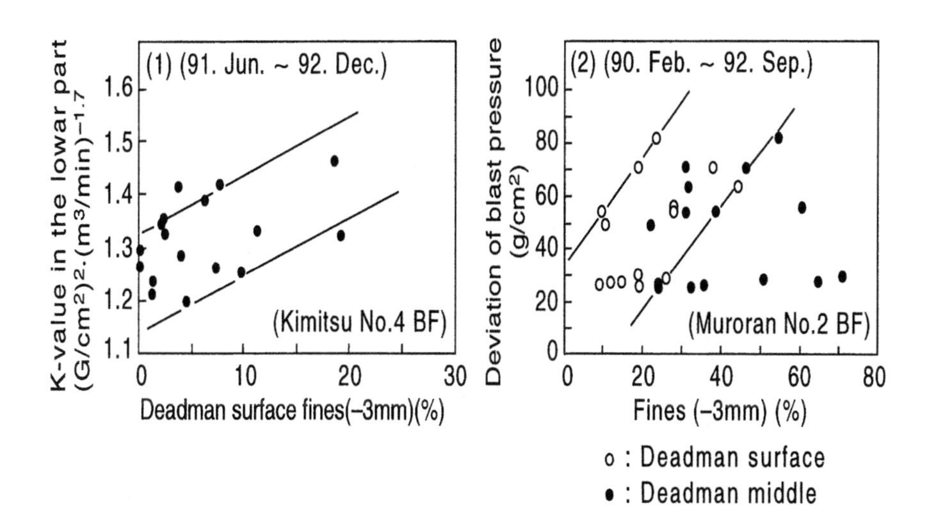

Fig.9-13 Influence of fines (-3mm) on permeability (Kimitsu 4 BF and Muroran 2 BF)[34]

shaft to the belly and about max. 25mass% (-3mm) to 40mass% (-5mm) inside the raceway. Based on the data acquired from the operating furnace, the effect of the existing area and amount of coke fines on the blast furnace operation was investigated. The results obtained are as described below.

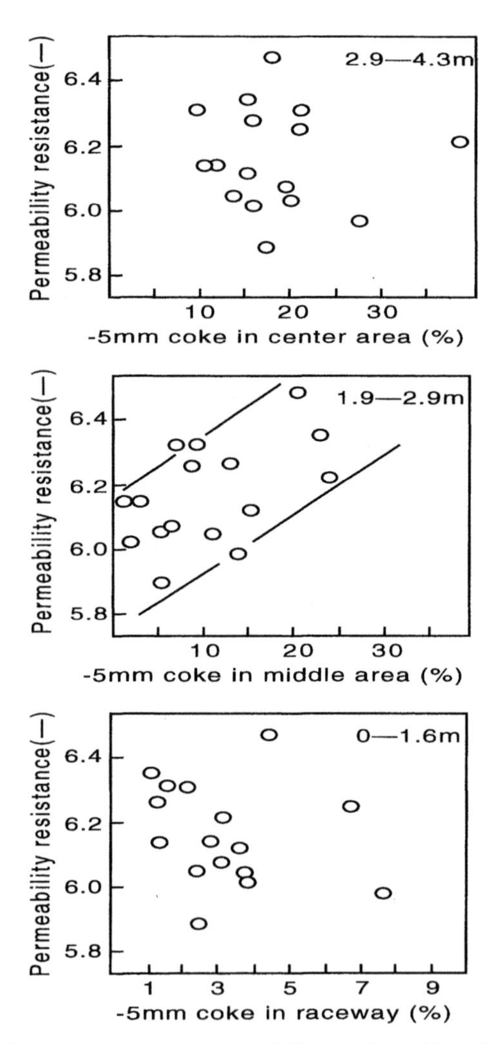

Fig.9-14 Relationship between amount of fine coke (-5mm) and permeability resistance (Kobe 3 BF)[35)]

(1) CRS and burden descent

As shown in Chapter 8 (Figs.8-4 and 8-5), the amount of coke fines at the lower shaft to the belly increased with the lowering of coke hot strength (CSR; Coke Strength after reaction) in the case of a CSR change test operation at Kimitsu 3 BF and Kashima 1 BF. Figures 9-11 and 9-12[32-33] show a relation between CSR and burden descent (slip), from which it is determined that the fines generated and accumulated in this area induce an abnormal burden descending.

(2) Effect of coke fines on pearmeability

Generally, the increase in the amount of fines makes the permeation resistance at the lower part of the furnace larger and causes a variation in blast pressure. In regard to the position in the radial direction at the tuyere level affecting the operation, the following analytical results have been obtained.

Figure 9-13[34] shows the analytical results regarding Kimitsu 4 BF and Muroran 2 BF. It indicates that the fines at the deadman surface control the permeability and the effect of fines at the middle area of the furnace is small. Figure 9-14[35] shows the analytical results at Kobe 3 BF, from which it is seen that the fines at the middle area (1.6 to 2.9m) of the deadman control the permeability and the effect of fines at the central area of the furnace and inside the raceway is small. A similar analytical example is seen also regarding Kokura 2 BF.[36]

In any case, since the definition of the deadman surface and middle area differs according to the blast furnaces, the fines at the middle area of deadman, excluding the inside of the raceway and the central area, may be considered to be a controlling factor.

(3) Fine accumulation area controlling permeability

There is no example of an analysis made regarding the operating furnace as to which area's fines are controlling the permeability in the blast furnace, the dripping zone at the lower part in the furnace or the inside of the raceway, which are the main generation areas of coke fines in the blast furnace. According to a model experimental result shown in Fig.9-15,[37] the permeability at the lower part of the furnace is found to have an even stronger correlation with dust concentration at the bosh than that inside the deadman. It can be said from this result that it is imperative to control the generation of fines at the dripping zone at the lower part of the furnace. But there is still a need to continue research on this subject.

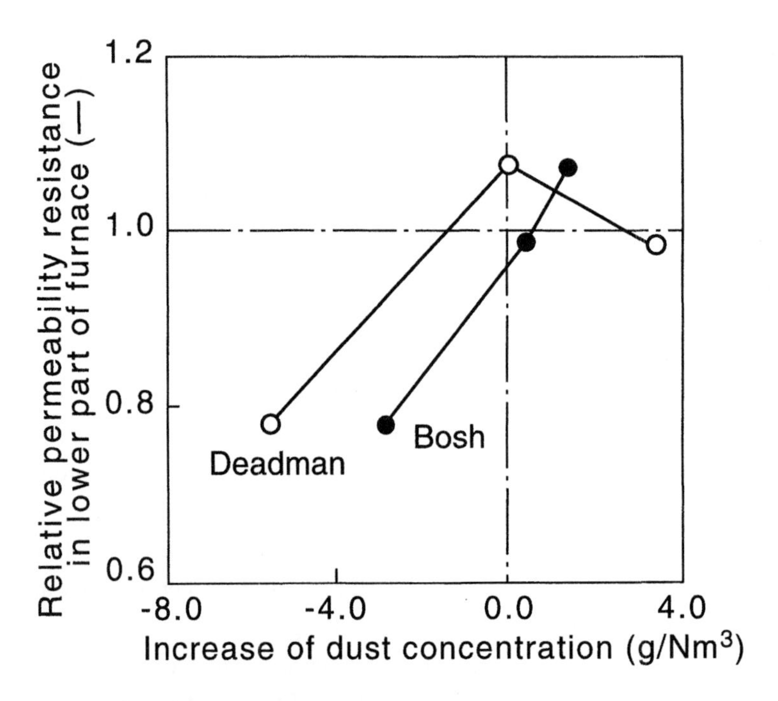

Fig.9-15 Relation between dust increment and permeability in lower part of furnace[36)]

As described in Chapter 8 (Figs.8-18 and 8-19), the traveling route of fines was examined by comparing the exposed temperature of lump with that of fine of coke sampled from the inside of the blast furnace. According to the results obtained then, fine generated inside the raceway flies as far as the bosh and accumulates there and fine generated at the dripping zone at the lower part of the furnace travels to the deadman together with the descending burden and accumulates there.[6,38)]

It can be assumed from the description in Chapter 8 (Fig.8-20) that the fine (-3mm) at the wall side of tuyere level, which is supposed to have been generated inside the raceway and has an exposed temperature of 80 to 100Å (1700 to 1800°C), exists at the middle area of the tuyere level and an area of the tuyere level +1.5m and the fine is presumed to fly as far as these areas, which has substantiated traveling of the fine generated inside the raceway to an area above and the middle area of the deadman. Also, according to the description in Chapter 8 (Fig.8-21), the fine (-5mm) supposed to have been generated inside the raceway and having an exposed temperature of 1800 to 2000°C exists even at the central area (6.5m) of the furnace, which indicates traveling of fine generated inside the raceway to the central area of the furnace. Further, a phenomenon that the fine generated at the dripping zone at the lower part of the furnace travels to the middle area of the deadman was seen also at the time of a CSR change test operation at Kimitsu 3 BF.[32)]

From the above literature information, it is seen that the permeability of the blast furnace is controlled by the generation of fine at the dripping zone at the lower part of the furnace and the fine generated at this area travels to the middle area of the deadman together with the descending burden and accumulates there. It would thus be important, for maintaining a proper permeability of the blast furnace, to control the fine generation at the dripping zone at the lower part of the furnace and, besides, to minimize the accumulation of fine in the middle area of the deadman.

9.2.2 Control of coke fine generation

Although a study on consumption of coke fines generated inside the blast furnace is

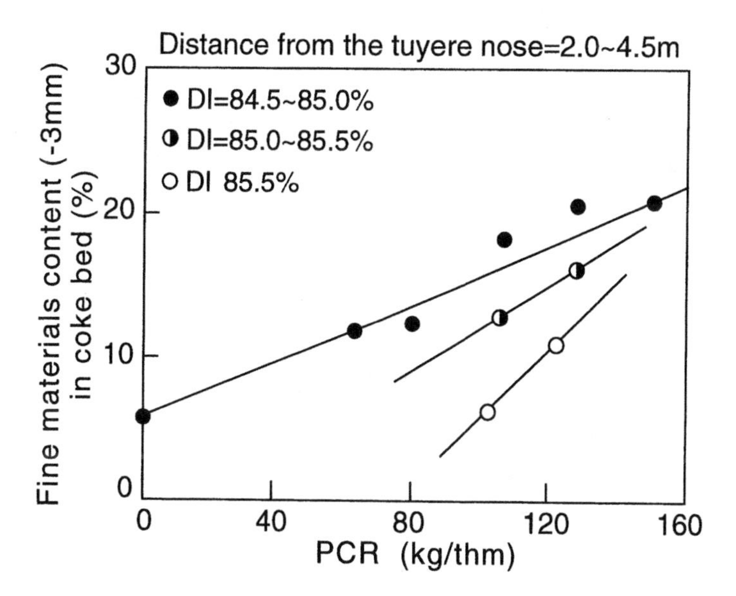

Fig.9-16 Effects of PC rate and cold strength of coke (DI) on fine content (-3mm) in deadman (Kimitsu 3 BF)[39)]

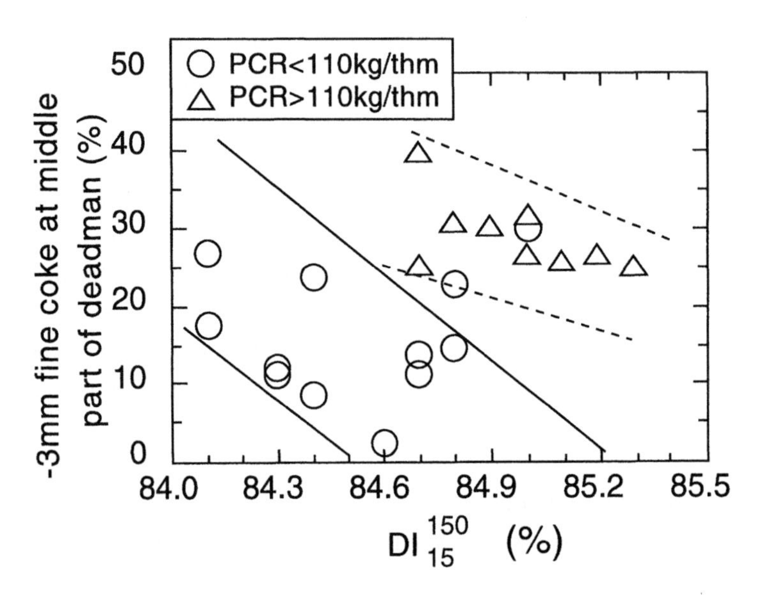

Fig.9-17 Relationship among cold strength of coke (DI), PCR and fine coke (-3mm) at middle part of deadman (Kakogawa 1 BF)[40]

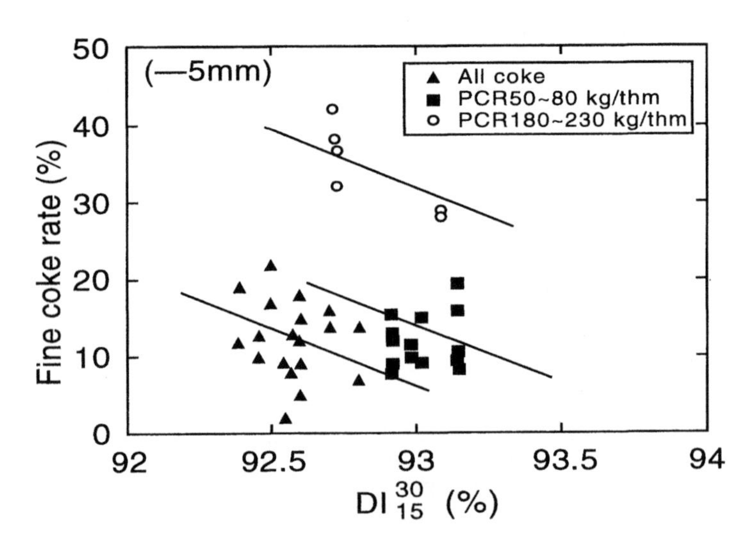

Fig.9-18 Relationship among cold strength of coke (DI), PCR and fine coke (-5mm) in deadman (Fukuyama 4 BF)[41]

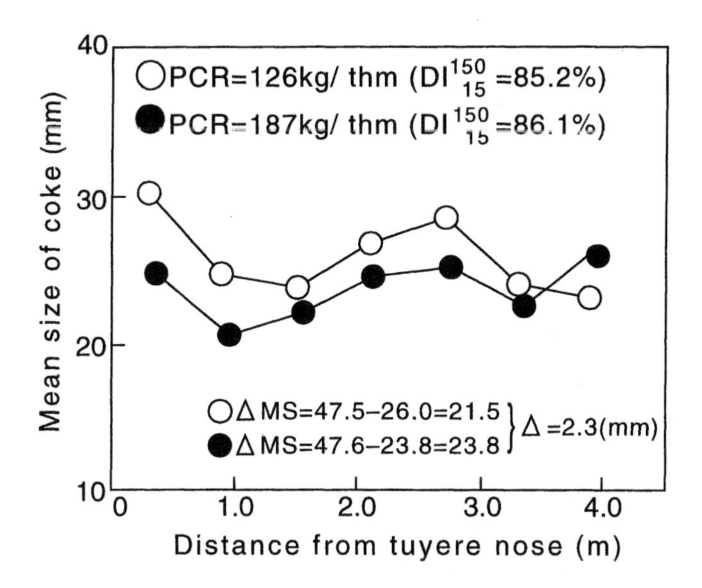

Fig.9-19 Change in mean size of coke sampled at tuyere level (Kimitsu 3 BF)[42]

a future project, consumption of coke fines in the actual blast furnace operation is hardly expected. Therefore, how to control the generation of coke fines is another important problem to be solved. In regard to this subject, various measures described on the succeeding pages have been taken at the operating furnace or research has been conducted by performing a model experiment, and thus from now on, a quantitative evaluation is expected.

(1) Improvement in cold strength (DI)

Measures for raising the cold strength (DI) have been adopted by almost all operating blast furnaces in the case of high rate PCI. The analytical results obtained from Kimitsu 3 BF shown in Fig.9-16,[39] Kakogawa 1 BF shown in Fig.9-17[40] and Fukuyama 4 BF shown in Fig.9-18[41] indicate that the increase in coke fines at the tuyere level, which becomes clear in the case of high rate PCI, can be controlled well with the improved DI.

There is no information from the operating furnace about the change in coke fines at the dripping zone at the lower part of the furnace in the case of high rate PCI. When the coke in an area 1.5m away from the wall at the tuyere level, which is shown in the particle size distribution of the coke sampled from the tuyere level at the time of the high rate PCI test operation performed at Kimitsu 3 BF in Fig.9-19,[42] is regarded as having dropped from an area just above the tuyere, it is indicated that when the PCI rate is 187kg/thm relative to 126kg/thm, the particle size is small despite DI having been raised. Thus, it is indicated indirectly that fine generation at the dripping zone at the lower part of the furnace is intensive at the time of high rate PCI. Whilst, as to the change in coke fines inside the raceway, the generation of coke fines at the deadman has been controlled successfully thanks to the improvement in DI at Kimitsu 3 BF, as shown in Chapter 8 (Fig.8-8). From the model experimental results shown in Chapter 8 (Fig.8-13), it can be seen that dust inside the deadman increases as it does in the operating furnace in the case of high rate PCI and it is supposed that the reason why the amount of dust at the bosh increases less than it does in the deadman is that a burden load in the model experiment is smaller than that on the operating furnace. The amount that dust increases at the bosh in the operating furnace should be assumed to be great.

Supposing, from the above results, that the coke fines increase at both area, that is, at the dripping zone at the lower part of the furnace and the inside of the raceway in high rate PCI operation, the measures should be taken into account. And the generation of coke fines at the inside of the raceway should be mainly controlled with the improvement in DI.

The increase in the amount of coke fines at the dripping zone at the lower part of the furnace in the case of high rate PCI might be attributable to; increase in burden load working on coke due to rise in Ore/Coke ratio, increase in residence time of coke in furnace and increase in amount of contact carburization with molten metal per coke particle. However, the increase of coke fines is not quantitatively analyzed yet.

As for the increase of coke fines inside the raceway, the analytical results described below are available. According to the results obtained by a mathematical model simulation, as the operation is transferred from all coke to PCI and further, the combustibility of PC has been improved, while the amount of coke reacting with O_2 decreases.[43] Consequently, dissipation of the surface part degraded at the lower part of the furnace due to the reaction with O_2 lessens and the coke fines increase.

Also, according to the examination results obtained by a single particle model of lump coke, there is no dissipation of the surface part degraded at the lower part of the furnace due to the reaction with O_2 and a thick degraded layer is formed on the coke particle surface inside the raceway because of the large amount of CO_2 gasification inside

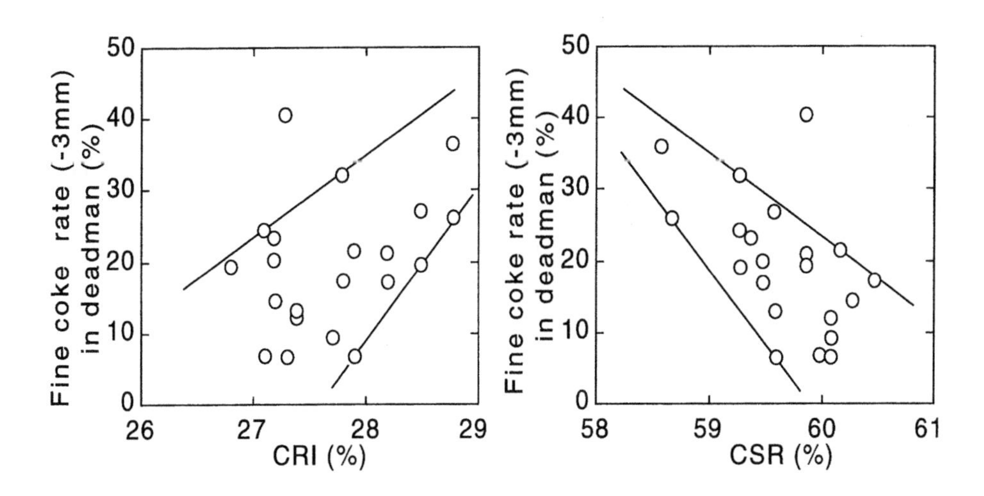

Fig.9-20 Relationship between coke qualities and coke fine (-3mm) in deadman Kure 1 BF)[44)]

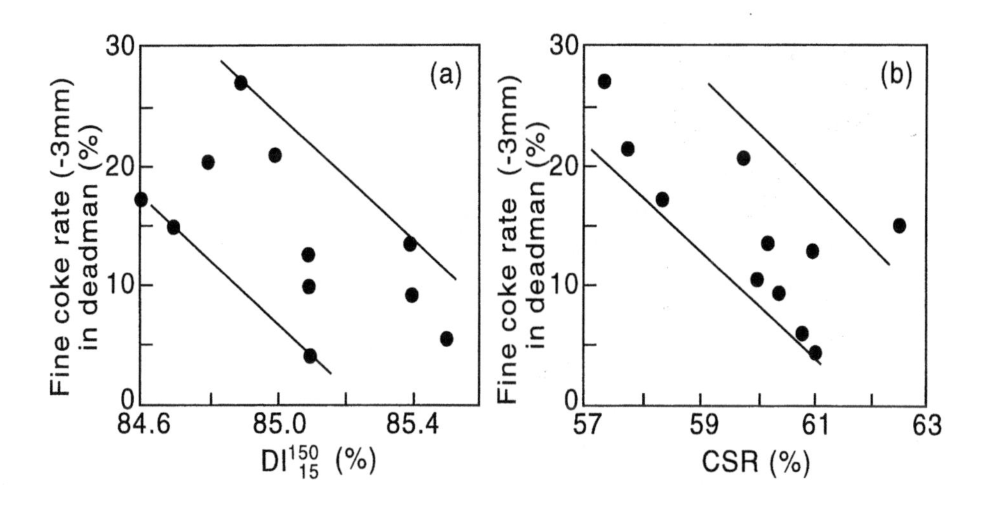

Fig.9-21 Relationship among cold strength (DI), hot strength (CSR) and fine content (-3mm) in deadman coke (Kimitsu 3 BF)[42)]

the raceway[37)] when the PCI rate is 203kg/thm, compared with a case where the rate is 118kg/thm. Thus, the coke fines increase.

(2) Improvement inCSR

Measures which have been very successful in raising the hot strength are being used in Europe and the United States. Also in Japan, as shown by the example of Kure 1 BF (Fig.9-20[44)]), the generation of coke fines at the tuyere level has been controlled substantially by the hot strength (CSR) improvement.

In regard to the separation of the effects of DI and CSR, the effective coefficients of both are compared by multiple regression analysis based on the operating furnace analytical results (Fig.9-21[42)]) of Kimitsu 3 BF. These effects become a coke fine ratio -12.8%/DI + 1% and a coke fine ratio -2.0%/CSR + 1%, and thus, DI has an effect which is about 6 times as much as CSR.[11)]

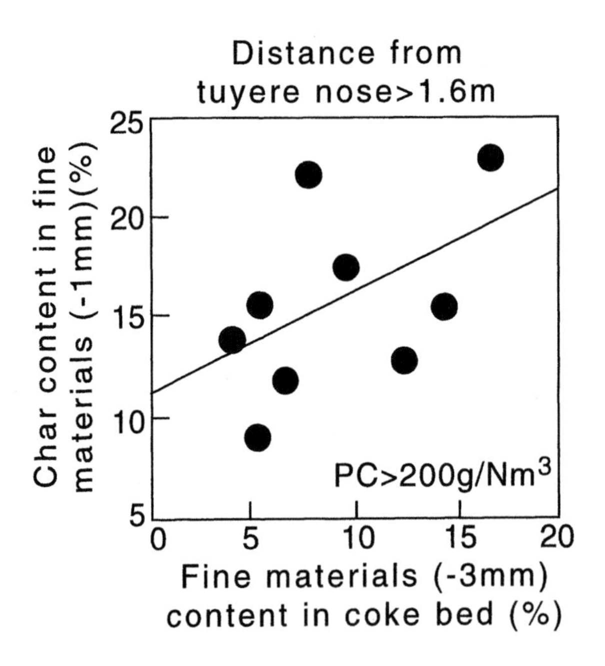

Fig.9-22 Relationship between fine content (3mm) in deadman coke and char content (-1mm) in fine (Kimitsu 4 BF)[39)]

(3) Improvement incombustion efficiency of PC

Figure 9-22[39] shows the result obtained by making an analysis at Kimitsu 3 BF. There is a positive correlation between the amounts of coke fines and unburnt char at the deadman surface, from which it is found that the amount of coke fines decreases owing to the improved combustibility of pulverized coal (control of generation of unburnt char). There are also the model experimental results indicating that when the combustibility of PC is raised by adopting double lances, the amount of coke fines discharged out of the furnace top becomes small.[43]

From the above results, it has been supposed that the reason why the amount of coke fines decreases is that the amount of gasification with CO_2 and H_2O lessens. But this is a problem which has to be studied more in the future to clarify the details of the mechanism causing such a decrease.

(4) Application of high reactivity of unburnt char

As described before, when the amount of unburnt char is increased, the reaction of lump coke at the lower part of the furnace is suppressed by the preferential gasification reaction of the char, and thus the amount of coke fines generated decreases.[26]

However, this measure is contrary to the measure for control of the generation of coke fines inside the raceway owing to the improved combustibility of PC and an optimum value is supposed to be for the amount of unburnt char generation.

This is the measure already adopted before proceeding with the high rate PCI, as described before. A small coke (about 20mm in size, 5 to 25mm, as an example) mixed into the iron ore layer is subjected to gasification reaction prior to a large coke of about 50mm in size being mixed therein, whereby the reaction of large coke at the lower part of the furnace is suppressed.[28] It can also be expected that the protective effect on lump coke will be further increased by enhancing the reactivity of the mixed small coke.[17] However, a proper amount of small coke to be mixed corresponding to the amount of generated unburnt char mentioned above should be available.

(5) Reactivity (CRI) of nuts coke

The rise in reactivity (CRI; Coke Reaction Index) is a phenomenon contrary to the improvement in hot strength (CSR) as obtained from Fig.9-20, and making its mechanism clear and developing the coke producing technology are problems to be solved in the

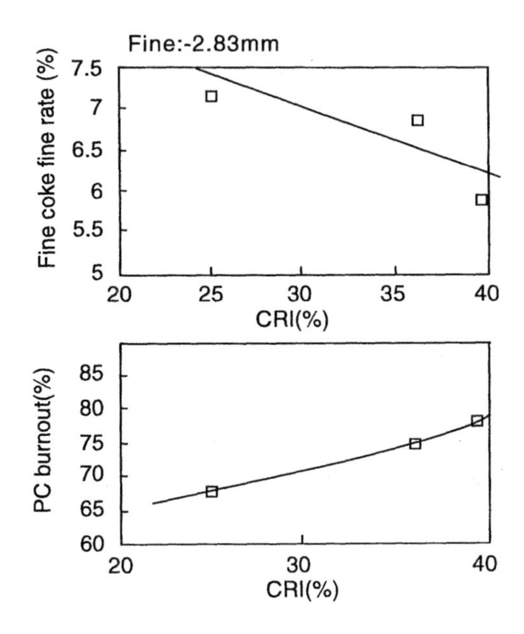

Fig.9-23 Effect of coke reactivity (CRI) on coke fine formation (-3mm) and PC combustibility in raceway[45]

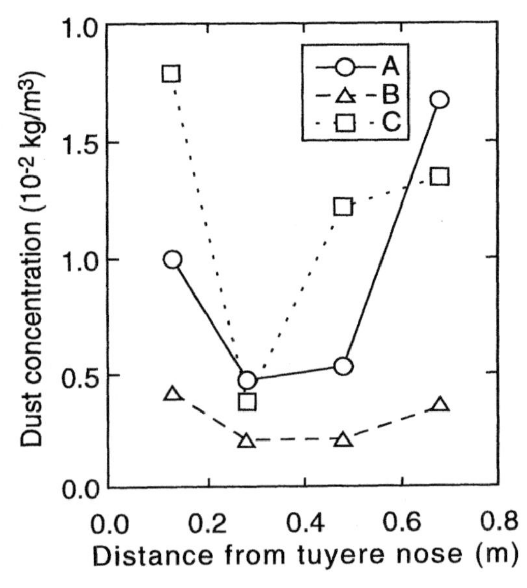

Fig.9-24 Distribution of dust concentration at 0.6m above tuyere[46]

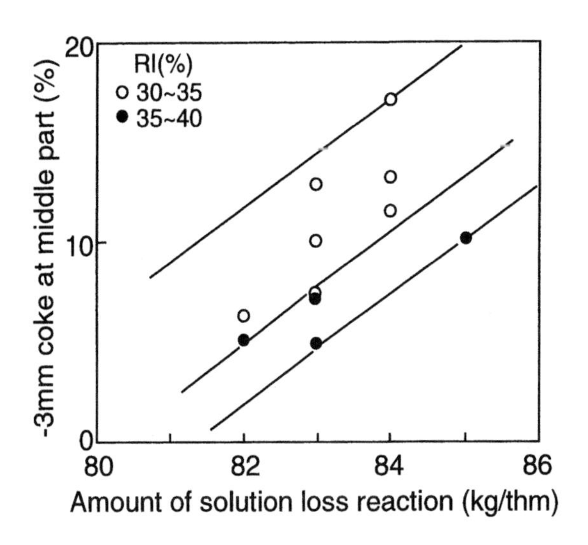

Fig.9-25 Relationship among amount of solution loss reaction, coke reactivity (RI) and fine coke (-3mm) in deadman (Kobe 3 BF)[47]

Table 9-2 Properties of coke used in experiment[46]

Coke No.	DI^{150}_{15} (-)	$I^{600}_{13.2}$ (-)	RI (-)	RSI (-)	Ash (%)	Porosity (%)
A	85.2	88.8	23.0	72.5	11.2	45.0
B	84.5	85.4	35.6	53.5	11.5	45.5
C	81.9	83.6	40.2	43.2	11.2	44.4

Table 9-3 Influence of CSR on size distribution of coke sampled at tuyere (Kashima 1 BF)[33)]

Post reaction strength (C S R)		60	50	45
Mean size (mm)		21.7	17.6	14.4
Size distribution (%)	+25mm	38.8	26.7	17.0
	15~25	17.8	17.7	19.8
	3~15	26.1	28.9	42.0
	−3	17.3	26.7	21.2
Degradation pattern				

future. When the rise in reactivity is compared with the improvement in DI or CSR, it may reduce the coke producing cost, and thus it should become a promising measure if it is actually materialized.

In the model experiment, the effect of suppressing the generation of coke fines at the tuyere level due to the rise in CRI has already been determined (Fig.9-23,[45)] Fig.9-24 and Table 9-2[46)]). Also in Chapter 8 (Fig.8-14), the same data are shown. Further, the same analytical results (Fig.9-25[47)]) have been obtained from the operating furnace operation. All of these results show that CRI has been raised about 40%.

However, as seen from Fig.9-20, the amount of fines in the deadman is increasing with the rise in CRI at a range where CRI is under 30%. In the case of a CSR change test operation performed at Kimitsu 3 BF, when CSR is 54% (CRI: 35%), the amount of fines at the tuyere level decreases, though the amount of coke fines at the belly increases.[32)]

Table 9-3[33)] shows the investigation results of coke sampled inside the furnace in a CSR change test operation performed at Kashima 1 BF, from which it was determined that when CSR is lowered as much as 45% (CRI: 40%), the particle size distribution of coke fines generated changes, the amount of -3mm fine decreases and the amount of fine particles of 3 to 15mm increases. And as shown in Fig.9-12, the descending burden does not deteriorate so much.

From these results, it is suggested that there is a minimum generation of coke fines in relation to CRI and at the same time, it is supposed that the effects of CRI at the dripping zone at the lower part of the furnace and inside of the raceway differ from each other.

It is also found from Fig.9-24 that degradation of B coke of which reactivity (RI is the same as CRI) and hot strength (RSI is the same as CSR) are both high is less than that of C coke of which RI is high and RSI is low. Accordingly, this measures should be

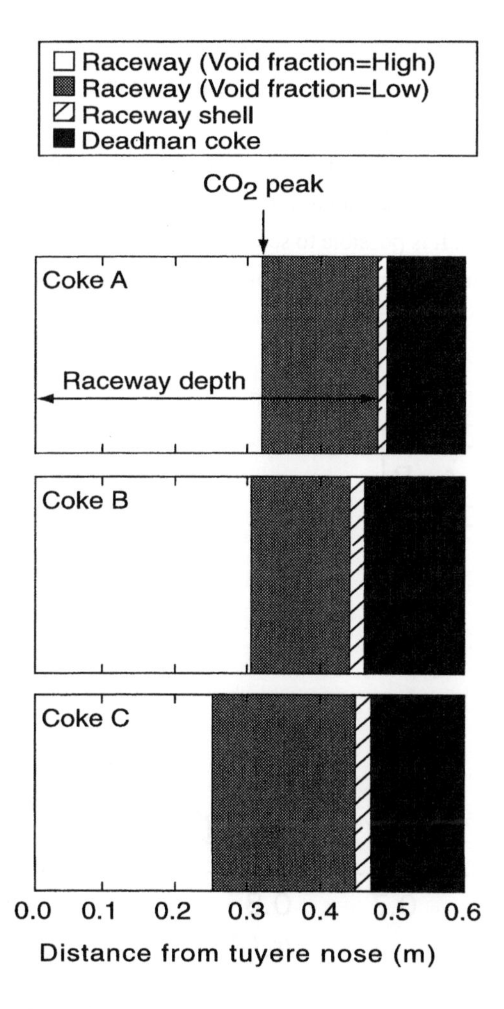

Fig.9-26 Effect of coke quality on raceway structure[46)]

studied further in the future, including its relation to the strength.

(6) Increase in coke reactivity

The effect of reactivity (RI) and strength (RSI) on the degradation behavior of coke in the raceway in high rate PCI was examined, using therefor a coke bed type combustion furnace incorporating a simulated raceway. The results (Fig.9-26[46)]) showed that the higher RI is, the higher the combustion (oxidation) rate of coke becomes, a maximum combustible point (CO_2 peak position) comes near to the tuyere side, raceway depth becomes small and at the same time, the temperature of combustion area is increased rapidly, which contributes to the increase in temperature of PC and improvement in combustibility. A trend similar to this is shown in Fig.9-23.

It is seen from the distribution of reaction degree in lumpy coke (Fig.9-27[46)]) that a reaction layer of B and C cokes each having high RI is limited to a surface, compared with A coke having low RI. It is possible to see that the degradation of A coke has become noticeable because of an increasing mechanical impact force due to a wide raceway

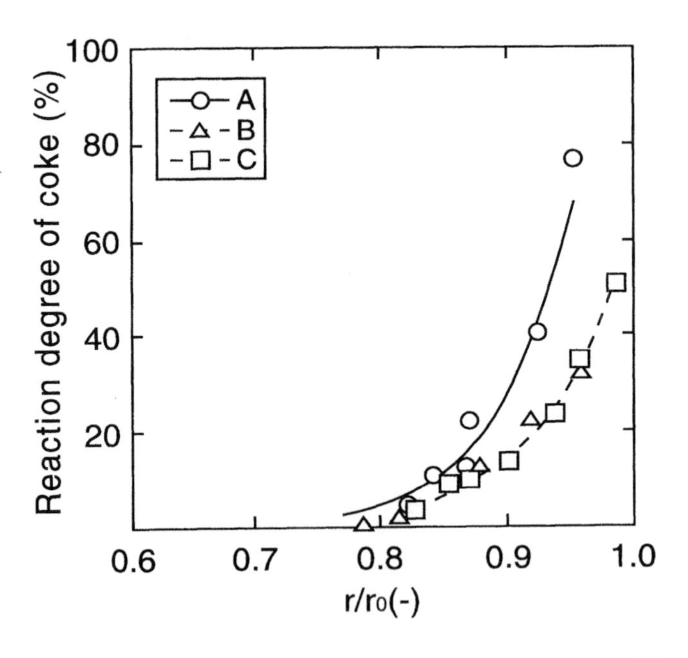

Fig.9-27 Distribution of reaction degree in lump coke sampled in raceway[46)]

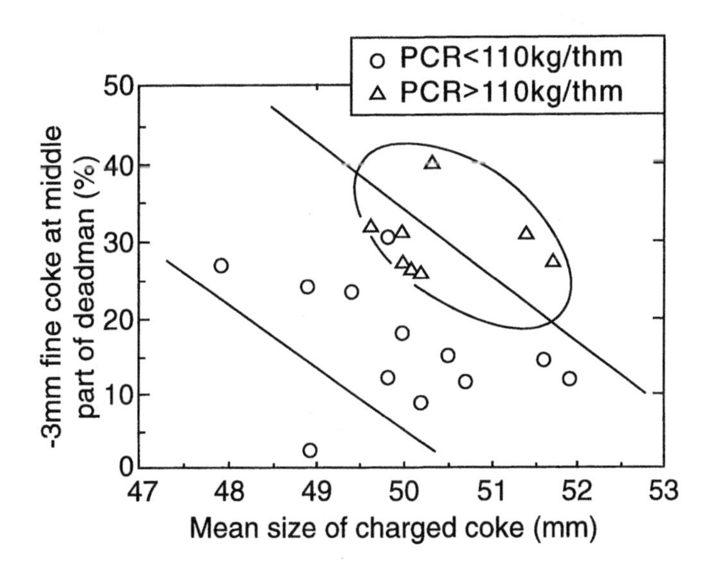

Fig.9-28 Relationship among average size of charged coke, PCR and fine coke ratio (-3mm) at middle part of deadman[13)]

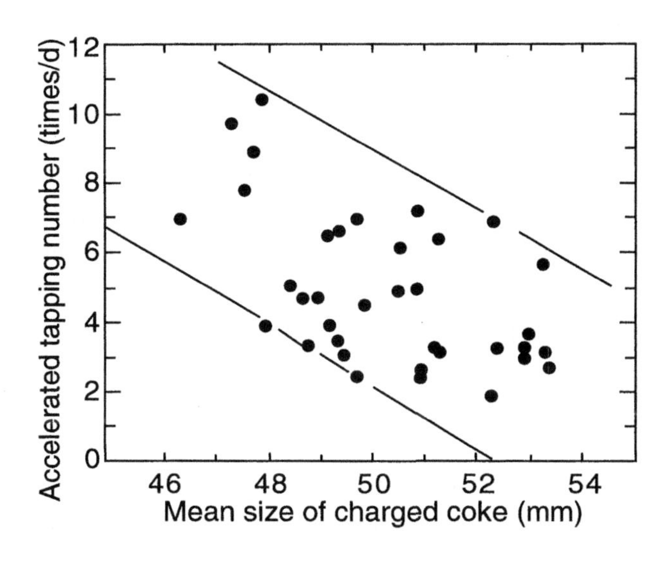

Fig.9-29 Relationship between mean size of charged coke and accelerated tapping number[12)]

space and a longer traveling distance, in addition to the above fact.

From the above description, it can be seen that coke required for high rate PCI should have the qualities of high strength and high reactivity .

Further, an examination was made on the coke reactivity (CRI) and other factors, using therefor an experimental apparatus by which heat transfer accompanied by reaction at the lower part of the blast furnace can be simulated.[37] The results have revealed that the concentration of dust in the deadman is lowered with the rise in CRI as shown in Chapter 8 (Fig.8-14). The same tendency is shown in Fig.9-23. However, as to the concentration of dust at the bosh, no correlation with CRI is seen, and thus it is controlled by another factor. The generation of fines at the bosh is not dependent on the coke properties and the result obtained shows that it is effective to thicken the coke layer in order to suppress the degradation according to the burden load, as shown in Chapter 8 (Fig.8-15). In regard to the cause of the degradation of coke at the bosh, much is expected to be found out from future research, together with determination of the factors controlling the aforementioned bulk breakage.

(7) Increase in coke size

There are few examples of analyses being made to determine the particle diameter of charging coke in the case of high rate PCI. At Kakogawa 1 BF, the amount of fine (-3mm) at the middle area of the deadman at the tuyere level has been decreased with an increase in the particle diameter of charging coke (Fig.9-28[13]) and the same effect as that obtained secured by improving DI has been obtained.

Also, as shown in Fig.9-29,[12] the increase in particle diameter of charging coke contributes also to the stability of the tapping operation. But there are lots of subjects which should be studied in the future according to the effect of particle diameter.

It is generally understood in any case, that the working rate of a coke oven decreases accompanied by high rate PCI and there is a tendency for the coke particle diameter to increase. It is hoped that future research will confirm these matters, including whether it is necessary to increase the particle diameter.

9.2.3 Enhancement of coke fines consumption

Little quantitative research on the consumption of coke fines has so far been conducted. In the model experiment or the analysis made using the operating furnace in regard to the consumption of coke fines, the effect brought about by addition of steam has been reported.

Figure9-30[32] shows the model experimental result obtained by suppressing the increase of coke fines by the addition of steam (from 6 to 60g/Nm^3) when CSR is lowered.

Figure 9-31[48] shows the analytical results obtained using the operating furnace at Kure 1 BF, which show that the amount of fines at the middle area of the deadman decreases with the addition of steam.

The fines are supposed to be consumed by; (1) gasification with CO_2 (H_2O), (2) contact with molten FeO and (3) carburization due to contact with molten metal, and thus there is a need in the future to understand the effects of these factors quantitatively.

9.2.4 Summary

(1) improving the cold and hot strengths (DI and CSR) are the measures taken at present to provide the coke with a quality suitable for high rate PCI. In addition, indirect measures are; (2) securing appropriate combustibility of pulverized coal and (3) mixing

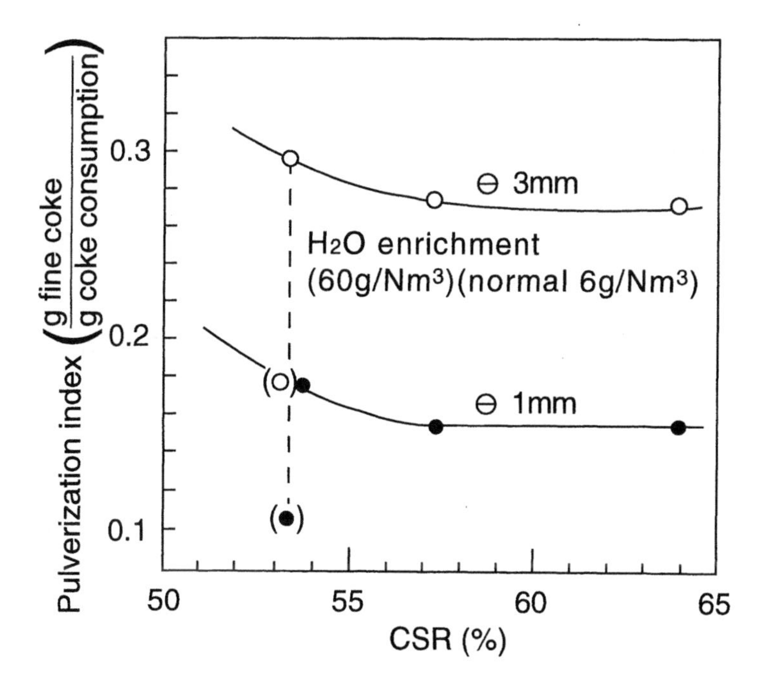

Fig.9-30 Effects of CSR and steam injection on coke fine formation in combustion zone [32]

small coke into an iron ore layer and raising reactivity of small coke.

Against such measures accompanied by the rise in producing cost of coke as improving DI and CSR, research on (4) sharp rise in reactivity (CRI) (greater than 40%) obtained by proper cold strength is going on. In the currently-available coke producing technology to this measures, there is no other way besides the formed coke. It is necessary to conduct a research further hereafter.

Again, for estimating the coke quality, there is a problem of whether DI, CSR and CRI are appropriate or not for doing this, and in regard to this, a search for a new estimating method is under way. In order to accomplish a steadily high rate PCI, it is indispensable to conduct future research on the following items.

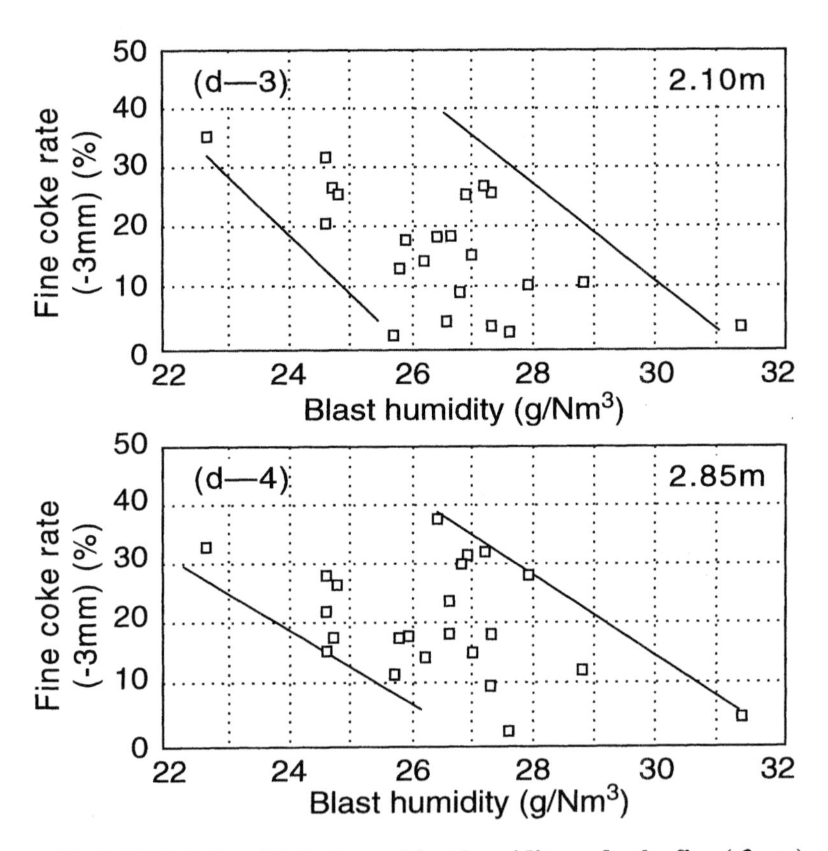

Fig.9-31 Relationship between blast humidity and coke fine (-3mm) (Kure 1 BF)[48)]

(1) Extraction of degradation controlling factors for producing coke hard to degrade at lower part of furnace and inside raceway and clarification of its control technology
(2) Presentation of new quality control index for producing hard-to-degrade coke
(3) Clarification and quantification of mechanism of how fines generated inside raceway travel to and accumulate on lower part of the furnace and the deadman, and mechanism of how fines generated at lower part of furnace travel to and accumulate on the deadman
(4) Clarification and quantification of mechanism of how fines generated inside the raceway and at lower part of the furnace are consumed.

REFERENCES

1) T.Kamijo, N.Takahashi, M.Shimizu, Y.Yoshida and R.Ito: Proc. 1st Int. Cong. Science & Tech. Ironmaking, (1994), Sendai, ISIJ, p.505.
2) K.Kakiuchi, S.Matsunaga, A.Sakamoto, H.Matsuoka, H.Ueno and K.Yamaguchi: CAMP-ISIJ, 7(1994), p.126.
3) T.Okuda, H.Nakaya, R.Ono, J.Yamagata, T.Goto, R.Ito and M.Mizuguchi: CAMP-I SIJ, 6(1993), p.111.
4) K.Mori, N.Takagaki, H.Inoue, H.Mitsufuji, S.Kishimoto and A.Sakai: CAMP-ISIJ, 8(1995), p.319.
5) K.H.Peters, H.Mohnkern and H.B.Lungen : Proc. 1st Int. Cong. Science & Tech. Ironmaking, (1994), Sendai, ISIJ, p.493.
6) Y.Yoshida, S.Kitayama, S.Ishiwaki, M.Isobe and H.Miyagawa: CAMP-ISIJ, 6(1993), p.30.
7) M.Naito, Y.Hosotani, K.Yamaguchi, M.Higuchi, Y.Inoue and T.Haga: CAMP-ISIJ, 9(1996), p.627.
8) T.Sato, K.Takeda and H.Itaya: CAMP-ISIJ, 9(1996), p.620.
9) N.Kon-no, H.Shiota, N.Takamatsu, Y.Fujiwara and M.Naito: CAMP-ISIJ, 6(1993), p.18.
10) T.Sugiyama, M.Naito, S.Matsuzaki, H.Kumaoka and T.Nakayama: CAMP-ISIJ, 7(1994), p.46.
11) S.Amano, S.Matsunaga, K.Kakiuchi, H.Ueno, N.Kon-no and K.Yamaguchi: JSPS 54th Committee , No.54-2025, (1994).
12) Y.Yukubo: Private letter.

13) K.Kadoguchi, T.Goto, R.Ito, T.Yabata and M.Shimizu: R & D Kobe Steel Engineering Reports, 46, 1(1996), p.2.
14) A.Maki, A.Sakai, N.Takagaki, K.Mori, T.Ariyama, M.Sato and R.Murai: ISIJ Int., 36(1996), p.650.
15) T.Umezaki, T.Murai, K.Sato, M.Onishi, Y.Hatano and A.Ogawa: CAMP-ISIJ, 9(1996), p.735.
16) K.Katayama, H.Ito, M.Matsumura, A.Koike and M.Hoshi: CAMP-ISIJ, 10(1997), p.873.
17) M.Naito: CAMP-ISIJ, 10(1997), p.743.
18) K.Yamaguchi, K.Higuchi and Y.Hosotani: CAMP-ISIJ, 10 (1997), p.946.
19) A.Tayama, Y.Shimomura, K.Kushima, T.Nakata and K.Fujita: Proc. 39th Ironmaking Conf., Washington, AIME, (1980), p.390.
20) A.Mochizuki, T.Murai, Y.Kawaguchi and Y.Iwanaga: Tetsu-to-Hagane, 72(1986), p.1855.
21) M.Hoshi and T.Kawaguchi: CAMP-ISIJ, 9(1996), p.813.
22) T.Watanabe, A.Maki, S,Hamaya, H.Sato, N.Sakamoto, H.Noda and K.Ichikawa: CAMP-ISIJ, 10(1997), p.193.
23) S.Nagami, T.Murai, Y.Shimoda, S.Komatsu and T.Inada: CAMP-ISIJ, 6(1993), p.26.
24) J.Kiguchi, M.Shimizu, R.Ito and G.Hoshino: CAMP-ISIJ, 9 (1996), p.639.
25) K.Mori, A.Maki, A.Shimomura, H.Hayasaka, K.Oya, H.Noda, K.Ichikawa and N.Sakamoto: CAMP-ISIJ, 10(1997), p.874.
26) T.Inada, S.Uejo, K.Katayama and T.Yamamoto: CAMP-ISIJ, 9 (1996), p.22.
27) T.Kamijo, M.Shimizu, R.Ito and G.Hoshino: CAMP-ISIJ, 9 (1996), p.26.
28) Y.Abe, K.Okuda, K.Yamaguchi, K.Yamaguchi, N.Ishioka, T.Furukawa, Y.Shimomura and K.Ono: JSPS 54th Committee, No.54-1701,(1984).
29) A.Sato, M.Arizuka, J.Yamagata, H.Miyagawa, T.Goto and M.Mizuguchi: CAMP-ISIJ, 5(1992), p.1064.
30) H.Fukuyo, M.Hattori, K.Ishii, A.Maki, N.Sakamoto and T.Sumigama: CAMP-ISIJ, 6(1993), p.14.
31) Y.Hatano, T.Umezaki, S.Muramatsu, T.Kawaguchi and M.Hoshi: CAMP-ISIJ, 9(1996), p.631.
32) Y.Ishikawa, M.Kase, Y.Abe, K.Ono, M.Sugata and T.Nishi: Proc. 42nd Ironmaking Conf., Atlanta, AIME, (1983), p.357.
33) I.Kurashige, H.Nakamura, M.Hatano, Y.Iwanaga, H.Nomiyama, M.Kojima and

Y.Aminaga: JSPS 54th Committee, No.54-1689, (1984).

34) M.Ichida, K.Kunitomo, Y.Fujiwara, H.Kamiyama and Y.Morizane: CAMP-ISIJ, 6(1993), p.860.

35) S.Kitano and T.Kamijo: Private letter.

36) Y.Hatano and T.Inada: Private letter.

37) K.Yamaguchi, T.Uno, T.Yamamoto, H.Ueno, N.Kon-no and S.Matsuzaki: Tetsu-to-hagane, 82(1996), p.641.

38) H.Shimizu, K.Sato, M.Kojima, Y.Aminaga, H.Nakamura and Y.Iwanaga: Tetsu-to-hagane, 72(1986), p.195.

39) K.Yamaguchi, T.Sugiyama, T.Yamamoto, H.Ueno and S.Matsunaga: CAMP-ISIJ, 8(1995), p.2.

40) M.Shimizu, A.Kasai, T.Kamijo, H.Iwakiri, R.Ito and M.Atushi: JSPS 54th Committee, No.54-2037 (1995).

41) A.Sakai and M.Sato: Private letter.

42) S.Amano: Private letter.

43) A.Shimomura and M.Sato: Private letter.

44) Y.Tomita: Private letter.

45) K.Takeda and H.Kokubu: Private letter.

46) A.Kasai, K.Miyagawa, T.Kamijo, J.Kiguchi and M.Shimizu: Tetsu-to-Hagane, 83(1997), p.239.

47) M.Shimizu, R.Ito and G.Hoshino: CAMP-ISIJ, 8(1995), p.6.

48) K.Tanaka: Private letter.Key word: ironmaking; blast furnace process; pulverized coal injection; unburnt char; sinter; reducibility; reduction degradation; softening and melt-down properties; coke; degradation; fine generation; reactivity.

CHAPTER 10

Upper limit of PCR

Pulverized coal (PC) is injected at high rates into the blast furnaces to reduce the fuel cost and extend the life of coke ovens. When the pulverized coal rate (PCR) exceeded 180 kg/thm, however, most of recent blast furnaces have become unstable, and their fuel rate has risen to 500 kg/thm or more. In this chapter, the overall coke-to-pulverized coal replacement ratio was evaluated, the conditions that control the phenomena taking place in the blast furnace injected with high rate PC were clarified, and the maximum possible rate of pulverized coal injection (PCI) was discussed.

10.1 Replacement ratio of coke to PC

Estimating the coke-to-pulverized coal replacement ratio accurately would be important in optimizing the heat balance of the blast furnace(BF), preventing the fluctuation of heat level of furnace from the change of coal type, and evaluating the production cost of the blast furnace. It is a matter of common knowledge that the coal having lower calorific value has a large effect on the replacement ratio.[1] The actual replacement ratio in actual blast furnaces was investigated, the effects of coal types and blast furnace operating conditions on the replacement ratio were discussed, and the replacement ratio was evaluated by studying the pulverized coal consumption mechanism using a design of operation model of BF.

10.1.1 Actual replacement ratio of blast furnace

The practical coke-to-pulverized coal replacement ratios of the blast furnaces of the Japanese steel makers are shown in Fig. 10-1.[1][2] When the coke rate(CR) is corrected for the blast furnace operating conditions, the overall replacement ratio ΔCR/ΔPCR is about 0.85. The actual replacement ratio varies with the coal type and PCR. At a PCR of 170 kg/thm, the actual replacement ratio is 0.85 (no correction) for NKK, 0.8 to 0.85 (after correction) for Nippon Steel, 0.95 to 1.00 (without correction) and 0.7 to 0.9 (after correction for Sumitomo Metal Industries, 0.9 (without correction) for Kawasaki Steel, 0.95 to 1.0 (without correction) for Kobe Steel, 0.98 (without correction) and 0.85 (after correction) for Nisshin Steel, and 0.93 for Nakayama Steel Works.

In the example of Company A, the correction factors for coke rate are 0.1kg/thm per blast temperature °C, 0.8 kg/thm per the blast humidity g/Nm^3, 5.0 kg/thm per coke ash content %, 0.15 kg/thm per slag volume kg/thm, 0.3 kg/thm per silicon content in hot metal %, -1.0 kg/thm per sinter ratio %, -0.5 kg/thm per pellet ratio %, and 0.7 kg/thm per sinter FeO content %.

10.1.2 Effects of PC properties and operation conditions on replacement ratio

The replacement ratio is considered to be affected significantly by pulverized coal properties (e.g., volatile matter, total carbon, and calorific value). The operation condition of BF also could be changed by the replacement ratio to a great extent. Then, the relationship between the pulverized coal properties and operation condition of BF were investigated in actual BF.

(1) Coal type and component

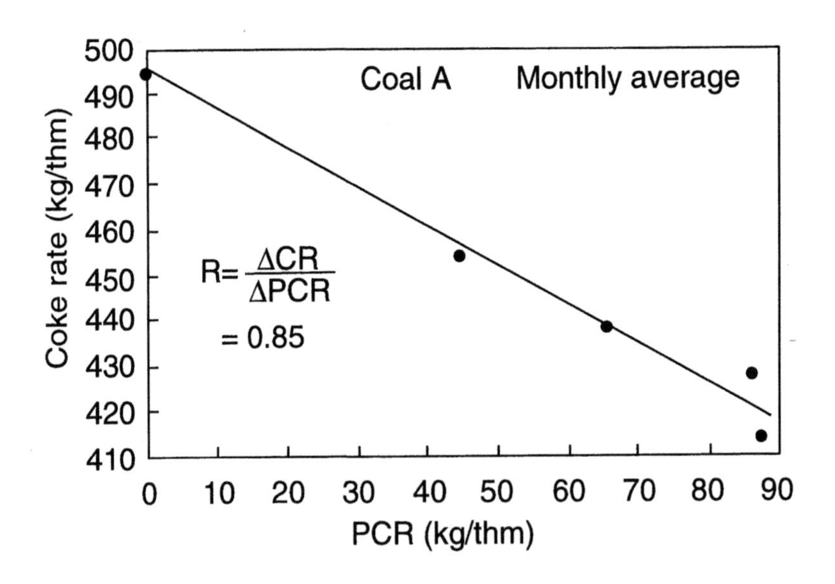

Fig.10-1 Relationship between CR and PCR[1)]

The effects of the lower calorific value and volatile matter (total carbon) content of pulverized coal on the actual replacement ratio are shown in Fig. 10-2a) [2] b) [3]. The actual replacement ratio is strongly correlated with these pulverized coal properties. This means that the replacement ratio mostly depends on the component of pulverized coal when the BF condition is stable.

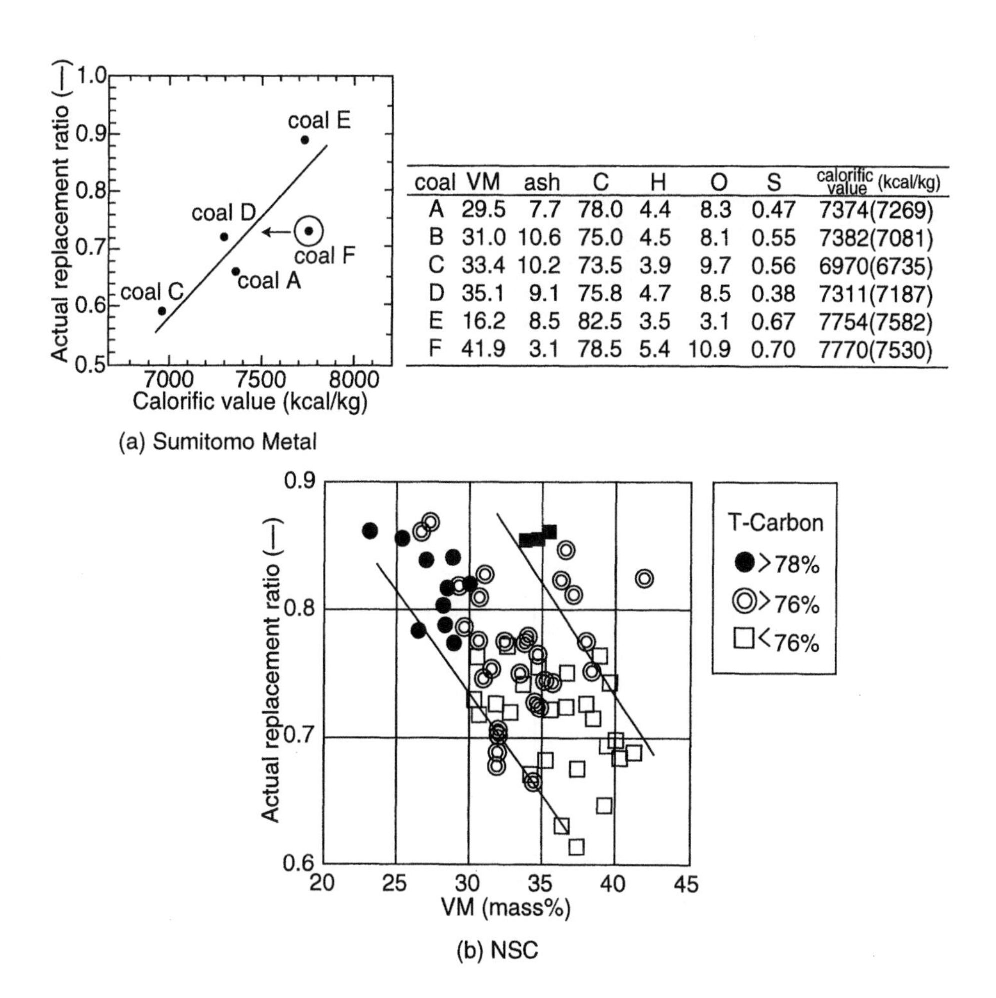

coal	VM	ash	C	H	O	S	calorific value (kcal/kg)
A	29.5	7.7	78.0	4.4	8.3	0.47	7374(7269)
B	31.0	10.6	75.0	4.5	8.1	0.55	7382(7081)
C	33.4	10.2	73.5	3.9	9.7	0.56	6970(6735)
D	35.1	9.1	75.8	4.7	8.5	0.38	7311(7187)
E	16.2	8.5	82.5	3.5	3.1	0.67	7754(7582)
F	41.9	3.1	78.5	5.4	10.9	0.70	7770(7530)

Fig.10-2 Relationship among calorific value, VM and coke replacement ratio[2,3]

(2) PCR

The replacement ratio decreases with increasing pulverized coal rate as shown in Fig. 10-3a) [4)]b) [3)]c) [5)]. This decreasing replacement ratio may be generally explained as follows. As the pulverized coal rate (PCR) increases, the combustibility of pulverized coal decreases in order to the decrease of the stoichiometric oxygen ratio. The decrease in the heat flux ratio increases the sensible heat of top gas, the heat loss and the fuel rate, which in turn reduces the replacement ratio. As the ore/coke (O/C) ratio increases, the burden distribution becomes inappropriate and lowers the shaft efficiency. For example, in the

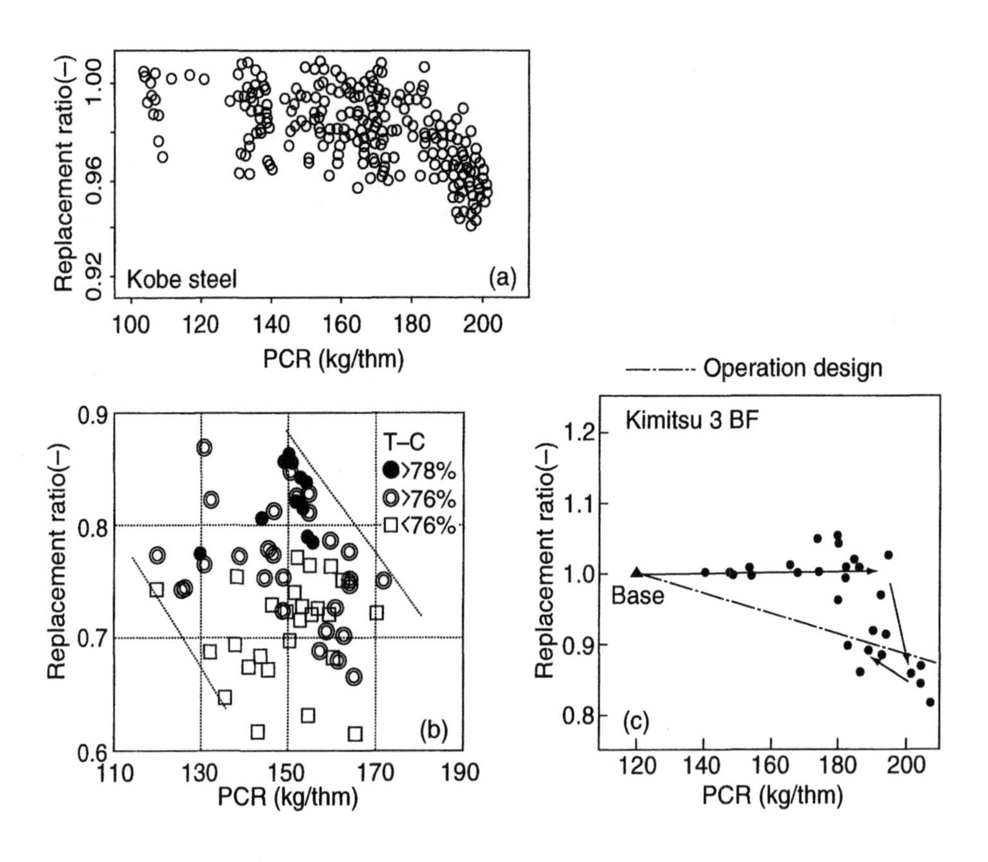

Fig.10-3 Relationship between PCR and coke replacement ratio[3,4,5)]

case of Kimitsu No. 2 blast furance[5] in Nippon Steel, the actual replacement ratio was maintained about 1.0 until 190kg/thm. However, the actual replacement ratio fell off when PCR was beyond 190 kg/thm. This phenomenon would be caused by the aggravation of the burden descent and the rapid drop of the shaft efficiency.

(3) Stoichiometric oxygen ratio (Excess air ratio)

When the stoichiometric oxygen ratio is defined as the ratio of the amount of input oxygen to required oxygen for complete combustion of pulverized coal, it decreases with increasing PCR as shown in Fig. 10-4 [3], when the oxygen concentration or the fuel rate was low. A low stoichiometric oxygen ratio reduces the combustibility of pulverized

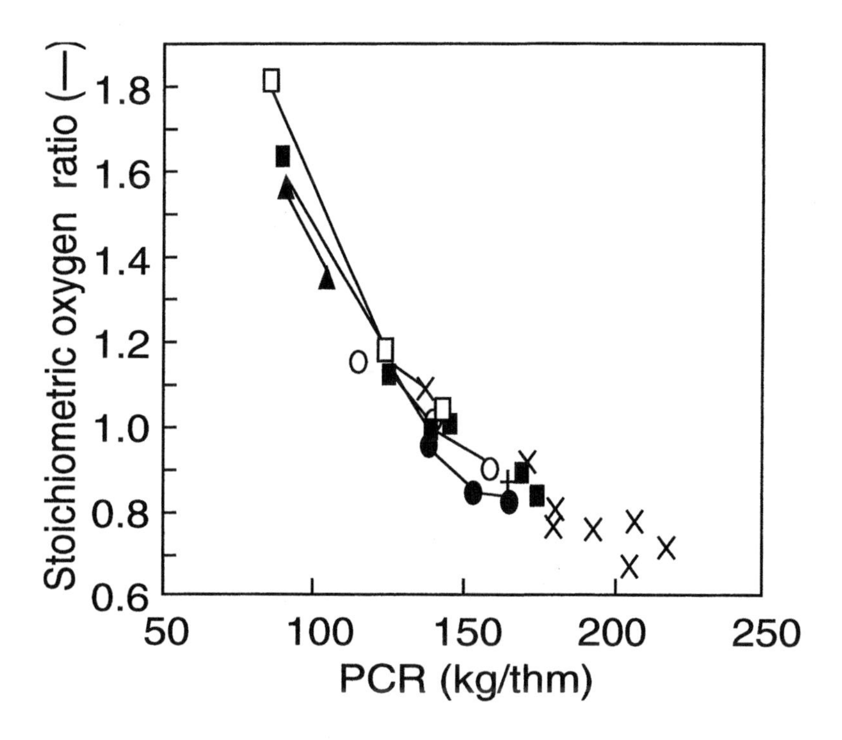

Fig.10-4 Relationship between PCR and stoichiometric oxygen ratio[3]

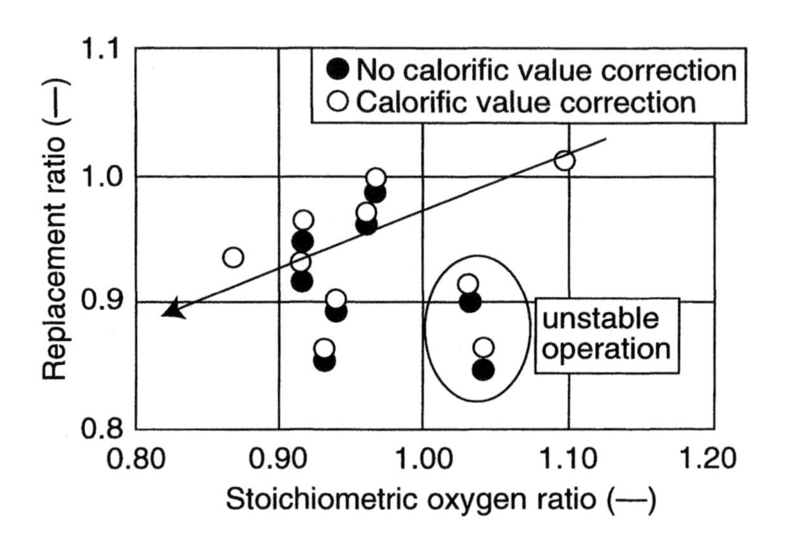

Fig.10-5 Relationship between stoichiometric oxygen ratio and coke replacement ratio[3)]

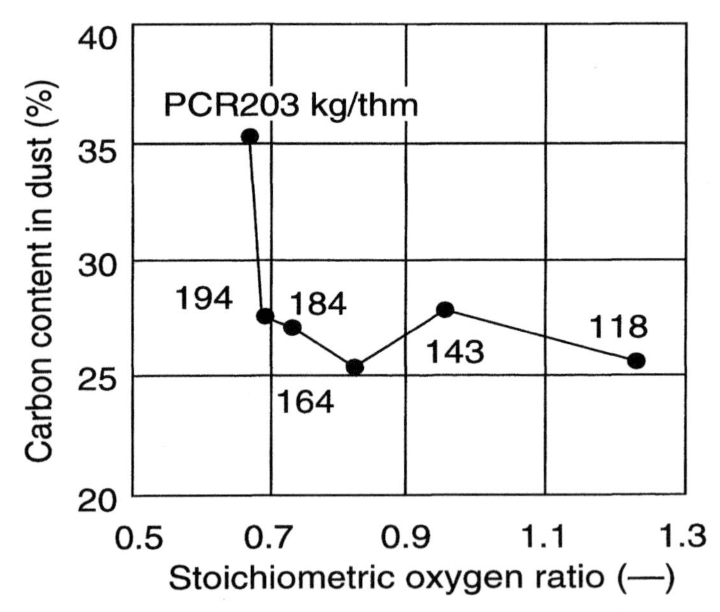

Fig.10-6 Relationship between Sstoichiometric oxygen ratio and percent carbon in dust[5)]

coal and lowers the replacement ratio as shown in Fig. 10-5 [3]. According to actual operating results, the minimum critical stoichiometric oxygen ratio was about 0.70 to 0.75 at which the blast furnace condition began to be unstable. In this case, the carbon content of dust and the unburnt char from the furnace top increased significantly, as shown in Fig. 10-6 [5].

(4) Theoretical flame temperature

The high rate PCI operation in a blast furnace is limited by the available blast temperature and oxygen enrichment capacities. The theoretical flame temperature(Tf,°C) at tuyere nose generally tends to decrease with increasing PCR as shown in Fig. 10-7 [6]. The actual replacement ratio decreases with decreasing theoretical flame temperature as shown in Fig. 10-8 [4]. This is probably attributable to the decreased combustibility of pulverized coal. Other factors may also be involved.

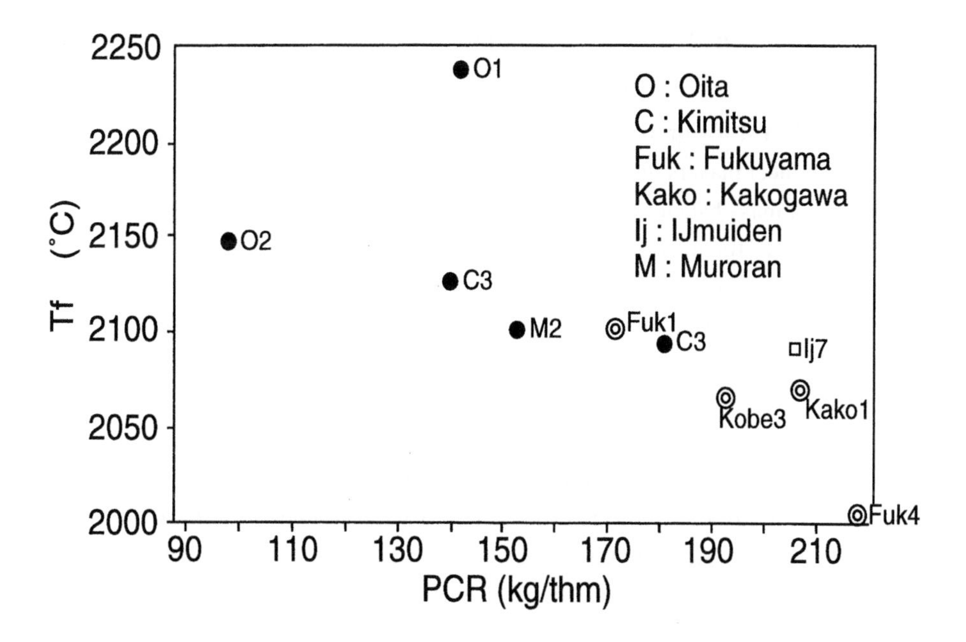

Fig.10-7 Relationship between PCR and theoretical flame temperature Tf [4].

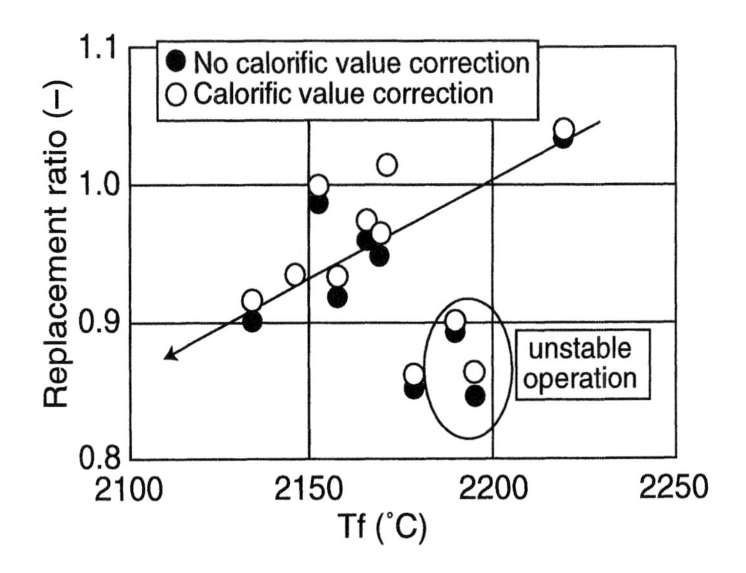

Fig.10-8 Relationship between flame temperature and coke replacement ratio[4].

10.1.3 Theoretical study on replacement ratio

In Japanese steel makers, using the calculation of the theoretical coke rate from the overall heat and mass balances, the theoretical coke replacement ratio is determined. Then, the effects of the pulverized coal injection, coal type and other factors on the operation condition of BF are predicted using those results. A steel company conducted a study based on the coke consumption mechanism throughout the blast furnace and tried to develop equations to estimate the replacement ratio of coke by pulverized coal injection.

(1) Estimation from total heat and mass balance

The effects of carbon, hydrogen, oxygen and volatile matter in the PC on the replacement ratio were studied. Assuming a shaft efficiency of 97% and a heat loss of 1.17GJ/thm, various components and calorific values of PC were investigated their effects on the replacement ratio at a pulverized coal rate of 170 kg/thm. The results are shown in Figs. 10-9 [4]and 10-10 [4], and might be summarized as follows:

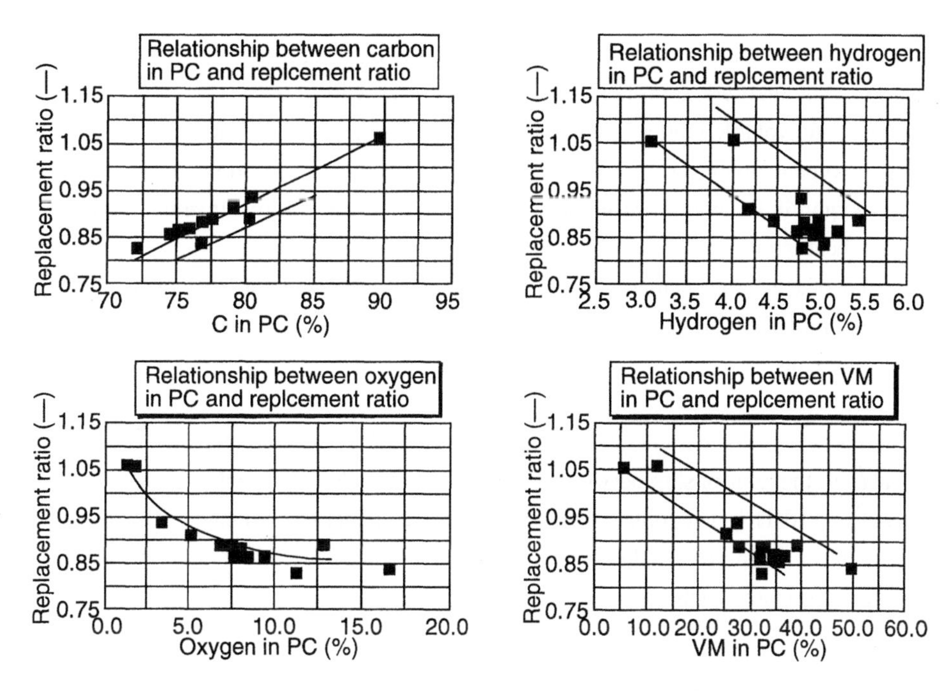

Fig.10-9 Relationship between chemical component in PC and coke replacement ratio[4].

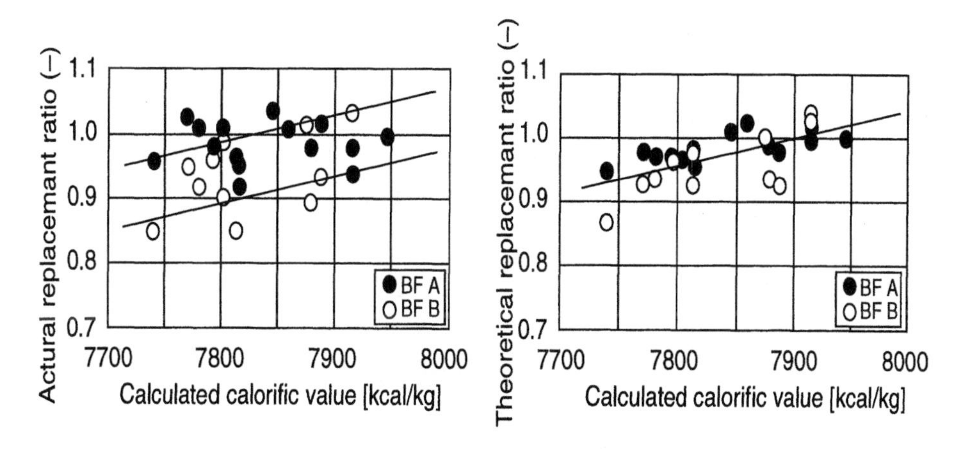

Fig.10-10 Relationship between calorie in PC and actual theoretical coke replacement ratio[4].

1) The theoretical replacement ratio can be practically represented by the carbon content of PC. The theoretical replacement ratio increased with the decrease of hydrogen, oxygen and volatile matter as a result of the increase of carbon content.
2) High oxygen content in coal decrease in the replacement ratio. It is considered that the oxygen in PC was working as cold oxygen in air, which result in the relatively lower heat input.
3) The plot of actual replacement ratio v.s. the calculated calorific value has large scattering, although the theoretical replacement ratio has good correlation with the calculated calorific value.

Similarly, the blast parameters until 200 kg/thm of PC were studied using three types of coals and the relationship between the PCR and the calculated replacement ratio are given in Fig. 10-11[2]. The replacement ratio decreased under the influence of the oxygen enrichment in the increasing PCR. Also the calculated calorific value and replacement ratio were correlated each other (refer to Fig. 10-2.).

Blast parameter

PCR(kg/thm)	0	50	100	150	200
Blast temperature(°C)	1100	1150	1200	1200	1200
Blast moisture(g/m3)	40	30	30	30	30
Oxygen enrichment(%)	0	0	2	4	6

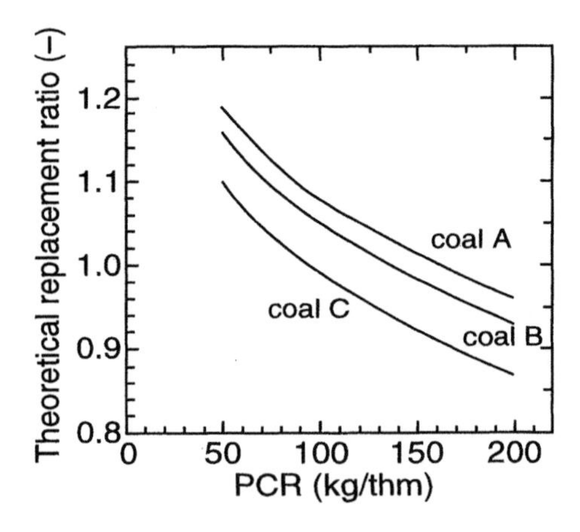

Fig.10-11 Relationship between PCR, type of coal and theoretical coke replacement ratio[2].

(2) Estimation considering carbon consumption

Murai et al.[1)] has presented the formulation of the replacement ratio based on the consumption mechanism of carbon in the BF. From their results, although the good correlation between the low calorific value in coal and actual replacement ratio was obtained, Some improvement of the formulation would be necessary. The difference from the conventional equation is to assume that the reaction of C and H in coal at combustion zone is defined until CO and H_2 formation. Furthermore, the heat and mass balances involved in FeO gas reduction, direct reduction, and hydrogen reduction are also taken into account as shown in Table.10-1[1)] and Fig.10-12.[1)]

The oxygen and heat balance equations were developed, and the equation of replacement ratio were tried to make up. The replacement ratio estimated by this equation was found to agree with the actual replacement ratio up to 100 kg/thm, as shown in Fig. 10-13[1)]. The deviation of the estimated replacement ratio from the actual replacement ratio increased as the PCR exceeded 150 kg/thm and the error was reached about 20% at 200 kg/thm(Fig. 10-14[1)]).

Furthermore, taking into account the heat loss in the high rate PCI, The modification of the heat balance equation was made by adding a heat loss term corresponding to the PCR. This correction reduced the deviation of the estimated replacement ratio from the actual replacement ratio to 10% as shown in Fig. 10-15[1)]. For future work, the modification

Table 10-1 Comparison of conventional and developed method for estimation of replacement ratio[1)].

	Conventional method Lower calorie	Developed method Effective calorie
Carbon in fuel	$C + O_2 = CO_2$ (complete combustion)	① Combustion to CO ② Indirect reduction ③ Direct reduction
Hydrogen in fuel	$H_2 + 1/2\ O_2 = H_2O$ (complete combustion)	Hydrogen reduction

Reduction reaction	Reduction efficiency	Calorific value
$FeO+CO=Fe+CO_2$	$\eta_{CO} \cdot Y$	Hi
$FeO+C=Fe+CO$	Direct reduction C:Cs	Hs
$FeO+H_2=Fe+H_2O$	$\eta_{H_2} \cdot Y$	Hh

$$C + \frac{1}{2}O_2 = CO$$

Heat of combustion to CO : Hc

Fig.10-12 Reactions taking into account in BF[1].

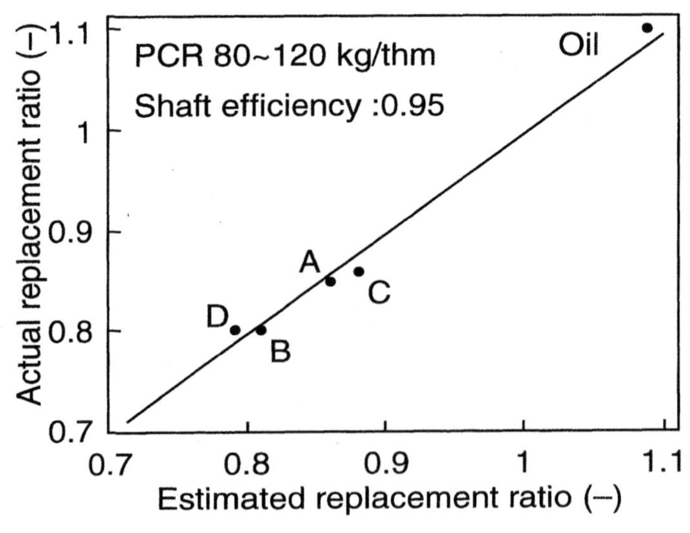

Fig.10-13 Relationship between estimated replacement ratio and actual value[1].

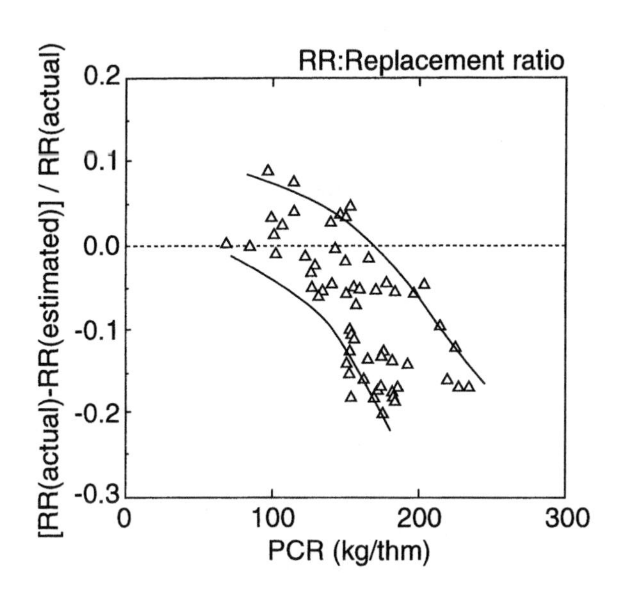

Fig.10-14 Deviation of estimated replacement ratio from actual in terms of PCR[1]

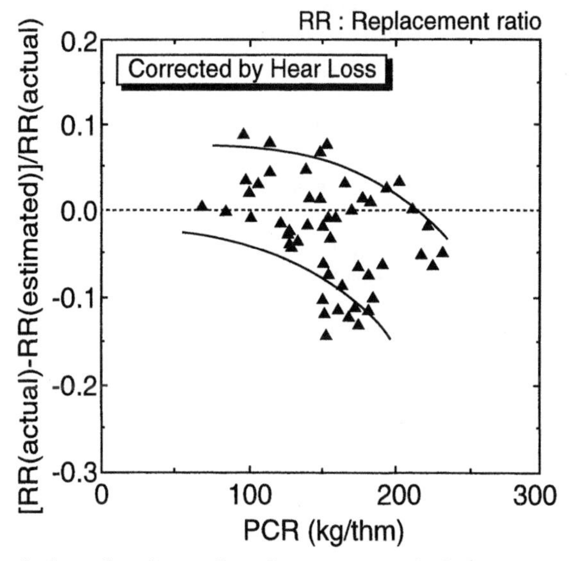

Fig.10-15 Deviation of estimated replacement ratio from actual in terms of PCR[1]

will be carried out to taking into account the decrease of heat flux ratio and the temperature of reduction equilibrium in addition of the unburnt char at high rate PCI.

10.1.4 Subjects for future study

The countermeasures for improving the coke-to-pulverized coal replacement ratio would contribute to the stabilization of blast furnace operation. The following procedures will be considered;

1) To improve the combustibility of pulverized coal, it is necessary to develop burners to

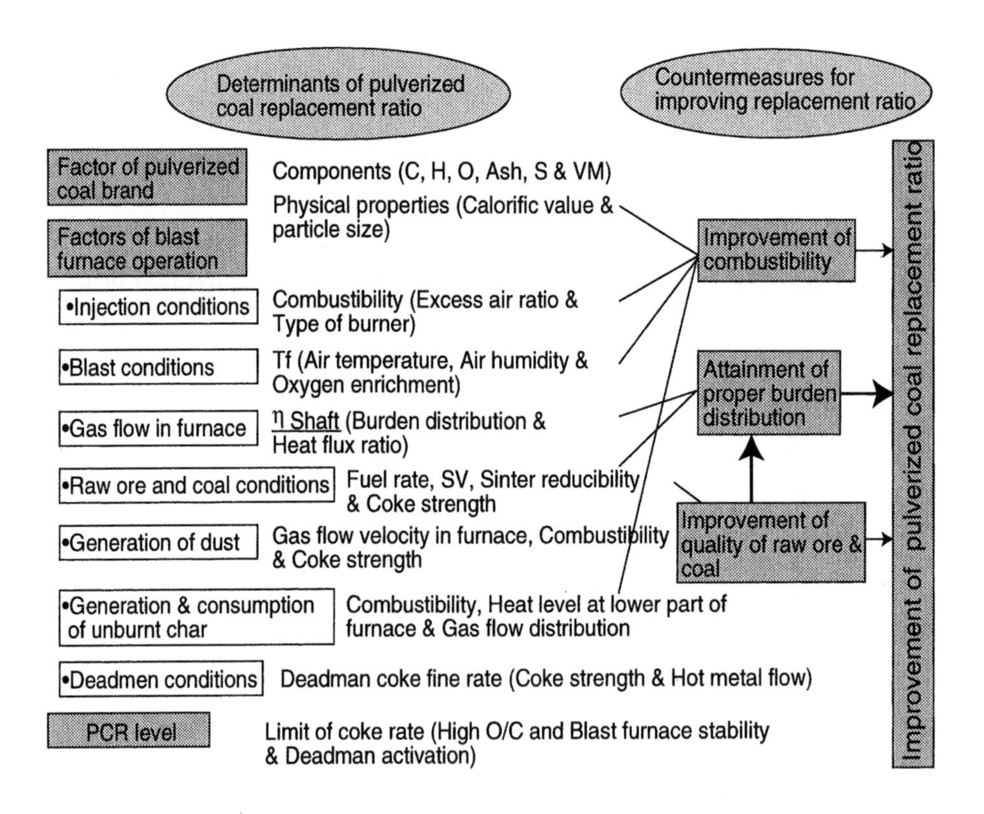

Fig.10-16 Summery about countermeasure of coke replacement ratio[3)]

be suitable for a lower stoichiometric oxygen ratio and to determine the blast condition such as a temperature at tuyere nose in multi-injection system. At the same time, it is unavoidable to evaluate the consumption mechanism of unburnt char to prevent the degradation of coke.

2) The improvement of the control of burden distribution is most important problem to accomplish the stable operation of blast furnace in a high rate PCI. As mentioned above, the stable burden distribution and improvement of η_{CO} at a high ore/coke ratio greatly affect the replacement ratio.

3) To improve the replacement ratio at a high rate PCI, it is also important to improve the reducibility and meltdown properties of ore to control a higher ore/coke ratio in the periphery together with maintaining the center flow of gas.

4) Research on coke is going on in terms of the quality of coke related to the degree of solution loss at low coke rate and the degradation in the raceway.

These countermeasures are summarized in Fig. 10-16[3).

10.2 Upper limit of PCR

The purpose of high rate PCI is to reduce the fuel cost by saving coke and to extend the life of coke oven by reducing the coke oven load. The operating results of blast furnaces injected with 200 kg/thm or more has been reported that the fuel rate increased up to 500 kg/thm in most of the blast furnaces, when PCR exceeded 180 kg/thm. The coke rate could not decrease in order to high fuel rate and the productivity dropped, and the advantage of PCI was halved. It is thus necessary to increase PCI without substantially raising the fuel rate.

From this standpoint, both of the high rate PCI and the low coke rate should be maintained, and it would be important to understand the limit of the replacement ratio, coke rate and PCI. That is, the maximum replacement ratio of coke by pulverized coal and the minimum coke rate in terms of the maximum PCR should be investigated. The following four conditions were addressed based on the past results of studies:

1) Limit of lowest stoichiometric oxygen ratio (combustion efficiency of coal in tuyere)
2) Limit of PCI in terms of limit of unburnt char consumption
3) Restriction of PCI from the viewpoint of the degradation of coke and the accumulation of fine coke in the lower part of furnace and the deadman.
4) Restriction of the profile of burden distribution (e.g., lower limit of coke layer thickness

in belly)

There would be other restrictions, such as gas flow distribution (limit of gas channeling) in the blast furnace, the molten slag flow in the lower part of the furnace and the local increase in the heat flux ratio near the wall. No quantitative relationships between those and PCR are clarified and remain as future subjects of study.

10.2.1 PCR based on minimum stoichiometric oxygen ratio

Okuno et al. [7] calculated the effect of the stoichiometric oxygen ratio (μ) on the pulverized coal rate and coke rate from the operating data of the Kimitsu No. 3 blast furnace at high rate PCI, using the model of blast furnace operation design. The calculated

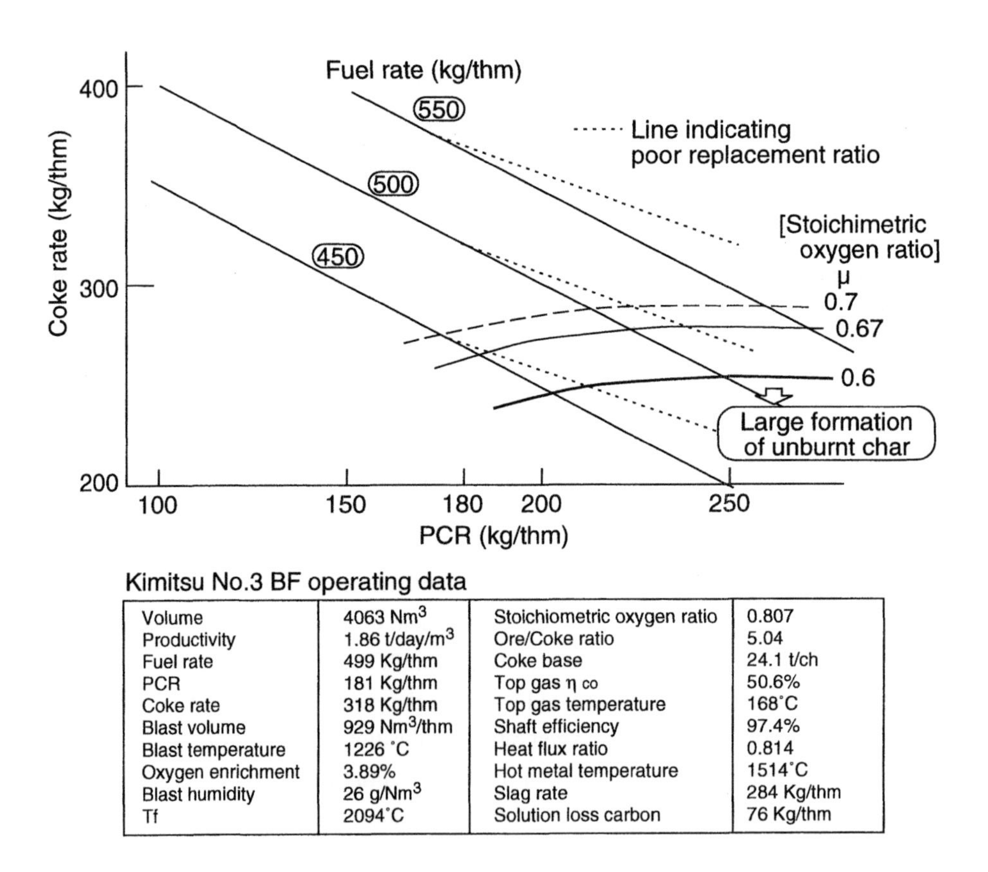

Kimitsu No.3 BF operating data

Volume	4063 Nm^3	Stoichiometric oxygen ratio	0.807
Productivity	1.86 t/day/m^3	Ore/Coke ratio	5.04
Fuel rate	499 Kg/thm	Coke base	24.1 t/ch
PCR	181 Kg/thm	Top gas η_{co}	50.6%
Coke rate	318 Kg/thm	Top gas temperature	168°C
Blast volume	929 Nm^3/thm	Shaft efficiency	97.4%
Blast temperature	1226 °C	Heat flux ratio	0.814
Oxygen enrichment	3.89%	Hot metal temperature	1514°C
Blast humidity	26 g/Nm^3	Slag rate	284 Kg/thm
Tf	2094°C	Solution loss carbon	76 Kg/thm

Fig.10-17 Relationship between PCR and CR, Stoichiometric oxygen ratio[7]

results are shown in Fig. 10-17 [7].

The lower limit of the stoichiometric oxygen ratio (μ) was reported to be 0.7 (PCR = 203 kg/thm) for the Kimitsu No. 3 blast furnace and 0.67 (PCR = 206 kg/thm) for the IJmuiden No. 7 furnace. In the case of the Kimitsu No. 3 blast furnace, the carbon content of dust and unburnt char in the top gas suddenly increased when the stoichiometric oxygen ratio was 0.7 or less, as described in section 10.1.2.(3). As the PCR was over 203 kg/thm and the stoichiometric oxygen ratio fell below 0.7, the carbon content in dust and unburnt char steeply increased. If the stoichiometric oxygen ratio of 0.67 in the IJmuiden No. 7 blast furnace could be taken as the lower limit of the successful operation of blast furnaces, the minimum coke rate would be in the range from 270 to 280 kg/thm, and the minimum PCR from 180 to 280 kg/thm.

The minimum coke rate, which vary with the fuel rate and the stoichiometric oxygen ratio, would be from 293 to 253 kg/thm in Fig.10-17. The recent high rate PCI operation in Japan was achieved by moving to the operation of high fuel rate. This result indicates that the stoichiometric oxygen ratio is a main factor on the restriction of high rate PCI. In any case, an stoichiometric oxygen ratio of less than unity means that oxygen is insufficient for the complete combustion of pulverized coal. Inevitably, the unburnt of pulverized coal generated must be consumed by the reaction with the molten FeO or gasification reaction with CO_2. Therefore, in order to make possible to further low stoichiometric oxygen ratio, it is necessary to increase in the combustion efficiency of pulverized coal at the nose of the tuyere and in the reaction efficiency between pulverized coal and molten FeO, so that the voidage of coke layer should be secured on the deadman surface. This means that the quality of coke is important in terms of the stoichiometric oxygen ratio as well.

10.2.2 Limit of PCR based on maximum consumption of unburnt char

The consumption of unburnt char by solution loss reaction was calculated from the assumption of balance of reactions concerned, and the consumption limit of unburnt char was estimated, when the upper limit of consumption was the amount of PCI. The lower limit of coke rate was calculated from the amount of carbon solution reaction regarding to the quality of coke used.

(1) Estimation of consumption rate of unburnt char

Sato et al. [9] reported that the PCR can be expressed as the sum of the combustion in

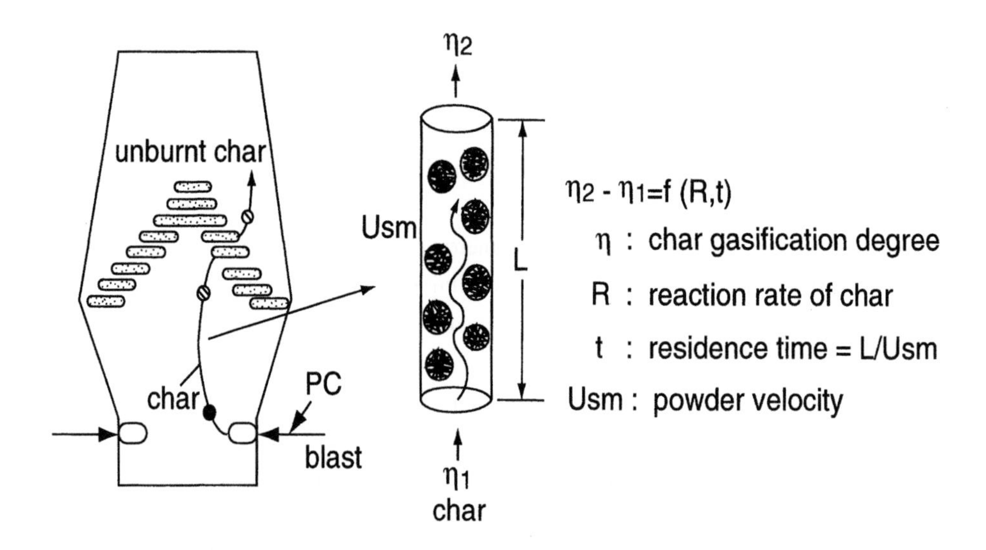

Fig.10-18 Calculation model for behavior of unburnt char in BF[9).

the raceway and the consumption of unburnt char in the blast furnace. The consumption of unburnt char is restricted by the mass balance of carbon in the blast furnace. For a high rate PCI, it is necessary to ensure high efficiency of combustion of PC so as to inhibit the fine generation of char not to exceed the limit of consumption of unburnt char in the blast furnace.

It was known that the unburnt char was consumed by the solution loss reaction preferentially in the furnace. It is important to clarify the mechanism of consumption of char in presence of coke (both of the fine coke and the coke in packed bed). Then, the investigation of the residual time of char in the furnace and the reaction kinetics of carbon solution on the limit of consumption of char was carried out, the upper limit of PCR and combustion efficiency to clarify the consumption limit of char was discussed. At first, experiments were carried out to obtain the mean powder velocity in the upward stream for the rate of reaction of unburnt char in the furnace. Subsequently, the upper limit of PCR was evaluated in terms of consumption limit of unburnt char in the blast furnace.

The transfer and consumption behavior of unburnt char in blast furnace are schematically shown in Fig. 10-18[9)].

To make an appropriate evaluation of the reaction degree(char gasification degree),

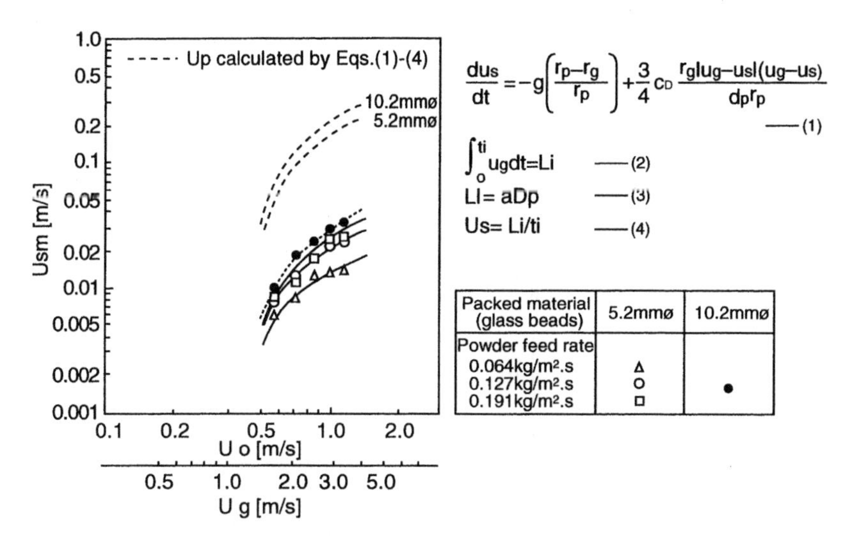

Fig.10-19 Velocity of powder obtained from residence time distribution in the packed bed[8].

η of unburnt char in the blast furnace, it is necessary to know the reaction rate R and residence time t, or average powder velocity Usm, accurately.

A cold-model experiment was carried out to formulate the mean powder velocity Usm of unburnt char. The results are given in Fig. 10-19[8]. From Fig. 10-19, the average ascending velocity of powder increases with increasing superficial velocity of gas. The average ascending velocity also increases with increasing feed rate of powder and the size of packed material. The ascending velocity Up obtained by a collision model is shown in Fig. 10-19 by the broken lines. The obtained ascending velocity is higher by an order of magnitude than that obtained from the residence time in the packed bed.

Furthermore, the behavior of fine in the packed bed was directly observed, and a model shown in Fig. 10-20[9] that involves the repeating an acceleration and a stop of powder caused by collision was developed. The model assumes that the mean velocity Usm of powder in the packed bed is determined by the both effect of the average local velocity of powder Us and the residence time td/L of powder in a stagnant area per unit length. This relationship is formulated as 1/Us + td/L = 1/Usm. From Fig. 10-21[8], the stagnant time plays a dominant role in determining the entire residence time. The 1/Us and td/L ratios are linearly correlated with the solid/gas ratio, as shown in Fig. 10-22[8]

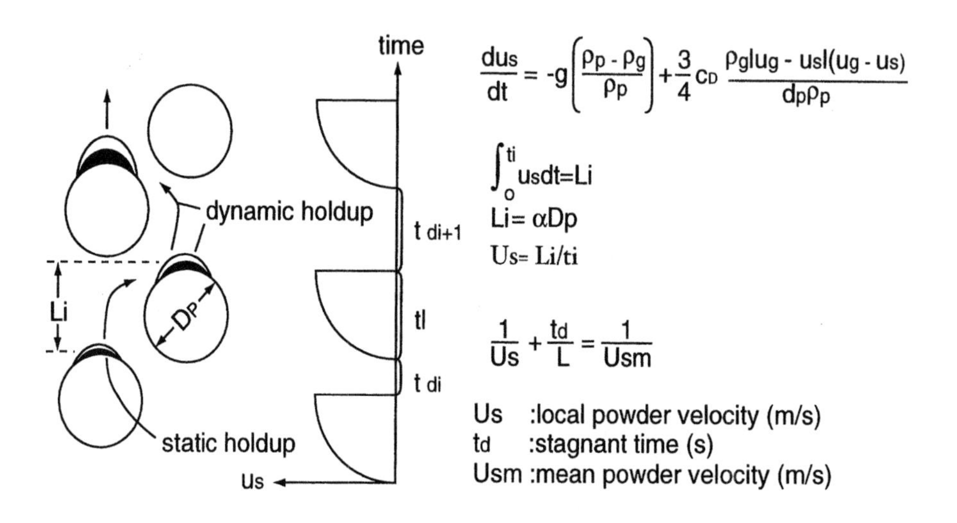

Fig.10-20 Powder behavior model in a packed bed[9).

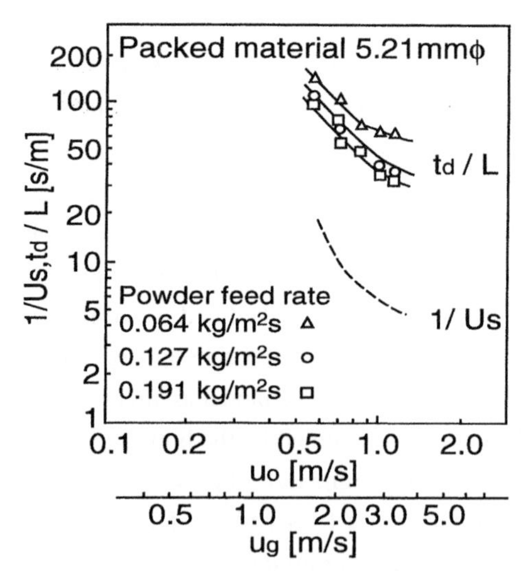

Fig.10-21 Comparison between stagnant time and ascending time of powder in the packed bed[9).

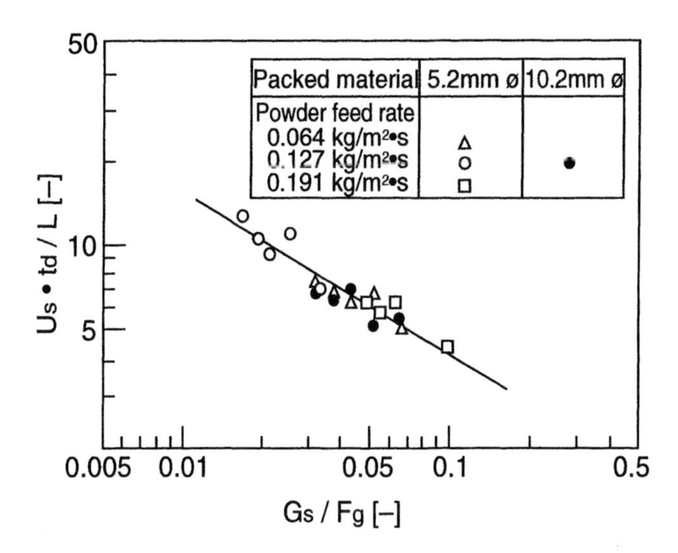

Fig.10-22 Influence of solid-gas ratio on velocity of powder and stagnant time[9].

and the empirical equation, Us*(td/L) = -4.39 -8.64*log(Gs/Fg) was obtained.

The mean velocity of ascending particles in the packed bed of coke in the lower part of the blast furnace was calculated by applying the above-mentioned equation to unburnt char and fine coke. The results are shown in Fig. 10-23[9]. As can be seen from Fig. 10-23, the ascending velocity of fine coke is considerably small, and its residence time is long. Figure 10-24 [9]shows that the increase in the solid/gas ratio or in the input of unburnt char increases the ascending velocity, so that the unburnt char is easy to be out of the reaction system.

Subsequently, the reaction rate of unburnt char in the blast furnace was obtained. The rates of solution loss reaction of char and fine coke in the packed bed of coke were calculated by using the mean velocity of particle Usm. A high-temperature reaction test furnace was used for the measurement. Figure 10-25[9] shows the relationship between the solid/gas ratio and the reaction rate at the superficial velocity of 1.2 m/s that close to the actual condition in the lower part of the blast furnace. As can be seen from Fig. 10-25, to obtain the same reaction rate, it is necessary to raise the combustion efficiency and to reduce the formation of unburnt char. Figure 10-26 [9]shows the effects of the solid/gas

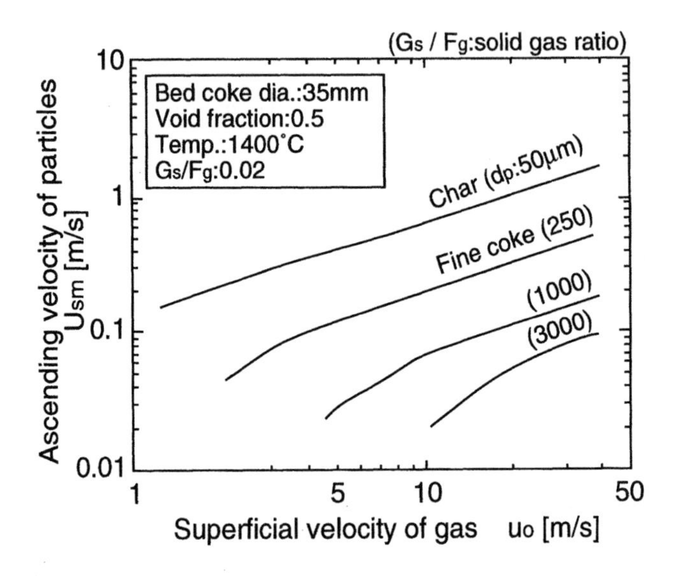

Fig.10-23 Ascending velocity of particles in packed bed[9].

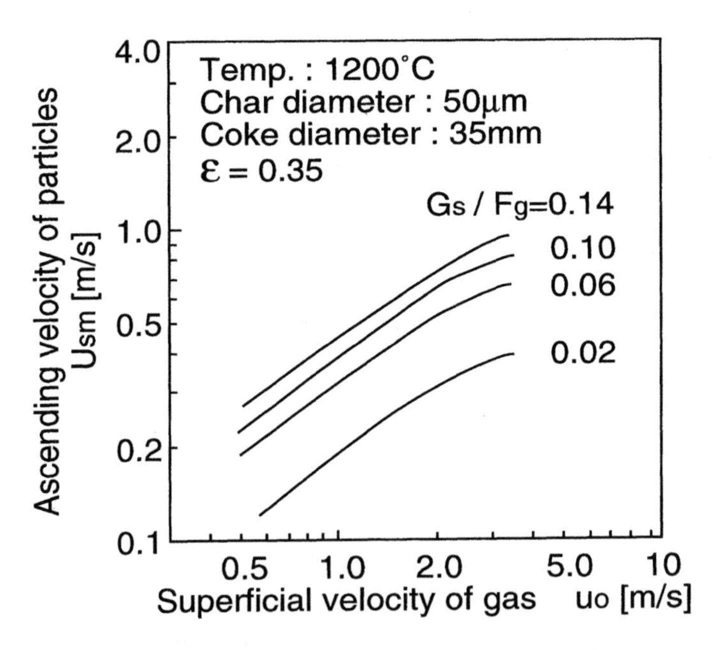

Fig.10-24 Calculated velocity of char in BF[9].

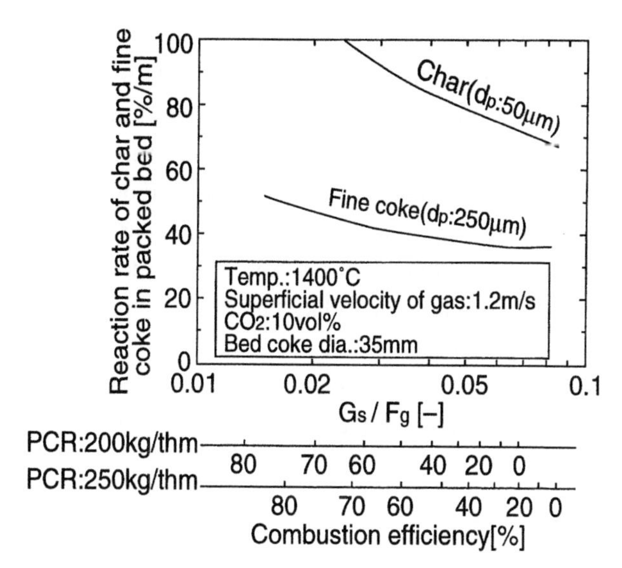

Fig.10-25 Relationship between solid-gas ratio and reaction rate[9].

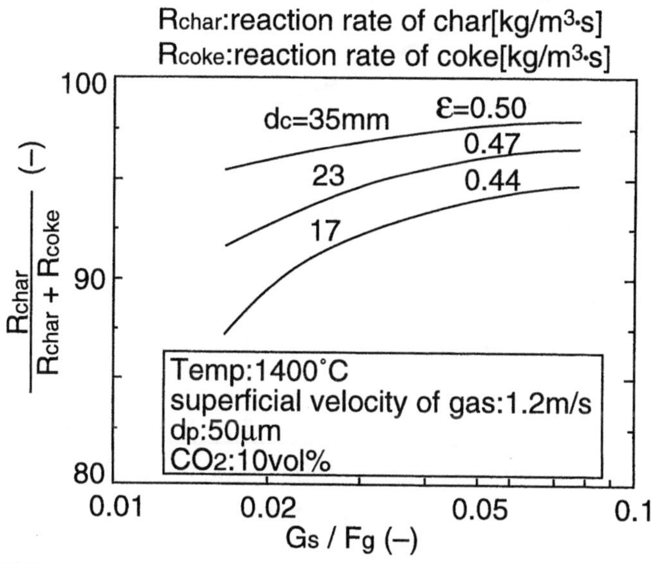

Fig.10-26 Effect of coke diameter and voidage of coke bed on reaction rate of char in packed bed[9].

ratio and the packed bed conditions of coke on the solution loss ratio (Rchar/(Rchar + Rcoke)). Assuming that the coke size is 35 mm in the lower part of the blast furnace, solution loss reaction of unburnt char is to be about 95% of the total solution loss reaction.

The limit of PCR in terms of the consumption limit of unburnt char in the blast furnace was estimated by using the equations discussed above. The results are shown in Fig. 10-27[9]. Assuming that the amount of solution loss carbon predicted from the Rist model is consumed, which is equivalent to unburnt char under the operating conditions of NKK Fukuyama No. 4 blast furnace, the eccentric double lances could inject 230 kg/thm of pulverized coal. To achieve a 250 kg/thm PCI, it is necessary to improve the combustion efficiency up to 80%.

Consequently, most of the solution loss reaction occurred in unburnt char by the above analysis that took into account the residence time of unburnt char in the packed

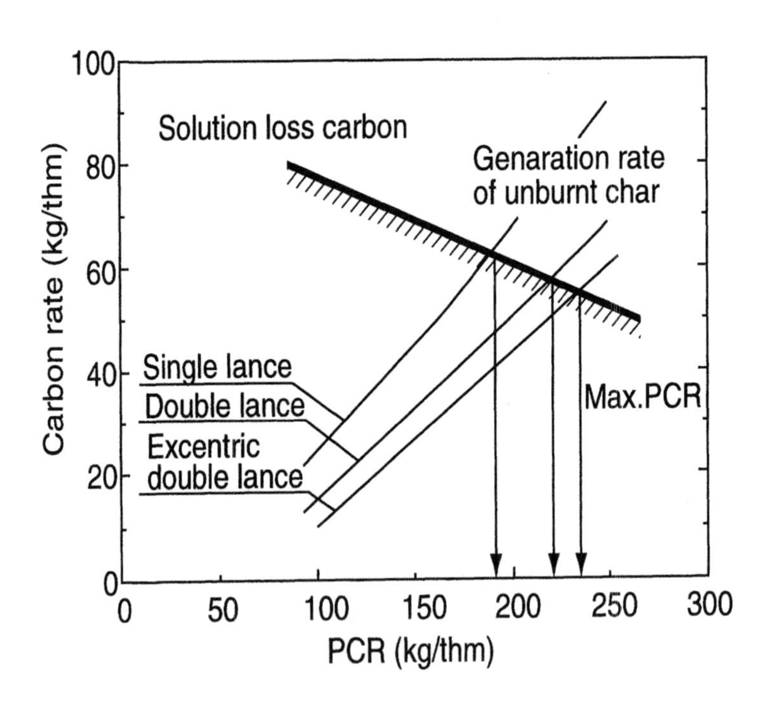

Fig.10-27 Estimation of maximum PCR[9].

bed of coke and the reaction rate of solution loss. Further, the limit of PCR is raised up to about 230 kg/thm by the use of eccentric double lances. To make PCR of 250 kg/thm possible, such lances to accomplish 80% combustion efficiency will be necessary..

(2) Reactions of unburnt char

The existence of unburnt char in the blast furnace related to the poor gas permeability in the blast furnace and the increase of heat loss in the lower part of furnace in the case of high rate PCI. Shimizu et al.[4] fundamentally studied the amount of PC combustion in the raceway and the reaction behavior of unburnt char which react with gases and FeO melts. The limit of PCR and the optimum conditions of PCI on the basis of the mass balance of each reaction products.

Fig. 10-28[4] shows the relationship between void fraction and dimensionless depth of raceway. The void fraction of the raceway is almost constant at 0.9 to 1.0 to the center of the raceway, and decreases toward the end of raceway linearly. From the voidage of

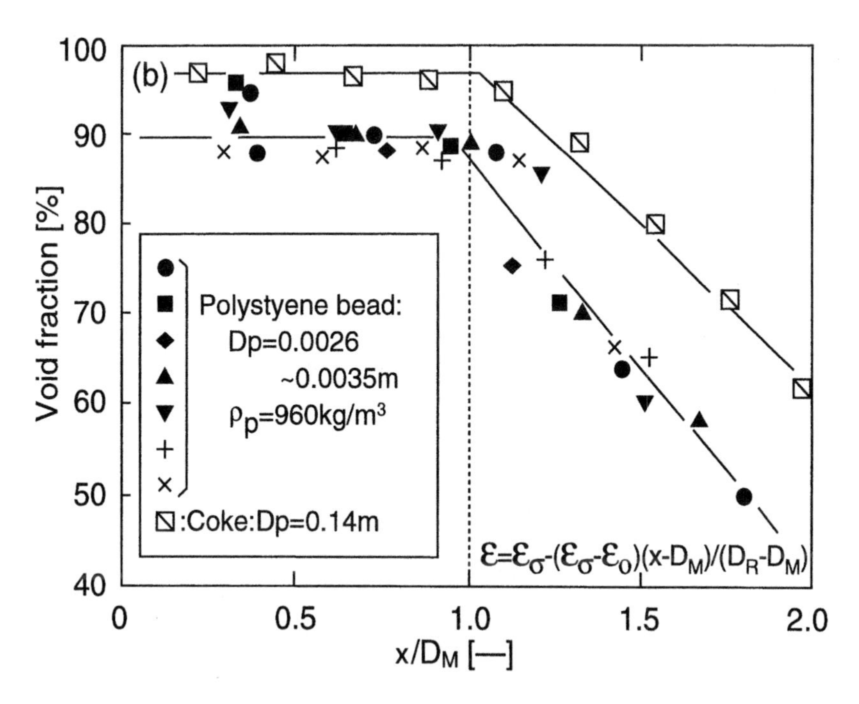

Fig.10-28 Relationship between void fraction and dimensionless distance normalized by x/D_M (D_M: distance from tuyuer nose to center of raceway)[4].

raceway, the effective reaction surface area of coke particles was estimated and combustion efficiency of PC was obtained by a raceway combustion model. The results are shown in Fig. 10-29[4]. The calculated results agree with the measured one. The combustion degree is about 68 to 73% when the PCR is 200 to 250 kg/thm and the oxygen enrichment rate is 3%. This means that unburnt pulverized coal flows into the blast furnace at about 60 to 75 kg/thm.

The amounts of solution loss reactions of unburnt char, which consists the reaction with the dripping slag and metal and gasification around the cohesive zone, were calculated.

In the case of the reaction of unburnt char with slag and metal, the unburnt char flowing out of the raceway are to contact with molten metal and/or slag in the dripping zone and to be consumed partially by carburization of metal and reaction with FeO melt. The amounts of each reactions were investigated by using the horizontal rotary reaction

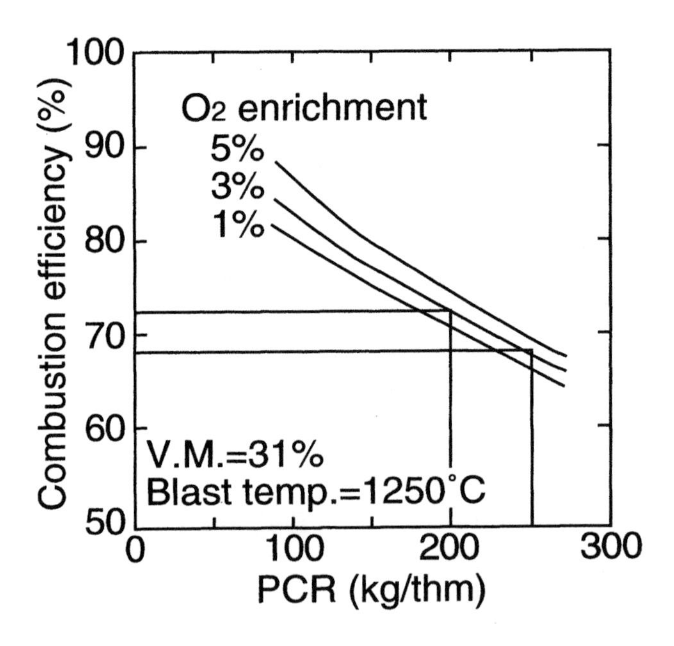

Fig.10-29 Combustion efficiency of PC in raceway calculated by mathematical model[4].

furnace illustrated in Fig. 10-30[4]. The results are shown in Fig. 10-31[4]. The carbon content of metal increases with increasing unburnt char, however, carburization of metal by graphite which simulated one with coke was small. The amount of unburnt char reacting with the metal in the actual blast furnace is estimated to be 1.5% or less when the combustion efficiency in the raceway is 70 or 90% as shown in Fig. 10-32[4]. The amount of unburnt char reacting with the molten slag is shown in Fig. 10-33[4]. The reaction of unburnt char with molten slag will finish in 3 min and the degree of reaction of unburnt char with the molten slag is almost constant and about 4%, regardless of the injection rate of unburnt char.

From these results, the total amount of unburnt char consumed in the dripping zone would be about 5% considering the both of the carburization of molten metal and reaction with FeO slag.

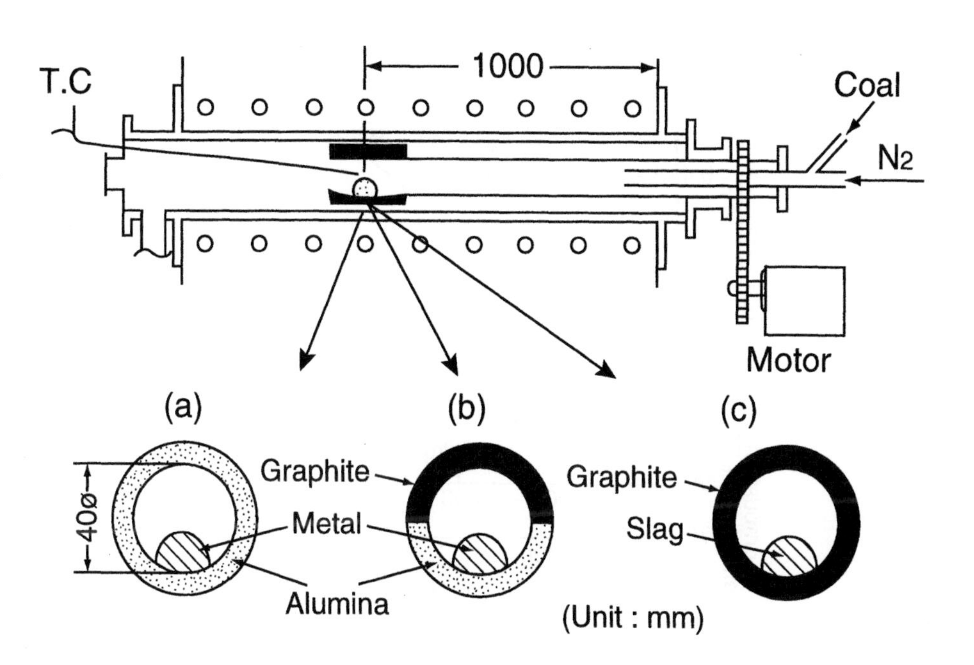

Fig.10-30 Illustration of horizontal rotary furnace[4].

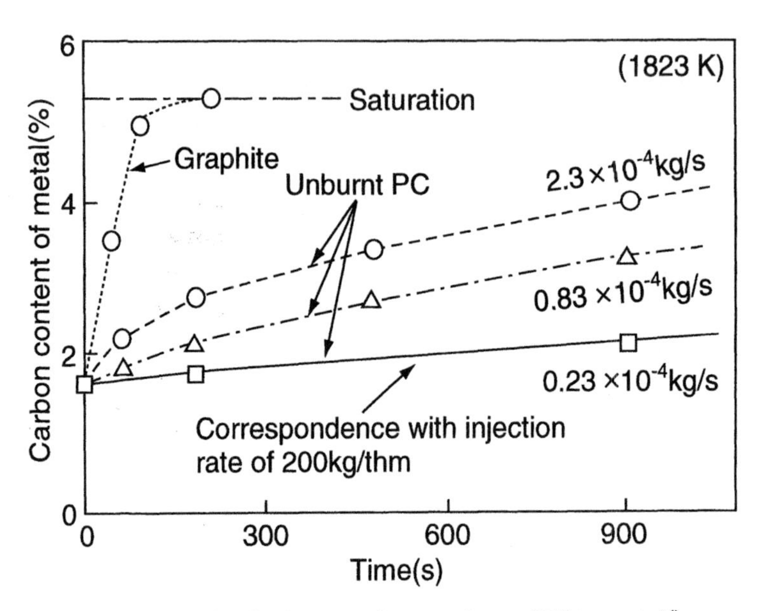

Fig.10-31 Dissolution rate from unburnt PC to metal[4].

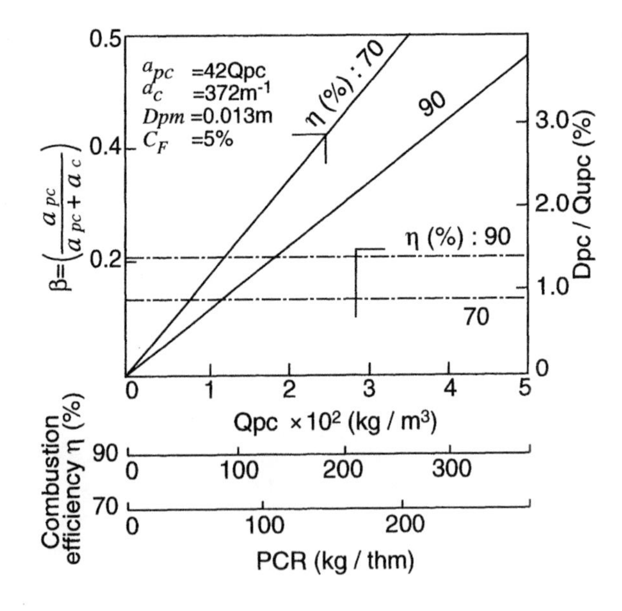

Fig.10-32 Dissolution rate of carbon from unburnt PC to metal[4].

The amount of gasification around cohesive zone was then determined. The reaction of unburnt char in the blast furnace would be the solution loss reaction by the CO_2 gas in the softening and cohesive zones, mainly. So the reaction of unburnt char should be considered together with the reaction of coke around the cohesive zone.

Figure 10-34 shows the relationship between the specific surface area of unburnt char and the reaction rate of unburnt char with CO_2 gas at 1273K. The reaction rate of unburnt char is about two to seven times higher than that of coke in same size. Assuming that the combustion efficiency of pulverized coal is 70 to 90%, the reaction rate of unburnt char is 4.1×10^{-4} kg/(kg-PC·s). The reaction rate of coke with the CO_2 gas is expressed as Fig. 10-35. The reaction rate gradually declines when the particle size exceeds 25 mm. Assuming that the size of coke is about 45 mm in the blast furnace, the reaction rate would be estimated as 5×10^{-6} kg/(kg-coke· s). The contribution of both reaction of char

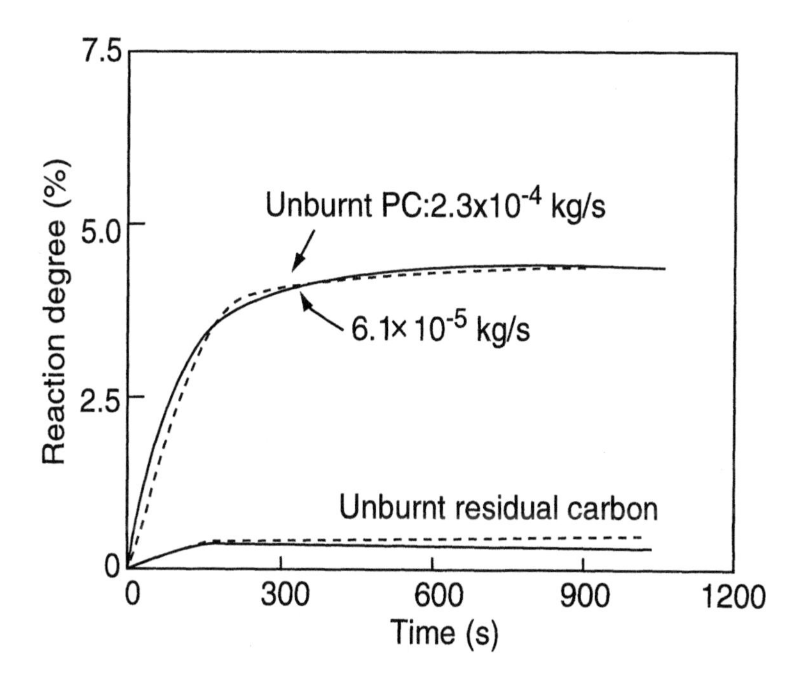

Fig.10-33 Reaction rate of unburnt PC with molten slag[4).

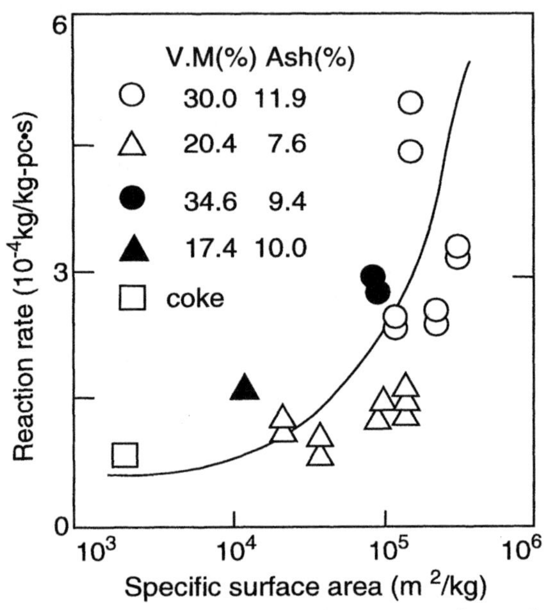

Fig.10-34 Relationship between specific surface area and reaction rate of char around cohesive zone[4].

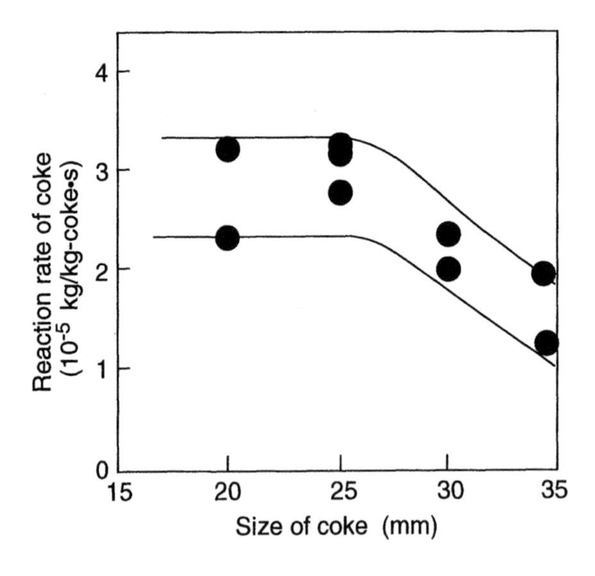

Fig.10-35 Effect of size of coke on reaction rate[4].

and coke to the carbon solution reaction in the blast furnace can be estimated from these values. When the coke rate is 300 kg/thm and the input of unburnt char is 100 kg/thm or more, the contribution of unburnt char to the solution loss reaction is 95% or more, because the specific surface area of unburnt char is quite larger than that of coke.

According to the above fundamental study, the consumption balance of unburnt char in the blast furnace was calculated from the reactions with the metal and slag in the dripping zone and from the gasification reaction around the softening and cohesive zones. The calculated results are given in Fig. 10-36[4]. The carbon content of char is expressed as the amount of residual carbon assumed to be reduced by 5% owing to the metal and slag reactions. The amount of solution loss carbon is obtained in an actual blast furnace. From Fig. 10-36, the limit of PCR is estimated to be 220 and 200 kg/thm, when the

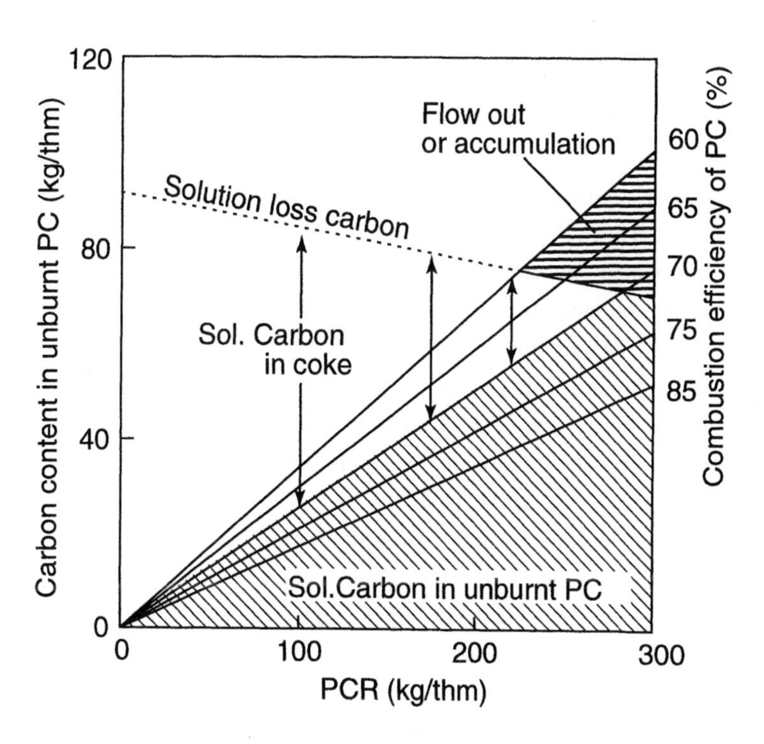

Fig.10-36 Mass balance of carbon consumed by solution loss reaction[4].

combustion efficiency in the raceway is 60 and 70%, respectively. In actual blast furnace operation, the combustion efficiency of PC was about 70% , when PCR was 200 kg/thm. This means that about 50 kg/thm of unburnt char flows into the blast furnace. All of this unburnt char should be consumed by the reactions in the blast furnace. To summarize, unburnt char has much greater specific surface area than coke, so that it is preferentially gasified with CO_2. If a maximum combustion efficiency of PC 7 could be 75% in the raceway, the possible PCR would reached to 300 kg/thm.

10.2.3 Coke rate and carbon solution loss reaction

It is a matter of common knowledge in high rate PCI operation that coke is attacked by reactions (mainly by the solution loss reaction) before it reaches to the tuyere nose. This attack of reaction makes coke brittle. The resultant fine coke accumulates on the surface of the deadman, inhibits the penetration of gas into the deadman and retards the heat transfer to the deadman. In general, the deadman has the role of buffer of heat variation in the lower part of the blast furnace and has the function of consuming fine coke by letting it react with dripping metal and slag. When the fine coke accumulated on the surface of the deadman exceeds a tolerable quantity, it is difficult to maintain a stable blast furnace operation. Specifically, as the increase of PCR decreases coke rate, the carbon consumed by solution loss reaction increases relatively and raises the degradation ratio of the coke, if the overall solution loss reaction occurred in the coke charged. Actually, the unburnt char is also consumed by solution loss reaction. The contribution of the unburnt char was mentioned in section 10.2.2 (1) and 10.2.2 (2).

Therefore, If coke degradation would proceed under low coke rate, the gas and liquid permeability in blast furnace would be aggravated seriously. Shimizu et al.[10)] took samples of coke from the center region of deadman during outage of blast and measured the fraction of -3mm coke. The relationship between -3mm coke and the ratio of solution loss carbon to coke rate, [SLC/CR] is shown in Fig. 10-37[10)]. The -3mm coke has a tendency to increases with increasing [SLC/CR] in the PCR less than 110kg/thm and keeps high percentage in high PCR(>110kg/thm). So, in high rate PCI operation, the fine coke generation is enhanced in the deadman and the gas and liquid permeability will be worse. Furthermore, the ratio of [SLC/CR] would be important and an upper limit could be mentioned as about 0.27 from Fig. 10-37.

Okuno, et al [6)] investigated the relationship between [SLC/CR] and PCR in the world's blast furnace (Fig.38(a)). The [SLC/CR] of foreign blast furnace and Ohita and Kobeís blast furnace which has low fuel rate was relatively high. This figure also shows the upper limit of [SLC/CR] from 0.27 to 0.3. Moreover, They developed a mathematical model to understand this result quantitatively. The calculated results are given in Fig. 10-38(b).

From Fig. 10-38(b), it is found that the coke rate less than 300 kg/thm is difficult in the fuel rate of 450 kg/thm and the PCR of 150 kg/thm or more. This result indicates that the high quality (DI and CSR) of coke is important in a low fuel rate ranged from 450 to 500kg/thm. It also indicates that the restriction occurred from coke degradation could be avoidable to a large extent if the fuel rate could increase. These estimations agreed with the actual results from blast furnace that the furnace condition became unstable, when the PCR increased in 150kg/thm during a low fuel rate operation about 480 kg/thm. Consequently, the operation of blast furnaces with a high rate PCR always calls into

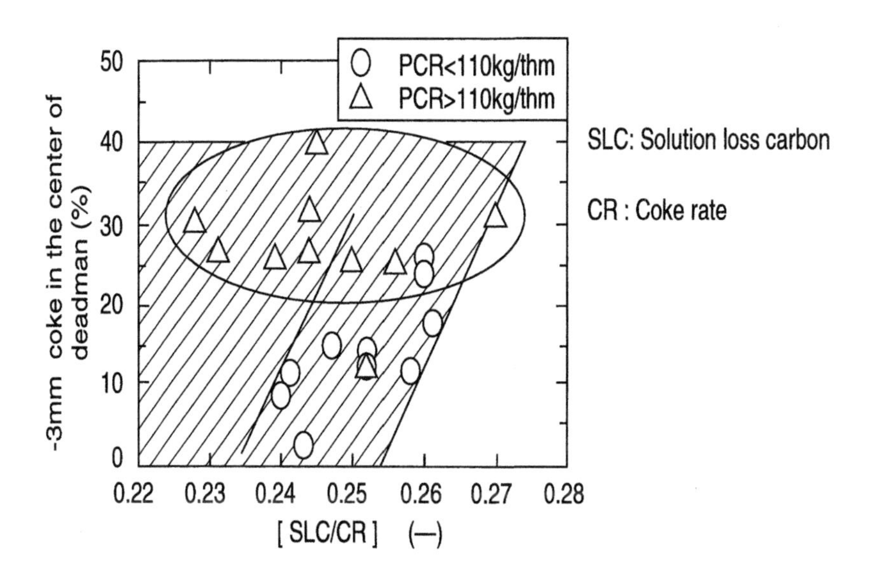

Fig.10-37 Relationship between [SLC/CR] and -3mm fine coke in deadman[10].

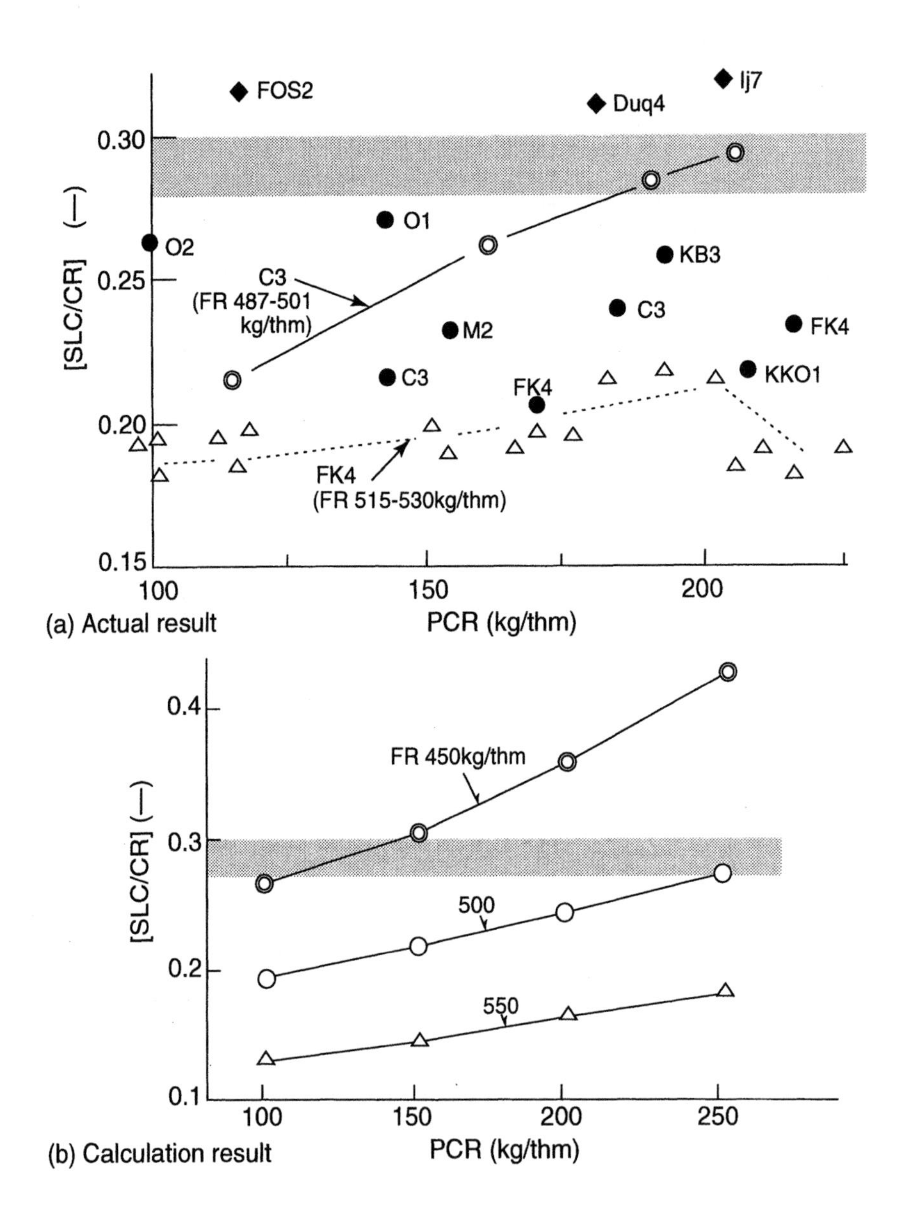

Fig.10-38 Relationship between PCR and [SLC/CR]. [6]
(a) :actual result, (b): calculation

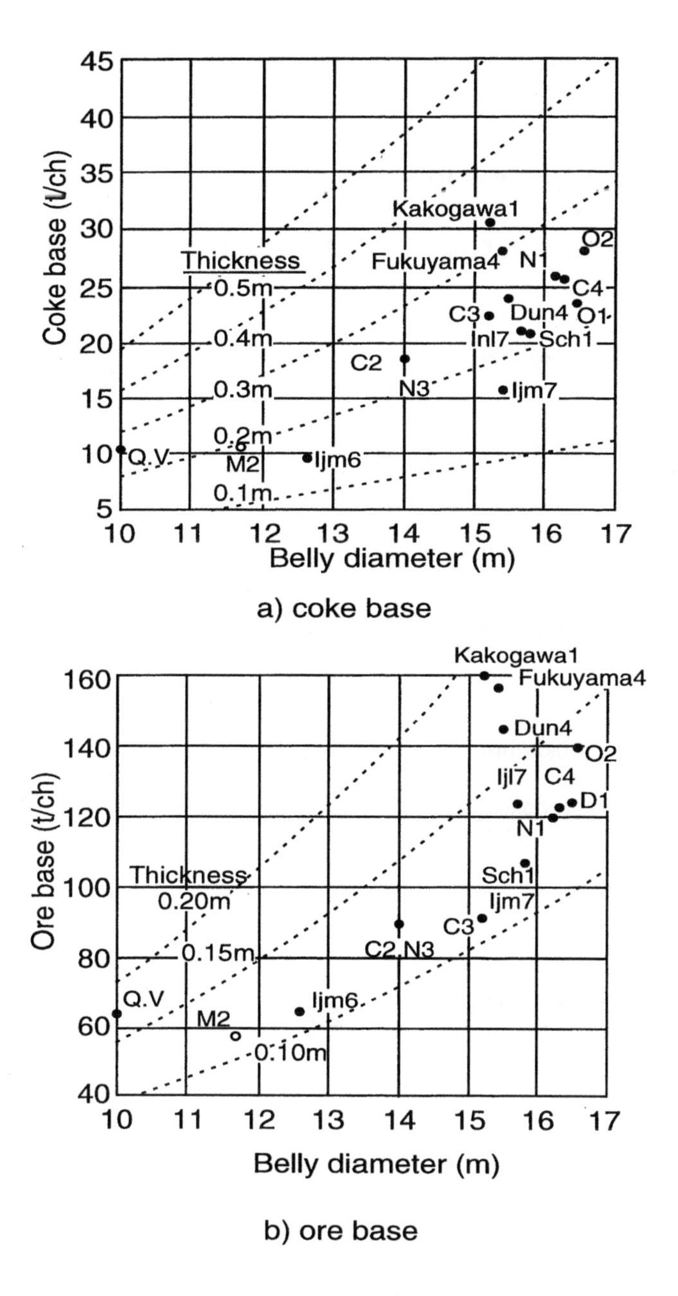

Fig.10-39 Relationship between diameter of belly and coke base, ore base. [7]
(a): coke base, (b): ore base

question whether to permit the high fuel rate operation or the usage of high quality coke and ore.

10.2.4 Coke rate and burden distribution

The Okuno, et al[7]. shows that when PCR exceeds 170 kg/thm, the ore/coke ratio is more than 5 and the function of coke slit which has a role of distributing of gas in the

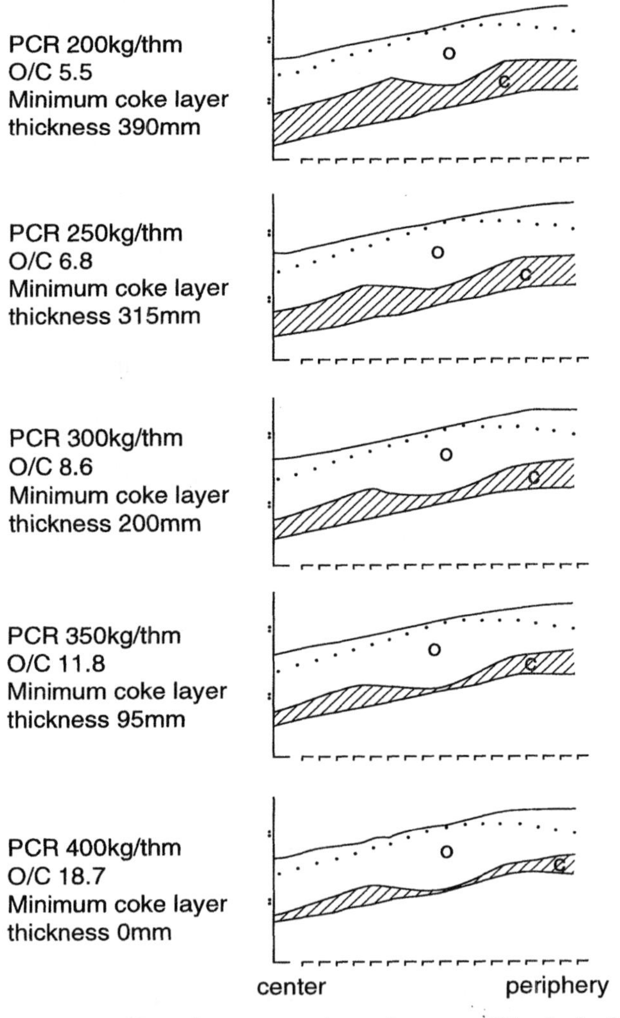

Fig.10-40 Burden profile at furnace top in various condition(calculation) [7].

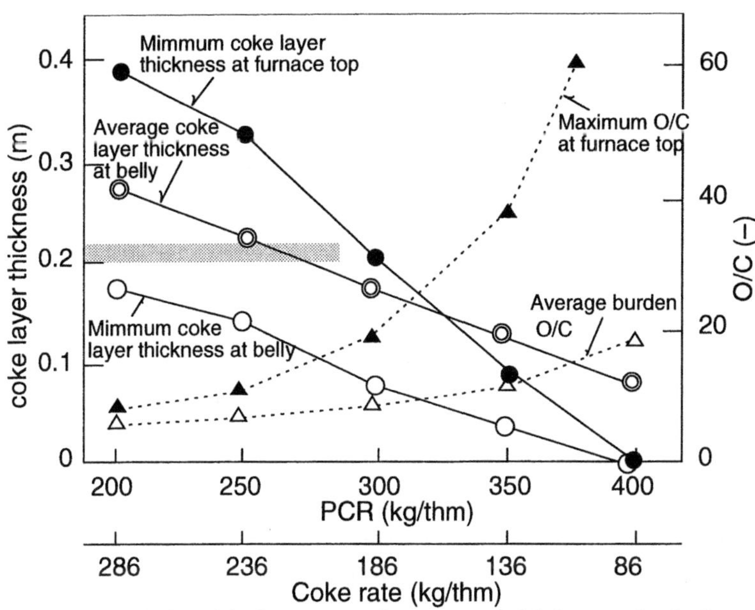

Fig.10-41 Relationship between coke rate and thickness of coke layer.[8]

shaft is expected to aggravate. The coke rate will have a lower limit when the coke layer thickness decreases to make the gas flow distribution nonuniform. In actual blast furnace operation, The coke base and ore base [t/charge] are plotted in terms of the diameter of belly when the minimum allowable thickness of coke layer was assumed to be from 200 to 220 mm (Fig.10-39).

Here, the profile of coke layer in radial direction and the gas flow distribution were calculated by the mathematical burden distribution model 'RABIT' using fuel rate of 487 kg/thm, ore/coke ratio of 5.56, coke base of 24.1 t/charge and productivity of 2.1 as initial conditions. Then, the minimum coke rate was discussed according to the results of calculation. In this case, a charging mode was selected that the coke layer thickness became relatively uniform in the radial direction. The minimum coke rate was studied in terms of the minimum thickness of coke layer, when the attention was focused on the thickness of coke layer become a minimum in the intermediate region by the collapse of coke layer.

Figure 10-40[7] shows the ore and coke layer profiles in the region of furnace top when the ore/coke ratio in the burden increased (the ore base being constant). Considering that the gas flow distribution in the radial direction was restricted by the minimum coke layer thickness in the intermediate region, the relationship between the minimum coke layer thickness and the ore/coke ratio in that position was obtained from Fig. 10-40 and shown in Fig. 10-41[8].

From Fig. 10-41, the coke rate at which the minimum coke layer thickness becomes '0' is 86 kg/thm. When the average coke layer thickness is 200 to 220 mm, which is considered to be the lower limit in the belly, the minimum coke layer thickness in the intermediate region is about 120 mm which is equivalent to the height of only two pieces of coke(average size). Since this thickness (120mm) is the minimum one to form coke slit, the average coke layer thickness of 200 to 220 mm in the belly would be a reasonable lower limit for blast furnace operation..

In general, the profile of coke layer is changed in various manners in accordance with the coke base, ore base and charging mode. It is difficult to make an accurate estimation of the minimum coke rate from only one example. If it was possible to assume that the thickness corresponding to the height of two pieces of coke maintained the function of the coke slit, the minimum coke rate would be calculated to be 220 to 230 kg/thm.

10.3 Summary

The maximum possible rate of PCI has been discussed in terms of the coke-to-pulverized coal replacement ratio and the blast furnace phenomena.

1) When the trends of blast furnaces in Japan (including the results estimated by mathematical models) were analyzed, the replacement ratio began to decrease from the PCR about 180 to 190 kg/thm, although the replacement ratio was related to the coal type and the fuel rate. It could be considered that the PCR about 180 to 190 kg/thm would be the maximum limit from the viewpoint of overal estimation of blast furnace operation including the cost.

2) The maximum possible rate of PCI was also discussed concerning the blast furnace phenomena, such as the minimum stoichiometric oxygen ratio, formation and consumption of unburnt char, solution loss carbon, and control of burden distribution.

   i) If the minimum stoichiometric oxygen ratio was from 0.6 to 0.7 observed in actual blast furnaces, the minimum coke rate was from 270 to 280 kg/thm, and the maximum PCR could be expected from 180 to 270 kg/thm.

   ii) The consumption limit of unburnt char by the solution loss reaction indicated that the maximum PCR of 230kg/thm would be possible in the case of eccentric double lances. On the other hand, from the viewpoint of combustion efficiency, when it was about 75% in the raceway, PCR could be 300 kg/thm.

iii) If the fuel rate was about 480 kg/thm, a coke rate of 285 kg/thm and a 200 kg/thm PCR would be the maximum limits for the stable operation of blast furnace from the standpoint of the amount of solution loss reaction per unit coke weight on the basis of the present coke quality.

iv) If the function of coke slit (220 mm coke layer thickness) was maintained, the minimum coke rate is 220 to 230 kg/thm in terms of the restriction of burden distribution. If the fuel rate was about 480 kg/thm, the maximum PCR would be 260 kg/thm.

Considering the problems such as the aggravation of raw materials quality, the load of coke oven and the demand to decrease of exhausted CO_2 in future, it would be impossible to correspond these problems with only present procedures. It will be required to increase PCR while decreasing the coke rate at a low fuel rate. In other words, it will be necessary to improve the combustion efficiency of PC, improve the control method of burden distribution and improve the quality of low-cost raw materials and fuels.

REFERENCES

1) R.Murai,M.Sato,T.Ariyama: Research Group of Pulverized Coal Combustion in Blast Furnace, Rep-40(1996), JSPS Ironmaking 54 Committee

2) K.Sunahara, T.Suzuki, T.Kamijou, T.Yamamoto, M.Fukui, A.Shinotake: Research Group of Pulverized Coal Combustion in Blast Furnace, Rep-41(1996), JSPS Ironmaking 54th Committee

3) T.Deno, A.Shinotake:Research Group of Pulverized Coal Combustion in Blast Furnace Rep-49(1996), JSPS Ironmaking 54th Committee

4) M.Shimizu, T.Kamijou: Research Group of Pulverized Coal Combustion in Blast Furnace Rep-45(1996), JSPS Ironmaking 54th Committee

5) S.Amano,S.Matsunaga,K.Kakiuchi,H.Ueno,N.Konno,K.Yamaguchi: JSPS

ironmaking 54th committee, No.54-2025(1994)

6) Y.Okuno, H.Ueno, K.Yamaguchi, K.Tamura, M.Nose, M.Nakayama, A.Matsui: JSPS ironmaking 54th committee, No.54-1942(1994)
7) Y.Okuno,Private letter
8) T.Deno,Private letter
9) M.Sato,R.Murai,T.Ariyama:Research Group of Pulverized Coal Combustion in Blast Furnace, Rep-48(1996), JSPS Ironmaking 54th Committee
10) M.Shimizu, A.Kasai, T.Kamijou, H.Iwakiri, R.Ito, M. Atsu: JSPS Ironmaking 54th Committee, No.54-2037(1995)

Index

D

E

F

G

H

M

N

O

P

Q

R

Printed and bound by CPI Group (UK) Ltd, Croydon, CR0 4YY

08/05/2025

01864783-0004